ANNUAL EDITIONS

Environment 13/14

Thirty-Second Edition

Editor

Richard Eathorne
Northern Michigan University

Richard Eathorne is Assistant Professor in the Department of Earth, Environmental, and Geographical Sciences at Northern Michigan University. His primary academic interest in the field of human geography with particular attention to the human–environment relationship helped create a new major at the university beginning in the 2011 academic year, *Environmental Studies & Sustainability*. This *Annual Editions: Environment 13/14* was a direct outgrowth of that effort. Professor Eathorne lived for a decade in Alaska where he spent most of his time living and learning the human–environment relationships of the Inupiaq Eskimos of the Unalakleet region of western Alaska and the Gwich'in of Arctic Village in the central Brooks Range mountains of northern Alaska. He has also traveled extensively throughout Central America and to the Galapagos Islands, the Ecuadorian rainforest, and highlands of Peru exploring the man–land relationships of those regions as well. He brings his human geography experiences to the teaching of introductory classes in human geography and environmental science as well as upper-level classes in economic geography, geography of Latin America, and geography of tourism.

Mc
Graw
Hill

Connect
Learn
Succeed™

The McGraw·Hill Companies

Mc Graw Hill — *Connect Learn Succeed*™

ANNUAL EDITIONS: ENVIRONMENT, THIRTY-SECOND EDITION

Published by McGraw-Hill, a business unit of The McGraw-Hill Companies, Inc., 1221 Avenue of the Americas, New York, NY 10020. Copyright © 2014 by The McGraw-Hill Companies, Inc. All rights reserved. Printed in the United States of America. Previous edition(s) © 2013, 2012, and 2011. No part of this publication may be reproduced or distributed in any form or by any means, or stored in a database or retrieval system, without the prior written consent of The McGraw-Hill Companies, Inc., including, but not limited to, in any network or other electronic storage or transmission, or broadcast for distance learning.

Some ancillaries, including electronic and print components, may not be available to customers outside the United States.

This book is printed on acid-free paper.

Annual Editions® is a registered trademark of The McGraw-Hill Companies, Inc. Annual Editions is published by the **Contemporary Learning Series** group within the McGraw-Hill Higher Education division.

1 2 3 4 5 6 7 8 9 0 QDB/QDB 1 0 9 8 7 6 5 4 3

ISBN: 978-0-07-351562-5
MHID: 0-07-351562-0
ISSN: 0272-9008 (print)
ISSN: 2159-1059 (online)

Acquisitions Editor: *Joan L. McNamara*
Marketing Director: *Adam Kloza*
Marketing Manager: *Nathan Edwards*
Senior Developmental Editor: *Jade Benedict*
Senior Project Manager: *Joyce Watters*
Buyer: *Nichole Birkenholz*
Cover Designer: *Studio Montage, St. Louis, MO*
Content Licensing Specialist: *DeAnne Dausener*
Media Project Manager: *Sridevi Palani*

Compositor: Laserwords Private Limited
Cover Image Credits: Courtesy of Richard Eathorne (inset and background)

www.mhhe.com

Editors/Academic Advisory Board

Members of the Academic Advisory Board are instrumental in the final selection of articles for each edition of ANNUAL EDITIONS. Their review of articles for content, level, and appropriateness provides critical direction to the editors and staff. We think that you will find their careful consideration well reflected in this volume.

ANNUAL EDITIONS: Environment 13/14
32nd Edition

EDITOR

Richard Eathorne
Northern Michigan University

ACADEMIC ADVISORY BOARD MEMBERS

Preface

Agent Smith: *"I'd like to share a revelation that I've made during my time here. It came to me when I tried to classify your species. I realized that you're not actually mammals. Every mammal on the planet instinctively develops a natural equilibrium with the surrounding environment, but you humans do not. You move to an area and you multiply, and multiply until every natural resource is consumed. The only way you can survive is to spread to another area. There is another organism that follows the same pattern. Do you know what it is? A virus. Human beings are a disease, a cancer of this planet. You are a plague. . . . You hear that Mr. Anderson? That is the sound of inevitability. . . . It is the sound of your death. . . . Goodbye, Mr. Anderson. . . . "*

Mr. Anderson: *My name . . . is Neo.*

From the film *The Matrix*

Harsh words to be sure. But not to worry. This is only a fictional quote from a fictional character, in a fictional film, intended for dramatic effect.

On the other hand, when a world renowned environmental scientist like E.O. Wilson (and many more are sharing Professor Wilson's observation) is quoted saying that *"the pattern of human population growth in the twentieth century was more bacterial than primate"* is it then perhaps time for us to wonder if we ourselves are living a fictional reality, mystical existence, believing we humans can continue to consume our Earth at the rate we are and live sustainably, forever.

Consuming the Earth is the theme of this *Annual Edition: Environment 13/14* volume. How we presently consume the earth—transform its raw resources to meet our appetites— as well as the stress we put upon the planet to attend to our waste may be perhaps our most critical environmental issue now and into the future. As the world's population grows, its societies and nations are destined to evolve socioeconomically as well. And with this evolution, global patterns of resource demands and consumption behaviors will change for billions of humans. Herein may lie our greatest future challenge in the history of civilization: How do we change old eating habits and encourage new generations to take on the healthy consumption lifestyles necessary for humanity to create a sustainable relationship with its planet?

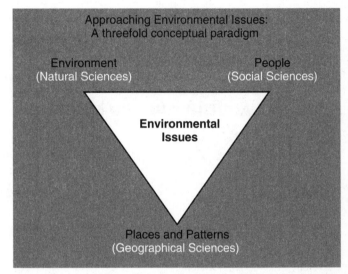

Figure 1 Addressing environmental issues requires a threefold approach that emphasizes the connections and linkages imbedded within any environmental problem. In addition, it is important to remember that the science applied (e.g., available technologies) and the people involved (e.g., cultural/ethnic, gender, religious factors) in a particular environmental problem can vary significantly by "place." Such place variability can significantly influence policy implementations, acceptance, appropriateness, and ultimately, success.

This volume of *Annual Editions: Environment 13/14* is constructed first and foremost upon the premise that approaching the *environmental consumption* issue requires a threefold paradigm that implies connections and linkages. Figure 1 illustrates this idea. Whenever we address environmental issues, there is general agreement that we cannot view an issue's environmental aspects independent of the people aspect. However, what is oftentimes overlooked, as well as underappreciated, is that any particular environmental issue must also be "placed" geographically. Only by doing so can we begin to uncover and understand better the variable global patterns of our environmental challenges as well as how those variable patterns (of wealth, access to resources, poverty, political instability, etc.) are connected to finding resolutions or obstructing our quest to succeed in implementing policy. For example, regarding the *environmental consumption* issue, we understand the role that earth's natural resources and ecosystem services play in our transforming those resources to meet human needs and desires (the natural sciences). We also are aware of how peoples, cultures, economies, and politics orchestrate the distribution and access of the materials we create from earth's resources and how those *environmental-resource-needs and desires* are managed and met (the social sciences). But *in dealing with the environmental impacts and challenges,* all this planetary resource transformation/ecosystem service consumption generates will vary considerably by *place.* For example, impoverished regions/nations will have different challenges to policy implementation compared to wealthy regions/nations. Rural agricultural economies will have different environmental priorities than urban manufacturing economic areas. Dispersed human populations inhabiting desert/dry shrubland biomes will have far less agency in discourse regarding fresh water management than will the concentrated populations that inhabit the earth's temperate broadleaf/mixed forests.

Secondly, the theme of this edition of *Environment 13/14,* **Consuming the Earth,** is intended to provide a platform for "active learning" regarding this issue and upon which the reader can consider connections with other issues and components of the Human–Earth relationship. In keeping with the spirit and intent of the entire McGraw-Hill Annual Editions series, the *Unit Overviews* as well as posted *Learning Outcomes* and *Critical Thinking Questions* all converge to encourage the reader to approach this text not for the consumption of more information but to actively engage in the assessment of the articles' theses and content with regard to their linkages (or lack of) with the classroom, textbook, and reader's real-world experience. This kind of proactive reader approach requires going beyond analytical "consumption reading" (knowledge accumulation/rote response). Critical-thinking environmental science reading seeks to generate questions and insights; help construct predictions and extrapolations; discover imbedded linkages and components of the environmental consumption issue that can lead to resolutions. It also encourages exploration of the *connections* between what the reader learns in class and what he or she sees outside of class. And finally, critical thinking of all environmental issues demands assessment of our purpose in not only environmental science but also all scientific disciplines.

Distinguished economist and noted conservationist, Barbara Ward (1914–1981), reflected well this need for critical thinking and understanding of connections in environmental science when she noted, *"For an increasing number of environmental issues, the difficulty is not to identify remedies. The remedies are well understood. The problem is to make them socially, economically, and politically acceptable. The solutions to environmental problems increasingly involve* [seeing the connections between] *human social systems* as well as *natural systems."* And the National Science Foundation's Advisory Committee for Environmental Research and Education 2009 report underscores further that to live sustainably on the Earth, *"greater priority must be given to advancing an integrated approach to Earth systems, and addressing the complexity of coupled natural and human systems from local to regional to global scales."*

The articles in this edition are compiled from a myriad of scientific, professional, and general audience as well as governmental publications. They were selected to provide illustrations of the multitude of concerns, views, interpretations and implications revolving around perhaps the most critical challenge to achieving *sustainability—changing planetary human consumption behaviors.* Several articles were selected precisely because of their earlier publication date. The intent is to allow readers to compare the "hypotheses" presented at that time of publication regarding future patterns of planetary consumption and compare the evidence available today. Were those predictions valid? If not, what "data" had not been included or available at the time? How might a better understanding of the reasons behind the accuracy/inaccuracy of these predictions help us in the construction of new hypotheses and new predictions leading to new knowledge for better planning for the future?

Finally, added to this volume is a new concluding section titled Afterword: Food for Thought. This *afterword* exercise asks students to define and reflect upon the two concepts of *sustainability* and *homeostasis.* My purpose and hope is the activity can serve as a sort of critical thinking springboard. A closer critical review of the two terms may help launch new discourse on the idea of *sustainability,* a concept some authors in this edition of *Environment* have suggested, albeit delicately, may becoming ineffectual as the guiding light to human survival.

Of course, embarking on any new directions of sustainability discourse and addressing the challenges that planetary consumption presents lie not in the exclusive realm of environmental science but will require insights and inputs from students and scholars of economics, geography, foreign policy, sociology, history, literature, and others. Therefore, I would like to express my gratitude to the McGraw-Hill Annual Editions Series and its entire editorial staff for providing a forum where scholars, authors, instructors, and students can bring together knowledge, understanding, and insights in an effort to resolve the challenges that lie ahead. Finally, feedback from readers, instructors, and students is crucial if we hope to maintain the *Annual Editions'* spirit and intent of fostering critical thinking, intellectual inquiry, and insight. Please use the postage-paid article rating form found on the last page to share your opinions and suggestions regarding not only articles selected but also the *Annual Editions: Environment 13/14* thematic organization, pedagogic structure, and construction of Critical Thinking Questions.

Richard Eathorne
Editor

The Annual Editions Series

VOLUMES AVAILABLE

Adolescent Psychology

Aging

American Foreign Policy

American Government

Anthropology

Archaeology

Assessment and Evaluation

Business Ethics

Child Growth and Development

Comparative Politics

Criminal Justice

Developing World

Drugs, Society, and Behavior

Dying, Death, and Bereavement

Early Childhood Education

Economics

Educating Children with Exceptionalities

Education

Educational Psychology

Entrepreneurship

Environment

The Family

Gender

Geography

Global Issues

Health

Homeland Security

Human Development

Human Resources

Human Sexualities

International Business

Management

Marketing

Mass Media

Microbiology

Multicultural Education

Nursing

Nutrition

Physical Anthropology

Psychology

Race and Ethnic Relations

Social Problems

Sociology

State and Local Government

Sustainability

Technologies, Social Media, and Society

United States History, Volume 1

United States History, Volume 2

Urban Society

Violence and Terrorism

Western Civilization, Volume 1

World History, Volume 1

World History, Volume 2

World Politics

Contents

UNIT 1
The Nature of Our Consumption: *Ecophagy*

1. **The Psychological Roots of Resource Over-Consumption,**
   Nate Hagens, *Postcarbon.org* (from the book *Fleeing Vesuvius*), posted May 11, 2011, www.postcarbon.org/article/331819-the-psychological-roots-of-resource-over-consumption

   Author Nate Hagens believes humans have an innate **need for status and novelty** in their lives. But the way we pursue these needs, Hagens argues, is **not sustainable.** The essay explores some of the **underlying drivers** of **resource depletion** and **planetary consumption.**   **3**

2. **Why Do We Over-Consume?,** Darek Gondor, *Our World 2.0,* posted December 14, 2009, http://ourworld.unu.edu/en/why-do-we-over-consume/

   In this brief essay blog, Darek Gordon asks, "Why do we **over-consume?**" He suggests our predisposition to needing more stuff goes back to our premodern roots. However, the **abundance of the resources** necessary to meet those needs **continues to diminish** while competition increases. The solution, the author argues, will lie in our **cultural evolution.**   **12**

3. **The Gospel of Consumption: And the Better Future We Left behind,**
   Jeffrey Kaplan, *Orion Magazine,* May/June 2008.

   According to author, Jeffrey Kaplin, "If we want to **save the Earth,** we must also save ourselves from our selves." While our obsession with work (actually, overwork) contributed to productivity and material wealth in the past, we need time to maintain and nurture the **human relationships** necessary for **sustaining a healthy planet.**   **15**

4. **Do We Consume Too Much?,** Mark Sagoff, *The Atlantic Online,* June 1977, www.theatlantic.com/past/docs/issues/97jun/consume.htm

   Author Mark Sagoff asked in 1997, "Do we consume too much?" Discourse about the future of our planet (in 1997) was dominated by people who believed an expanding world economy will **use up our natural resources** and others who saw no reasons to **limit economic growth.** The author concluded "neither side has it right."   **20**

5. **How Much Should a Person Consume?,** Ramachandra Guha, *GLOBAL DIALOGUE,* vol. 4, No. 1, Winter 2002; www.worlddialogue.org/content.php?id=180 (from AE: Environment 12/13, Article 38).

   Guha argues, "There are . . . more than 300 professional **environmental historians** in the United States . . . and not one has seriously studied the **global consequences** of the **consumer society . . . American Way of Life.**" The essay examines the answer to the title's question and concludes there are vast **inequalities of global consumption.**   **30**

The concepts in bold are developed in the article. For further expansion, please refer to the Topic Guide.

UNIT 2
The Human Factor: Environmental Virus or Symbiosis?

The concepts in bold are developed in the article. For further expansion, please refer to the Topic Guide.

UNIT 3
The Geopolitical-Economy of Planetary Consumption

UNIT 4
The Whole Earth Café: Bellying Up to the Trough

The concepts in bold are developed in the article. For further expansion, please refer to the Topic Guide.

The concepts in bold are developed in the article. For further expansion, please refer to the Topic Guide.

The concepts in bold are developed in the article. For further expansion, please refer to the Topic Guide.

UNIT 5
The Consumption-Sustainability Conundrum: Is There an Answer?

The concepts in bold are developed in the article. For further expansion, please refer to the Topic Guide.

The concepts in bold are developed in the article. For further expansion, please refer to the Topic Guide.

Correlation Guide

The *Annual Editions* series provides students with convenient, inexpensive access to current, carefully selected articles from the public press. **Annual Editions: Environment 13/14** is an easy-to-use reader that presents articles on important topics such as *consumption, economics, environmental ethics,* and many more. For more information on *Annual Editions* and other *McGraw-Hill Contemporary Learning Series* titles, visit www.mhhe.com/cls.

This convenient guide matches the units in **Annual Editions: Environment 1 13/14** with the corresponding chapters in three of our best-selling McGraw-Hill Environmental Science textbooks by Enger/Smith and Cunningham/Cunningham.

Annual Editions: Environment 13/14	Environmental Science: A Study of Interrelationships, 13/e by Enger/Smith	Environmental Science: A Global Concern, 11/e, by Cunningham/Cunningham	Principles of Environmental Science, 6/e, by Cunningham/Cunningham
Unit I: The Nature of Our Consumption: *Ecophagy*	**Chapter 1:** Environmental Relationships **Chapter 2:** Environmental Ethics **Chapter 8:** Energy and Civilization: Patterns of Consumption **Chapter 9:** Nonrenewable Energy Sources **Chapter 10:** Renewable Energy sources **Chapter 19:** Environmental Policy and Decision Making	**Chapter 1:** Understanding Our Environment **Chapter 6:** Population Biology **Chapter 7:** Human Populations **Chapter 8:** Environmental Health and Toxicology **Chapter 23:** Ecological Economics **Chapter 25:** What Then Shall We Do?	**Chapter 1:** Understanding Our Environment **Chapter 4:** Human Population **Chapter 6:** Environmental Conservation **Chapter 15:** Environmental policy and Sustainability
Unit 2: The Human Factor: Environmental Virus or Symbiosis	**Chapter 1:** Environmental Relationships **Chapter 2:** Environmental Ethics **Chapter 12:** Land-Use Planning **Chapter 7:** Populations: Characteristics and Issues **Chapter 8:** Energy and Civilization: Patterns of Consumption	**Chapter 6:** Population Biology **Chapter 7:** Human Populations **Chapter 8:** Environmental Health and Toxicology **Chapter 11:** Biodiversity: Preserving Species **Chapter 12:** Biodiversity: Preserving Landscapes **Chapter 13:** Restoration Ecology **Chapter 22:** Urbanization and Sustainable Cities	**Chapter 2:** Environmental Systems **Chapter 6:** Environmental Conservation **Chapter 14:** Economics and Urbanization
Unit 3: The Geopolitical-Economy of Planetary Consumption	**Chapter 1:** Environmental Relationships **Chapter 3:** Environmental Risk **Chapter 8:** Energy and Civilization: Patterns of Consumption **Chapter 11:** Biodiversity Issues **Chapter 12:** Land-Use Planning **Chapter 19:** Environmental Policy and Decision Making	**Chapter 4:** Evolutions, Biological Communities and Species Interactions **Chapter 7:** Human Populations **Chapter 8:** Environmental Health and Toxicology **Chapter 9:** Food and Hunger **Chapter 13:** Restoration Ecology **Chapter 17:** Water Use and Management **Chapter 21:** Solid, Toxic and Hazardous Waste **Chapter 23:** Ecological Economics **Chapter 24:** Environmental Policy, Law, and Planning	**Chapter 6:** Environmental Conservation **Chapter 8:** Environmental Health and Toxicology **Chapter 15:** Environmental policy and Sustainability **Chapter 12:** Energy **Chapter 10:** Water: Resources and Pollution
Unit 4: The Whole Earth Café: Bellying Up to the Trough	**Chapter 8:** Energy and Civilization: Patterns of Consumption **Chapter 9:** Nonrenewable Energy Sources **Chapter 11:** Biodiversity Issues **Chapter 12:** Land-Use Planning **Chapter 13:** Soil and Its Uses **Chapter 15:** Water Management **Chapter 14:** Agricultural Methods and Pest Management	**Chapter 8:** Environmental Health and Toxicology **Chapter 9:** Food and Hunger **Chapter 10:** Farming: Conventional and Sustainable Practices **Chapter 11:** Biodiversity: Preserving Species **Chapter 12:** Biodiversity: Preserving Landscapes **Chapter 16:** Air Pollution **Chapter 19:** Conventional Energy **Chapter 17:** Water Use and Management **Chapter 18:** Water Pollution	**Chapter 5:** Biomes and Biodiversity **Chapter 8:** Environmental Health and Toxicology **Chapter 6:** Environmental Conservation **Chapter 13:** solid and Hazardous Waste **Chapter 10:** Water: Resources and Pollution **Chapter 12:** Energy **Chapter 7:** Food and Agriculture

Correlation Guide

Unit 5: The Consumption-sustainability Conundrum: Is There an Answer?	**Chapter 1:** Environmental Relationships **Chapter 3:** Environmental Risk **Chapter 2:** Environmental Ethics **Chapter 8:** Energy and Civilization: Patterns of Consumption **Chapter 19:** Environmental Policy and Decision-Making	**Chapter 1:** Understanding Our Environment **Chapter 7:** Human Populations **Chapter 10:** Farming: Conventional and Sustainable Practices **Chapter 13:** Restoration Ecology **Chapter 17:** Water Use and Management **Chapter 20:** Sustainable energy **Chapter 22:** Urbanization and Sustainable Cities **Chapter 23:** Ecological Economics **Chapter 25:** What Then Shall We Do?	**Chapter 6:** Environmental Conservation **Chapter 8:** Environmental Health and Toxicology **Chapter 9:** Air: Climate and Pollutions **Chapter 10:** Water: Resources and Pollution **Chapter 15:** Environmental policy and Sustainability **Chapter 12:** Energy

Topic Guide

This topic guide suggests how the selections in this book relate to the subjects covered in your course. You may want to use the topics listed on these pages to search the Web more easily.

On the following pages a number of websites have been gathered specifically for this book. They are arranged to reflect the units of this Annual Editions reader. You can link to these sites by going to www.mhhe.com/cls

All the articles that relate to each topic are listed below the bold-faced term.

Agriculture
17. Radically Rethinking Agriculture for the 21st Century
18. The Cheeseburger Footprint
19. The New Geopolitics of Food
21. Rethinking the Meat-Guzzler
25. Water Footprints of Nations: Water Use by People as a Function of Their Consumption Pattern
31. Rich Countries Launch Great Land Grab to Safeguard Food Supply
33. Development at the Urban Fringe and Beyond: Impacts on Agriculture and Rural Land

Behavior
1. The Psychological Roots of Resource Over-Consumption
2. Why Do We Over-Consume?
3. The Gospel of Consumption: And the Better Future We Left behind
4. Do We Consume Too Much?
5. How Much Should a Person Consume?
8. Consumption and Consumerism
10. The Human Factor
13. The Competitive Exclusion Principle
41. Consuming Passions: Everything That Can Be Done to Bring the Age of Heroic Consumption to Its Close Should Be Done

Biodiversity
34. The End of a Myth
36. Economic Report into Biodiversity Crisis Reveals Price of Consuming the Planet

Consumption
3. The Gospel of Consumption: And the Better Future We Left behind
6. Consumption, Not Population Is Our Main Environmental Threat
8. Consumption and Consumerism
33. Development at the Urban Fringe and Beyond: Impacts on Agriculture and Rural Land
39. Collaborative Consumption: Shifting the Consumer Mindset
41. Consuming Passions: Everything That Can Be Done to Bring the Age of Heroic Consumption to Its Close Should Be Done

Cultural evolution
1. The Psychological Roots of Resource Over-Consumption
4. Do We Consume Too Much?
40. Toward a New Consciousness: Values to Sustain Human and Natural Communities

Demography/Population
8. Consumption and Consumerism
9. People and the Planet: Executive Summary
10. The Human Factor
11. Global Aging and the Crisis of the 2020s
12. The New Population Bomb: The Four Megatrends That Will Change the World
21. Rethinking the Meat-Guzzler
37. Putting People in the Map: Anthropogenic Biomes of the World

Economics
14. Of the 1%, by the 1%, for the 1%
15. The New Economy of Nature
17. Radically Rethinking Agriculture for the 21st Century
19. The New Geopolitics of Food
28. The Myth of Mountaintop Removal Mining

29. The Efficiency Dilemma
30. Jevons' Paradox and the Perils of Efficient Energy Use
33. Development at the Urban Fringe and Beyond: Impacts on Agriculture and Rural Land
36. Economic Report into Biodiversity Crisis Reveals Price of Consuming the Planet

Ecosystem services
9. People and the Planet: Executive Summary
13. The Competitive Exclusion Principle
35. Land-Use Choices: Balancing Human Needs and Ecosystem Function

Ecophagy
Preface

Energy
7. The Issue: Natural Resources, What Are They?
17. Radically Rethinking Agriculture for the 21st Century
21. Rethinking the Meat-Guzzler
27. Eating Fossil Fuels
28. The Myth of Mountaintop Removal Mining
29. The Efficiency Dilemma
30. Jevons' Paradox and the Perils of Efficient Energy Use
33. Development at the Urban Fringe and Beyond: Impacts on Agriculture and Rural Land

Environmental decision-making
28. The Myth of Mountaintop Removal Mining
32. Global Urbanization: Can Ecologists Identify a Sustainable Way Forward?
33. Development at the Urban Fringe and Beyond: Impacts on Agriculture and Rural Land
35. Land-Use Choices: Balancing Human Needs and Ecosystem Function

Environmental stress
6. Consumption, Not Population Is Our Main Environmental Threat
15. The New Economy of Nature
18. The Cheeseburger Footprint
21. Rethinking the Meat-Guzzler
23. The World's Water Challenge
28. The Myth of Mountaintop Removal Mining
33. Development at the Urban Fringe and Beyond: Impacts on Agriculture and Rural Land
34. The End of a Myth
35. Land-Use Choices: Balancing Human Needs and Ecosystem Function
36. Economic Report into Biodiversity Crisis Reveals Price of Consuming the Planet

Environmental values
1. The Psychological Roots of Resource Over-Consumption
3. The Gospel of Consumption: And the Better Future We Left behind
8. Consumption and Consumerism
13. The Competitive Exclusion Principle
40. Toward a New Consciousness: Values to Sustain Human and Natural Communities
41. Consuming Passions: Everything That Can Be Done to Bring the Age of Heroic Consumption to Its Close Should Be Done

Internet References

The following Internet sites have been selected to support the articles found in this reader. These sites were available at the time of publication. However, because websites often change their structure and content, the information listed may no longer be available. We invite you to visit www.mhhe.com/cls for easy access to these sites.

Annual Editions: Environment 13/14

General Sources

About: Geography
http://geography.about.com

This website, created by the About network, contains hyperlinks to many specific areas of geography, including cartography, population, country facts, historic maps, physical geography, topographic maps, and many others.

The Association of American Geographers (AAG)
www.aag.org

Surf this site of the Association of American Geographers to learn about AAG projects and publications, careers in geography, and information about related organizations.

Britannica's Internet Guide
www.britannica.com

This site presents extensive links to material on world geography and culture, encompassing material on wildlife, human lifestyles, and the environment.

Center for Science in the Public Interest
www.cspinet.org

Browse this site, a strong advocate for nutrition and health, food safety, and sound science for information on these topics.

Earth Science Enterprise
www.earth.nasa.gov

Information about NASA's Mission to Planet Earth program and its Science of the Earth program and its Science of the Earth System can be found here. Surf to learn about satellites, El Niño, and even "strategic visions" of interest to environmentalists.

Economics: Complete Guide to Economic Resources on the Web
www.economics.miningco.com

This resource "mines the Net" for information on economic subjects. It includes a large number of links and online articles from economics magazines and journals.

EnviroLink
www.envirolink.org

One of the world's largest environmental information clearing houses, EnviroLink is a grassroots nonprofit organization that unites organizations' volunteers around the world and provides up-to-date information and resources.

EPA
www.epa.gov/wastes/conserve/index.htm

This is the EPA's website concerning recycling.

Global Climate Change, NASA's Eyes on Earth
www.climate.nasa.gov

This site is a remarkably informative and graphical discussion about climate change.

Library of Congress
www.loc.gov

Examine this extensive website to learn about resource tools, library services/resources, exhibitions, and databases in many different subfields of environmental studies.

National Geographic Society
www.nationalgeographic.com

Links to National Geographic's huge archive are provided here. There is a great deal of material related to the atmosphere, the oceans, and other environmental topics.

The New York Times
www.nytimes.com

Browsing through the archives of the New York Times will provide a wide array of articles and information related to the different subfields of the environment.

RealClimate
www.realclimate.org

Climate Science from Climate Scientists—now that makes sense.

Research and Reference (Library of Congress)
http://lcweb.loc.gov/rr

This research and reference of the Library of Congress will lead to invaluable information on different countries. It provides links to numerous publications, bibliographies, and guides in area studies that can be of great help to environmentalists.

SocioSite: Sociological Subject Areas
www.pscw.uva.nl/sociosite/TOPICS

This huge sociological site form the University of Amsterdam provides many discussion and references of interest to students of the environment, such as the links to information on ecology and consumerism.

Solstice: Documents and Databases
http://solstice.crest.org/index.html

In this online source for sustainable energy information, the Center for Renewable Energy and sustainable Technology (CREST) offers documents and databases on renewable energy, energy efficiency, and sustainable living. The site also offers related websites, case studies, and policy issues.

United States Geological Survey
www.usgs.gov

This site and its many links are replete with information and resources in environmental studies from explanations of El Niño to discussion of concerns about water resources.

United States Global Change Research Program
www.usgcrp.gov

This government program supports research on the interaction of natural and human-induced changes the global environment and their implication for study. Find details on the atmosphere, climate change, global carbon and water cycles, ecosystems, and land use plus human contributions and responses.

Internet References

Union of Concerned Scientists
www.ucsusa.org

The site Citizens and Scientists for practical solution to environmental problems

Wikipedia
www.wikipedia.org

This free encyclopedia has millions of articles contributed collaboratively using Wiki software. It is a remarkable resource created by users the world over.

UNIT 1: The Nature of Our Consumption: *Ecophagy*

Alliance for Global Sustainability (AGS)
www.global-sustainability.org

The AGS is a cooperative venture seeking solutions to today's urgent and complex environmental problems. Research teams from four universities study large-scale, multidisciplinary environmental problems that the world's ecosystems, economies, and societies face.

Collaborative Consumption Hub
http://collaborativeconsumption.com/

This site has information about the reinvention of old market behaviors—renting, lending, swapping, bartering, gifting—through technology, taking place on a scale and in ways never possible before.

The Dismal Scientist
www.dismal.com

Often referred to as the "best free lunch on the Web," this is an excellent site with many interactive features. It provides access to economic data, briefings on the current state of the economy, and original articles on economic issues.

Global Footprint Network
www.footprintnetwork.org/en/index.php/GFN/

Programs at Global Footprint Network are designed to influence decision makers at all levels of society and to create a critical mass of powerful institutions using the Footprint to put an end to ecological overshoot and restore balance to our economies.

United States Information Agency (USIA)
www.america.gov

USIA's home page provides definitions, related documentation, and discussions of topics of concern to students of global issues. The site addresses today's Hot Topics as well as ongoing issues that form the foundation of the field.

World Population and Demographic Data
http://geography.about.com/cs/worldpopulation/

On this site, you will find information about world population and demographic data for all the countries of the world.

UNIT 2: The Human Factor: Environmental Virus or Symbiosis?

Can Cities Save the Future?
www.huduser.org/publications/econdev/habitat/prep2.html

This press release about the second session of the Preparatory Committee for Habitat II is an excellent discussion of the question of global urbanization.

Earth Renewal
www.earthrenewal.org/global_economics.htm

This site lists various excellent sites that address economic environmental policies and solutions.

Geography and Socioeconomic Development
www.ksg.harvard.edu/cid/andes/Documents/Background%20Papers/Geography&Socioeconomic%20Development.pdf

John I. Gallup wrote this 19-page background paper examining the state of the Andean region. He explains the strong and pervasive effects geography has on economic and social development.

Global Footprint Network
www.footprintnetwork.org/en/index.php/GFN/

This website has a link to Africa Factbook 2009, an electronic set of some of the most up-to-date charts and recent statistics about human and environmental trends in Africa. The Factbook is compiled by The Global Footprint Network, the Swiss Agency for Development, and other sponsors.

Global Trends Project
www.globaltrendsproject.org

The Center for Strategic and International Studies explores the coming global trends and challenges of the new millennium. Read its summary report at this website. Also access Enterprises for the Environment, Global Information Infrastructure Commission, and Americas at this site.

Graphs Comparing Countries
http://humandevelopment.bu.edu/use_existing_index/start_comp_graph.cfm

This site allows you to compare the statistics of various countries and nation-states using a visual tool.

IISDnet
www.iisd.org

The International Institute for Sustainable Development, a Canadian organization, presents information through gateways entitled Business, Climate Change, Measurement and Assessment, and Natural Resources. IISD Linkages is its multimedia resource for environment and development policymakers.

Linkages on Environmental Issues and Development
www.iisd.ca/linkages

Linkages is a site provided by the International Institute for Sustainable Development. It is designed to be an electronic clearing house for information on past and upcoming international meetings related to both environmental issues and economic development in the developing world.

People & Planet
www.peopleandplanet.org

People & Planet is an organization of student groups at universities and colleges across the United kingdom. Organized in 1969 by students at Oxford University, it is now an independent pressure group campaigning on world poverty, human rights, and the environment.

Population Action International
www.populationaction.org

According to its mission statement, Population Action International is dedicated to advancing policies and programs that slow population growth in order to enhance the quality of life for all people.

Internet References

Population Reference Bureau
www.prb.org

This site provides data on the world population and census information.

SocioSite: Sociological Subject Areas
www.pscw.uva.nl/sociosite/TOPICS

This huge site provides many references of interest to those interested in global issues, such as links to information on ecology and the impact of consumerism.

UNIT 3: The Geopolitical-Economy of Planetary Consumption

National Geographic Society
www.nationalgeographic.com

This site provides links to National Geographic's huge archive of maps, articles, and other documents. There is a great deal of material related to social and cultural topics that will be of great value to those interested in the study of cultural pluralism.

Penn Library: Resources by Subject
www.library.upenn.edu/cgi-bin/res/sr.cgi

This vast site is rich in links to information about subjects of interest to students of global issues. Its extensive population and demography resources address such concerns as migration, family planning, and health and nutrition in various world regions.

Sustainable Development.Org
www.sustainabledevelopment.org

Extensive links at this site lead to such sustainable development categories as agriculture, energy, environment, finance, health, microenterprise, public policy, and technologies.

United Nations
www.unsystem.org

Visit this official website locator for the United Nations System of Organizations to get a sense of the scope of international environmental inquiry today. Various UN organizations concern themselves with everything from maritime law to habitat protection to agriculture.

United Nations Environment Programme (UNEP)
www.unep.ch

Consult this home page of UNEP for links to critical topics of concern to environmentalists, including desertification, migratory species, and the impact of trade on the environment. The site will direct you to useful databases and global resource information.

World Health Organization (WHO)
www.who.ch

The WHO's objective, according to its website, is the attainment of the highest possible level of health by all peoples. *Health,* as defined in the WHO constitution, is a state of complete physical, mental, and social well-being and not merely the absence of disease or infirmity.

World Resources Institute (WRI)
www.wri.org

The World Resources Institute is committed to change for a sustainable world and believes that change in human behavior is urgently needed to halt the accelerating rate of environmental deterioration in some areas. It sponsors not only the general website above but also The Environmental

Information Portal (www.earthtrends.wri.org), which provides a rich database on interaction between human disease, pollution, and large-scale environmental, development, and demographic issues.

WWW Virtual Library: Demography & Population Studies
http://demography.anu.edu.au/VirtualLibrary

A definitive guide to demography and population studies can be found at this site. It contains a multitude of important links to information about global poverty and hunger.

UNIT 4: The Whole Earth Café: Bellying Up to the Trough

Alternative Energy Institute (AEI)
www.altenergy.org

The AEI will continue to monitor the transition from today's energy forms to the future in a "surprising journey of twists and turns." This site is the beginning of an incredible journey.

Alternative Energy Institute, Inc.
www.altenergy.org

On this site created by a nonprofit organization, discover how the use of conventional fuels affects the environment. Also learn about research work on new forms of energy.

EnviroLink
www.envirolink.org

One of the world's largest environmental information clearing houses, EnviroLink is a grassroots nonprofit organization that unites organizations' volunteers around the world and provides up-to-date information and resources. Many links to fresh water issues are found here.

Endangered Species
www.endangeredspecie.com

This site provides a wealth of information on endangered species anywhere in the world. Links providing data on the causes, interesting facts, law issues, case studies, and other issues on endangered species are available.

Friends of the Earth
www.foe.co.uk/index.html

Friends of the Earth, a nonprofit organization based in the United Kingdom, pursues a number of campaigns to protect the Earth and its living creatures. This site has links to many important environmental sites, covering such broad topics as ozone depletion, soil erosion, and biodiversity.

Freshwater Society
www.freshwater.org

The mission of the Freshwater Society is to promote the conservation, protection, and restoration of all freshwater resources.

Greenpeace
www.greenpeace.org

Greenpeace is an international NGO (nongovernmental organization) that is devoted to environmental protection.

The Hunger Project
www.thp.org

Browse through the site of this nonprofit organization whose goal is the sustainable end to global hunger through leadership at all levels of society. The Hunger Project contends that the persistence of hunger is at the heart of the major security issues threatening our planet.

Internet References

National Geographic Society

www.nationalgeographic.com

National Geographic is developing a strategic approach to freshwater issues with the goal of motivating people across the world to care about and conserve fresh water. Links to National Geographic's huge archive are provided here. There is a great deal of material related to fresh water.

Natural Resources Defense Council

http://nrdc.org

The Natural Resources Defense Council (NRDC) uses law, science, and the support of more than 1 million members and activists to protect the planet's wildlife, plants, water, soils, and other resources. The site provides abundant information on global issues and political responses.

Smithsonian Institution

www.si.edu

Looking through this site, which will provide access to many of the enormous resources of the Smithsonian, offers a sense of the biological diversity that is threatened by humans' unsound environmental policies and practices.

The World's Water

www.worldwater.org

For more than a decade, the biennial report The World's Water has provided key data and expert insights into our most pressing freshwater issues.

United Nations Environment Program (UNEP)

www.enep.ch

Consult this home page of UNEP for links to critical topics of concern related to fresh water.

World Wildlife Federation (WWF)

www.wwf.org

This home page of the WWF leads to an extensive array of information links about endangered species, wildlife management and preservation, and more. It provides many suggestions on how to take an active part in protecting the biosphere.

UNIT 5: The Consumption-Sustainability Conundrum: Is There an Answer?

Alliance for Global Sustainability (AGS)

http://globalsustainability.org

The AGS is a cooperative venture seeking solutions to today's urgent and complex environmental problems. Research teams from four universities study large-scale, multidisciplinary environmental problems that the world's ecosystems, economies, and societies face.

The Earth Institute at Columbia University

www.earth.columbia.edu

The Earth Institute at Columbia University, led by Professor Jeffery D. Sachs, is dedicated to addressing a number of complex issues related to sustainable development and the needs of the world's poor.

Earth Pledge Foundation

www.earthpledge.org

The Earth Pledge Foundation promotes the principles and practices of sustainable development—the need to balance the desire for economic growth with the necessity of environmental protection.

Energy Justice Network

www.energyjustice.net/peak

Find information regarding this organization that advocates for "a clean energy, zero-emission, zero-waste future for all."

EnviroLink

http://envirolink.org

EnviorLink is committed to promoting a sustainable society by connecting individuals and organizations through the use of the World Wide Web.

Global Footprint Network (GFN)

www.footprintnetwork.org

The major focus of the GFN is to encourage the use of the Ecological Footprint as an accounting tool that measures how impact on nature.

Going Green

www.goinggreen.com

Browse this site that provides suggestions for simple behavior changes that support sustainability.

The Green Guide

http://environment.nationalgeographic.com/environment/green-guide/

The National Geographic site provides assorted information on sustainable living.

Grass-roots.org

www.grass-roots.org

This site describes innovative grassroots programs in the United States that have helped people better their communities.

National Center for Electronics Recycling

www.electronicsrecycling.org/Public/default.aspx

This site is dedicated to the development and enhancement of a national infrastructure for the recycling of used electronics in the United States.

Resources

www.etown.edu/vl

Surf this site and its extensive links to learn about specific countries and regions, to research various think tanks and international organizations, and to study such vital topics as international law, development, the international economy, human rights, and peacekeeping.

Sustainable Communities Online

www.sustainable.org

The site focuses primarily on buildings but also provides links to other useful sustainability sites.

Terrestrial Sciences

www.cgd.ucar.edu/tss

The Terrestrial Sciences Section (TSS) is part of the Climate and Global Dynamics (CGD) Division at the National Center for Atmospheric Research (NCAR) in Boulder, Colorado. Scientists in the section study land-atmosphere interactions, in particular surface forcing of the atmosphere, through model development, application and observational analyses. Here, you'll find a link of VEMAP, The Vegetation/Ecosystem Modeling and Analysis Project.

YouSustain

www.yousustain.com/footprint/actions

A calculator at this site allows you to calculate how change in 24 behaviors will affect the environment.

World Map

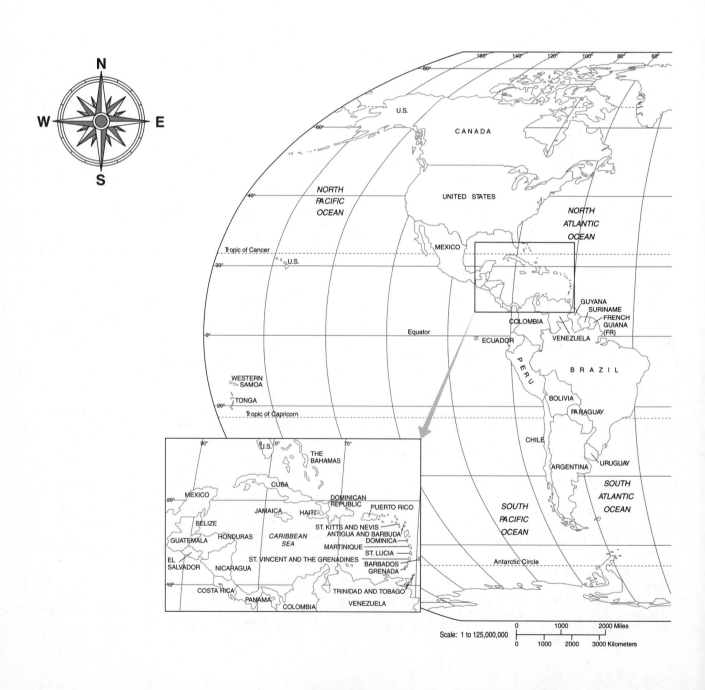

N
W   E
S

160° 140° 120° 100° 80° 60°
80°

U.S.

CANADA

60°

NORTH PACIFIC OCEAN

UNITED STATES

40°

NORTH ATLANTIC OCEAN

Tropic of Cancer

20°
U.S.

MEXICO

GUYANA
SURINAME
FRENCH GUIANA (FR)
COLOMBIA
VENEZUELA

Equator
0°
ECUADOR

P E R U

B R A Z I L

WESTERN SAMOA

BOLIVIA

TONGA
20°

PARAGUAY

Tropic of Capricorn

CHILE

SOUTH ATLANTIC OCEAN

ARGENTINA   URUGUAY

SOUTH PACIFIC OCEAN

Antarctic Circle

Inset

90°   0°   70°
U.S.

THE BAHAMAS

CUBA

20°
MEXICO

DOMINICAN REPUBLIC
PUERTO RICO

JAMAICA   HAITI

BELIZE

ST. KITTS AND NEVIS
ANTIGUA AND BARBUDA
DOMINICA

GUATEMALA   HONDURAS   CARIBBEAN SEA

MARTINIQUE
ST. LUCIA

EL SALVADOR   NICARAGUA   ST. VINCENT AND THE GRENADINES

BARBADOS
GRENADA

10°

COSTA RICA   PANAMA   TRINIDAD AND TOBAGO

COLOMBIA   VENEZUELA

Scale: 1 to 125,000,000

0   1000   2000 Miles

0   1000   2000   3000 Kilometers

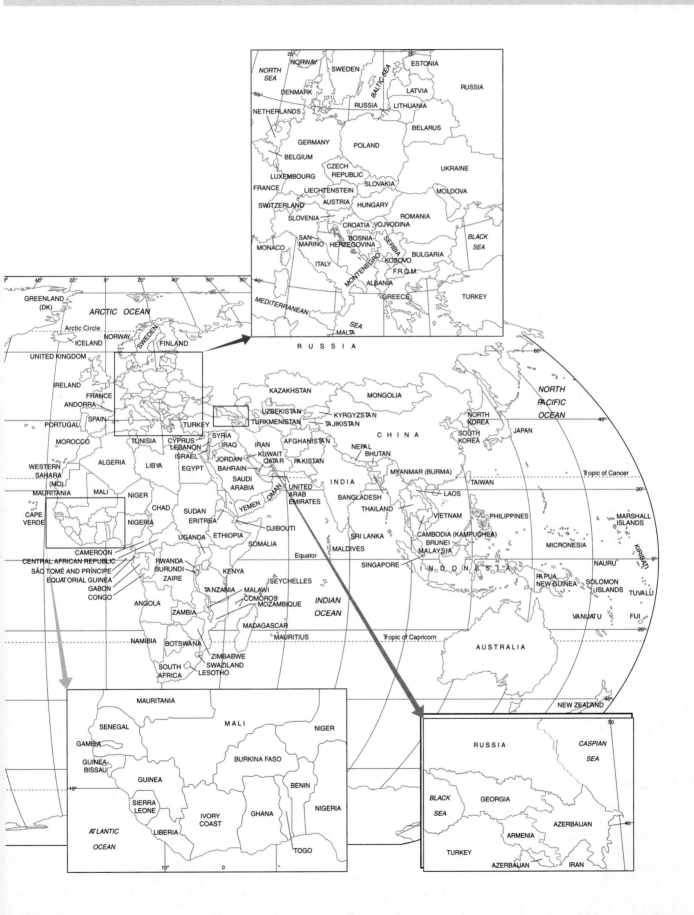

UNIT 1

The Nature of Our Consumption: *Ecophagy*

Unit Selections

Learning Outcomes

After reading this Unit, you will be able to:

- Describe the psychological roots of resource overconsumption and underlying drivers of resource depletion and planetary consumption.

- Describe the relationship between humans' predisposition for accumulating material possessions and how cultural evolution is argued to be able to change this behavior.

- Discuss how the American work ethic and erosion of human relationship values may be contributing to our current environmental consumption problems.

- Explain why there are such vast inequalities in global consumption and what that inequality may mean for environmental sustainability.

- Assess the question: "Do we consume too much?"

- Outline the reasons that global population growth itself is not our main environmental threat.

- Describe the current state of our existing natural resources and how the emergence of middle-class nations may change global patterns of resource use, extraction, and efficiency.

- Appraise the validity of the argument that human consumption and consumption patterns are the fundamental causes underlying all our present day environmental problems.

Student Website

www.mhhe.com/cls

Internet References

Alliance for Global Sustainability (AGS)
www.global-sustainability.org

Collaborative Consumption Hub
http://collaborativeconsumption.com/

The Dismal Scientist
www.dismal.com

Global Footprint Network
http://footprints@footprintnetwork.org

U.S. Information Agency (USIA)
www.america.gov

World Population and Demographic Data
http://geography.about.com/cs/worldpopulation/

The sun, the moon, and the stars would have disappeared long ago . . . had they happened to be within the reach of predatory human hands.

—Havelock Ellis, *the Dance of Life,* 1923

"Oh, how we consume our planet! How ravenous, how insatiable our appetite seems to be!" Such a sentiment seems to be shared by increasing numbers of Earth's scholars, scientists, philosophers, political and religious leaders, indigenous peoples, and global citizens. In this edition's exploration of **Consuming the Earth,** we open with examining the nature of our consumption in Unit I. And with that examination, I would like to introduce the term *ecophagy,* coined by Robert Freitas, a Senior Research Fellow, one of four researchers at the nonprofit foundation Institute for Molecular Manufacturing in Palo Alto, California. According to Freitas, *ecophagy* essentially refers to the consumption of an ecosystem. In the paper[1] that Freitas first introduced the word, he wrote,

Perhaps the earliest-recognized and best known danger of molecular nanotechnology is the risk that self-replicating nanorobots capable of functioning autonomously in the natural environment could quickly convert that natural environment (e.g. biomass) into replicas of themselves (e.g. nanomass) on a global basis, a scenario usually referred to as the "grey goo problem" but perhaps more properly termed global ecophagy.

The term *ecophagy,* therefore, I think describes well the current nature of our environmental consumption. And I think Articles 1 through 8 in Unit I, as well as the remaining Articles 9 though 44 in Units II through V lend substantial support to such a description. Not only is humanity gobbling up with impunity everything served at the Whole Earth Café—water, plants, animals, species, land, fossil fuels, and ecosystems—but also the line out the door keeps getting longer and longer as more and more hungry humans demand a seat at the buffet.

Fortunately, the growing global focus (or fad?) on our finding a sustainable relationship with Planet Earth appears to suggest that we may have to begin eating more sensibly. We are now recognizing, reluctantly at times (kicking and screaming at other times as we are hauled out of the Café), that if we continue our consumption misbehavior, the Whole Earth Café may very well be forced to shut its doors. And if the evidence is correct—both scientific and based on our personal daily observations—this recognition of a need to change our environmental dietary habits is not occurring a moment too soon. Earth has grown to more than 7 billion Whole Earth Café patrons. But while our species' birthrates are falling, before this era of population growth levels in 2050, there is a good chance habitat Earth will be home to more than 9 billion humans. Add to this the fact that we are seeing more and more of our offspring living to adulthood and seeing adults living longer lives, rates of earth resource consumption are destined to rise, and spread, as well. That's the *biological* aspect of humanity's consumption nature.

© McGraw-Hill Companies, Inc. Mark Dierker, photographer

On the *sociological* side of those consumption demographics we see that the percentage of humans who have achieved a decent standard of living is higher now than it has ever been. And the gap between the world's poorest and richest is now filling with a broad middle income group that had not existed half a century ago. But is that good news? Could it be possible that within such a "silver lining" may lie hidden an ominous cloud? The ten articles included in Unit I encourage the reader to seriously, and critically, asses the potential catastrophic impact that this 21st century "middle-income group" of the human species may have on the planet should their Earth consumption behavior be anything like that of their 20th century ancestors.

Article 1, the "Psychological Roots of Resource Over-Consumption," explores the question: Are humans hardwired to be resource over-consumers? Author Nate Hagens suggests that we are. He believes humans have an innate need for status and novelty. However, providing the material "stuff" necessary to meet such needs requires resource consumption and possible resource depletions and runs contrary to the spirit of sustainability.

In Article 2, "Why Do We Over Consume?" Derek Gondor thinks, as does Nate Hagens in Article 1, that is the result of our predisposition to "needing more stuff" and that this behavior dates as far back as our premodern roots. Again, the author's thesis contends that our original abundance of resources that allowed such copious consumption continues to diminish as global competition grows. But Gondor believes that cultural evolution may provide the solution.

Writer and activist Jeffrey Kaplan argues in Article 3, "The Gospel of Consumption," that our obsession with work (and overwork) and its associated high levels of productivity are linked directly to our compulsion to amass material wealth. Furthermore, this association is distracting us from making time to nurture the human relationships (nonmaterial production) necessary for sustaining a healthy planet.

A little more than 15 years ago, writer Mark Sagoff examined the now frequently asked question, "Do we consume too much?" in his essay by the same title (Article 4). He believed at that time the debate surrounding the answer to the question—between resource optimists and resource pessimists—was missing the real point and that neither side had it right. Sagoff argued that the "world has the wealth and resources to provide everyone the opportunity to live a decent life. . . . We consume too much when consumption becomes an end in itself and makes us lose affection and reverence for the natural world." Have the observations made in his article been validated 15 years later?

That there are vast inequalities in global consumption is well recognized. Also, the association between material production/consumption in industrialized countries and carbon emissions climate change is clearly documented. Given this information, the question of any conscientious global citizen would have to be then, "How much should a person consume?" In Article 5, Ramachandra Guha explores such a question but does not dictate a specific answer. In his analysis, he presents once again the evidence of how the industrialized North lays excessive claim to the "environmental space" of the South and also supports the fact that developing countries'/emerging economies' peoples (ecosystem people) would also like to enhance their own resource consumption to obtain some of the more positive fruits of the Industrial Revolution. Can everybody possibly enjoy the fruit? Probably. Can everybody in the world own a car? According to the author, never. Then how much should we consume? Far, far less.

Article 6, "Consumption, Not Population Is Our Main Environment Threat," is a brief essay by Fred Pearce inviting the reader to begin to reflect on and consider the possibility that human material consumption behavior, not human population growth, may be our greatest environmental threat. The author argues that by nearly every measure, a small proportion of the world's population consumes the majority of the world's resources, and as such, is responsible for the majority of its pollution (when we talk about consumption, we are talking also about carbon emissions). Who is this "small proportion"? Why are they (we?) eating us out of house and home? More critically, however, is to ask,

"Is this minority of the human population serving as a role model for the majority of its population? Or will the larger population of humans, as they evolve into middle class, instinctually become mega-omnivores like their predecessors?"

Article 7, "The Issue: Natural Resources, What Are They?" was compiled by the World Resource Forum and presents a brief review of our current state of natural resources, describes our current patterns of global use, resource extractions, resource efficiency, and concludes with some possible scenarios of future resource extraction. The article is included because it offers a graphic/cartographic visualization of the resources we consume and the inequitable global patterns of that consumption.

Finally, Article 8, "Consumption and Consumerism," by Anup Shah reports on a similar global consumption resource pattern scenario of extreme variability. Although the global wealth gap (consumption gap) is closing and more humans are enjoying better "living conditions," it will become even more important for us to be clear regarding our definition of "better living conditions." Are we referring to *higher standards of living* (increased material production/consumption) or *improved quality of life* (ensuring basic sustenance, self-esteem, and freedom). Consumption (associated with *higher standards of living*), according to the United Nations, is a leading cause of environmental degradation. Nonconsumptive behaviors (reflected in *improvements in quality of life*) such as education, democracy, freedom from tyranny, empowerment of women, ensuring human rights, and the like have yet to demonstrate any associative or causal relationship to environmental degradation.

In conclusion, Unit I encourages the reader to consider this: It may be a timeworn cliché, but during our evolution as a species, our transformation of the Earth, our ascension to civilization, and our metamorphosis from merely humans to humanity, have we lost sight of the forest for the trees?

Note

1. Robert Freitas (2000) *Some Limits to Global Ecophagy by Biovorous Nanoreplicators, with Public Policy Recommendations.*

The Psychological Roots of Resource Over-Consumption

Nate Hagens

The Psychological Roots of Resource Over-Consumption

Humans have an innate need for status and for novelty in their lives. Unfortunately, the modern world has adopted very energy—and resource—intensive ways of meeting those needs. Other ways are going to have to be found as part of the move to a more sustainable world.

Most people associate the word "sustainability" with changes to the supply side of our modern way of life such as using energy from solar flows rather than fossil fuels, recycling, green tech and greater efficiency. In this essay, however, I will focus on the demand-side drivers that explain why we continue to seek and consume more stuff.

When addressing 'demand-side drivers', we must begin at the source: the human brain. The various layers and mechanisms of our brain have been built on top of each other via millions and millions of iterations, keeping intact what 'worked' and adding via changes and mutations what helped the pre-human, pre-mammal organism to incrementally advance. Brain structures that functioned poorly in ancient environments are no longer around. Everyone reading this page is descended from the best of the best at both surviving and procreating which, in an environment of privation and danger where most 'iterations' of our evolution happened, meant acquiring necessary resources, achieving status and possessing brains finely tuned to natural dangers and opportunities.

This essay outlines two fundamental ways in which the evolutionarily derived reward pathways of our brains are influencing our modern overconsumption. First, financial wealth accumulation and the accompanying conspicuous consumption are generally regarded as the signals of modern success for our species. This gives the rest of us environmental cues to compete for more and more stuff as a proxy of our status and achievement. A second and more subtle driver is that we are easily hijacked by and habituated to novel stimuli. As we shall see, the prevalence of novelty today eventually demands higher and higher levels of neural stimulation, which often need increased consumption to satisfy. Thus it is this combination of pursuit of social status and the plethora of *novel activities* that underlies our large appetite for resource throughput.

Status

Evolution has honed and culled 'what worked' by combining the substrate of life with eons' worth of iterations. Modern biological research has focused on the concept of 'relative fitness', a term for describing those adaptations that are successful in propelling genes, or suites of genes, into the next generation and that will have out-competed those that were deleterious or did not keep up with environmental change. Though absolute fitness mattered to the individual organisms while they were alive, looking back it was 'relative fitness' that shaped the bodies and brains of the creatures on the planet today.

Status, both in humans and other species, has historically been a signaling mechanism that minimised the costs of competition, whether for reproductive opportunities or for material resources. If you place ten chickens in an enclosure there will ensue a series of fights until a pecking order is established. Each bird quickly learns who it can and cannot beat and a status hierarchy is created, thus making future fights (and wastes of energy) less common. Physical competition is costly behaviour that requires energy and entails risk of injury. Status is one way to determine who one can profitably challenge and who one cannot. In our ancestral environment, those men (and women) that successfully moved up the social hierarchy improved their mating and resource prospects. Those at the bottom of the status rung did not only possess fewer mating opportunities but many did not mate at all. Status among our ancestors was probably linked to those attributes providing consistent benefits to the tribe: hunting prowess, strength, leadership ability, storytelling skills etc. In modern humans, status is defined by what our modern cultures dictate. As we are living through an era of massive energy gain from fossil fuels, pure physical prowess has been replaced by digital wealth, fast cars, political connections, etc.

It follows that the larger a culture's resource subsidy (natural wealth), the more opportunity there is for 'status badges' uncorrelated with basic needs such as strength, intelligence, adaptability, stamina, etc. Though 'what' defines status may be culturally derived, status hierarchies themselves are part of our evolved nature. Ancestral hominids at the bottom of the mating pecking order, ceteris paribus, are not our ancestors. Similarly, many of our ancestors had orders of magnitude more descendants than others. For example, scientists recently discovered an odd geographical preponderance for a particular Y chromosome mutation which turns out to be originally descended from Genghis Khan. Given the 16 million odd male descendants alive today with this Y marker, Mr. Khan is theorised to have had 800,000 times the reproductive success than the average male alive on the planet in 1200 A.D. This does not imply that we are all pillagers and conquerors—only that various phenotypic expressions have had ample opportunity to become hardwired in our evolutionary past.[1]

Mating success is a key driver in the natural world. This is all studied and documented by evolutionary research into the theory of "sexual selection", which Charles Darwin once summarised as the effects of the "*struggle between the individuals of one sex, generally the males, for the possession of the other sex.*"[2] Biologists have shown that a primary way to reliably demonstrate one's 'quality' during courtship is to display a high-cost signal—e.g. a heavy and colourful peacock's tail, an energy-expending bird-song concert, or a $100,000 sports car.[3] These costly "handicap" signals are evolutionarily stable indicators of their producer's quality, because cheap signals are too easy for low-quality imitators to fake.[4]

In this sense 'waste' was an evolutionary selection! Think of three major drawbacks to a male peacock of growing such a hugely ornate tail:

1. the energy, vitamins and minerals needed to go into the creation of the tail could have been used for other survival/reproductive needs,
2. the tail makes the bird more likely to be spotted by a predator,
3. If spotted, the cumbersome tail makes escape from a predator less likely.

Overall, though, these negative "fitness hits" must have been outweighed by the drab female peahen's preference for males with larger, more ornate tails. With this filter, we can understand the rationale and prevalence of Veblen goods (named after the 19th-century economist who coined the term 'conspicuous consumption')—a group of commodities that people increasingly prefer to buy as their price gets higher because the greater price confers greater status. This biological precept of signalling theory is alive and well in the human culture.

Novelty

Modern man evolved from earlier hominids under conditions of privation and scarcity at least until about 10,000 years ago. The period since then has been too short a time to make a significant change to millions of years of prior neural sculpture. Nature made the brain's survival systems incredibly efficient. The brain is based on about 40% of all our available genes and consumes over 20% of our calorific intake. Incremental changes in how our brains recognise, process and react to the world around us either contributed to our survival and thus were carried forward, or died out.

Some changes affected *salience,* the ability to notice what is important, different or unusual. Salience recognition is part of what's called the mesolimbic dopamine reward pathway. This pathway is a system of neurons integral to survival efficiency, helping us to instantly decide what in the environment should command our attention. Historically, immediate feedback on what is 'new' was critical to both avoiding danger and procuring food. Because most of what happens around us each day is predictable, processing every detail of a familiar habitat wastes brain energy. Such activity would also slow down our mental computer so that what are now minor distractions could prove deadly. Thus our ancestors living on the African savanna paid little attention to the stable mountains on the horizon but were quick to detect any movement in the bush, on the plains, or at the riverbank. Those more able to detect and process 'novel cues' were more likely to obtain rewards needed to survive and pass on their suites of genes. Indeed, modern experimental removal of the (dopamine) receptor genes in animals causes them to reduce exploratory behaviour, a key variable related to inclusive fitness in animal biology.[5]

We are instinctually geared for individual survival—being both reward-driven, and curious. It was these two core traits that the father of economics himself, Adam Smith, predicted in *The Wealth of Nations* would be the drivers of world economic growth. According to Smith, uniting the twin economic engines of self-interest (which he termed self-love) and curiosity was ambition—"the competitive human drive for social betterment". About 70 years later, after reading Adam Smith's *Theory of Moral Sentiments,* Charles Darwin recognised the parallel between the pursuit of wealth in human societies and the competition for resources that occurred among animal species. Our market system of allocating resources and 'status' can therefore be seen as the natural social culmination for an intelligent species finding an abundance of resources.

But, as we shall soon see, the revered Scottish philosopher could not have envisioned heli-skiing, Starbucks, slot machines, Facebook, email and many other stimulating and pleasurable objects and activities that people engage in today and to which they so easily become accustomed.

The Mesolimbic Dopaminergic Reward System

Americans find prosperity almost everywhere, but not happiness. For them desire for well-being has become a restless burning passion which increases with satisfaction. To start with emigration was a necessity for them: now it is a sort of gamble, and they enjoy the sensations as much as the profit.

Alexis de Tocqueville, Democracy in America 1831

Traditional drug abuse happens because natural selection has shaped behaviour-regulation mechanisms that function via chemical transmitters in our brains.[6] Addicts can become habituated to the feelings they get from cocaine, heroin or alcohol, and they need to increase their consumption over time to get the same neurotransmitter highs. This same neural reward architecture is present in all of us when considering our ecological footprints: we become habituated via a positive feedback loop to the 'chemical sensations' we receive from shopping, keeping up with the Joneses (conspicuous consumption), pursuing more stock profits, and myriad other stimulating activities that a surplus of cheap energy has provided.

An explosion of neuroscience and brain-imaging research tells us that drugs of abuse activate the brain's dopamine reward system that regulates our ability to feel pleasure and be motivated for "more". When we have a great experience—a glance from a pretty girl, a lovemaking romp in the woods, a plate of fresh sushi, hitting 777 on a one-eyed bandit, catching a lunker pike, watching a sunset, hearing a great guitar riff etc.—our brain experiences a surge in the level of the neurotransmitter dopamine. We feel warm, 'in the zone' and happy. After a while, the extra dopamine gets flushed out of our system and we return to our baseline level. We go about our lives, looking forward to the next pleasurable experience. But the previous experience has been logged into our brain's limbic system, which, in addition to being a centre for pleasure and emotion, holds our memory and motivation circuitry.[7] We now begin to look forward to encores of such heady stimuli and are easily persuaded towards activities that promise such a chemical reprise. These desires have their beginnings outside our conscious awareness. Recent brain-imaging research shows that drug and sexual cues as brief as 33 milliseconds can activate the dopamine circuitry, even if a person is not conscious of the cues. Perhaps there are artistically shaped sexual images hidden in advertisements for whiskey after all. . .

Historically, this entire system evolved from the biological imperative of survival. Food meant survival, sex meant survival (of genes or suites of genes), and additional stockpiles of both provided success relative to others, both within and between species. There was a discrete payoff to waiting hours for some movement in the brush that signaled 'food', or the sound of a particular bird that circled a tree with a beehive full of honey, etc. Our pattern recognition system on the Pleistocene would have been a grab-bag of various environmental stimuli that 'excited' our brains towards action that correlated with resources (typically food). In sum, the brain's reward pathways record both the actual experience of pleasure as well as ensuring that the behaviours that led to it are remembered and repeated. *Irrespective of whether they are 'good' for the organism in the current context—they 'feel' good, which is the mechanism our brain has left us as a heritage of natural selection.*

The (Very Important) Mechanism of Habituation

Habituation—getting used to something—and subsequent substance abuse and addiction develops because of the way we learn. Learning depends crucially on the discrepancy between the prediction and occurrence of a reward. A reward that is fully predicted does not contribute to learning.[8] The important implication of this is that learning advances only to the extent to which something is unpredicted and slows progressively as a stimuli becomes more predictable.[9] *As such, unexpected reward is a core driver in how we learn, how we experience life, and how we consume resources.*

Dopamine activation has been linked with addictive, impulsive activity in numerous species. Dopamine is released within the brain not only to rewarding stimuli but also to those events that *predict* rewards. It has long been known that two groups of neurons, in the ventral tegmental and the substantia nigra pars compacta areas, and the dopamine they release, are critical for reinforcing certain kinds of behaviour. Neuroscientist Wolfram Schultz measured the activity of these dopamine neurons while thirsty monkeys waited for a tone which was followed by a squirt of fruit juice into their mouths. After a series of fixed, steady amounts of juice, the volume of juice was suddenly doubled. The rate of neuron firing went from about 3 per second to 80 per second. But after several trials, after the monkeys had become habituated to this new level of reward, their dopamine firing rate returned to the baseline rate of 3 firings per second after the squirt of juice. The monkeys had become habituated to the coming reward! The opposite happened when the reward was reduced without warning. The firing rate dropped dramatically, but eventually returned to the baseline rate of 3 firings per second.[10]

The first time we experience a drug or alcohol high, the amount of chemical we ingest often exceeds the levels of naturally occurring neurotransmitters in our bodies by an order of magnitude.[11] No matter how brief, that experience is stored in our neural homes for motivation and memory—the amygdala and hippocampus. Getting drunk with your friends, getting high on a ski-lift, removing the undergarments of a member of the opposite sex for the first time—all initially flood the brain with dopamine alongside a picture memory of the event chemically linked to the body's pleasurable response to it. As such we look forward to doing it again, not so much because we want to repeat the activity, but because we want to recreate that *'feeling'*.

But in a modern stimuli-laden culture, this process is easily hijacked. After each upward spike, dopamine levels again recede, eventually to below the baseline. The following spike doesn't go quite as high as the one before it. Over time, the rush becomes smaller, and the crash that follows becomes steeper. The brain has been fooled into thinking that achieving that high is equivalent to survival and therefore the 'consume' light remains on all the time. Eventually, the brain is forced to turn on a self-defence mechanism, reducing the production of dopamine altogether—thus weakening the pleasure circuits' intended function. At this point, an 'addicted' person is compelled to use the substance not to get high, but just to feel normal—since one's own body is producing little or no endogenous dopamine response. Such a person has reached a state of "anhedonia", or inability to feel pleasure via normal experiences. Being addicted also raises the risk of having depression; being depressed increases the risk of self-medicating, which then leads to addiction, etc. via positive feedback loops.

In sum, when exposed to novel stimuli, high levels of curiosity (dopamine) are generated, but it is the *unexpected reward* that causes their activation. If I order a fantastic array of sushi and the waiter brings me a toothpick and my check, I am going to have a plunge in dopamine levels which will create an immediate craving for food. It is this interplay between expected reward and reality that underlies much of our behavioural reactions. Ultimately, as it relates to resource consumption, repeated use of any dopamine-generating 'activity' eventually results in tolerance. Withdrawal results in lower levels of dopamine and continuous use is required to keep dopamine at normal levels, and even higher doses to get the 'high' levels of initial use. Consumers in rich nations are arguably reaching higher and higher levels of consumption tolerance. If there was such a thing as 'cultural anhedonia', we might be approaching it.

America and Addiction

It would be pretty hard to be addicted directly to oil; it's toxic, slimy and tastes really bad. But given the above background, we can see how it is possible to become addicted to the energy services that oil provides. Humans are naturally geared for individual survival—curious, reward-driven and self-absorbed—but modern technology has now become a vector for these cravings. Material wealth and the abundant choices available in contemporary US society are unique in human (or animal) experience; never before in the history of our species have so many enjoyed (used?) so much. Within a culture promoting 'more', it is no wonder we have so many addicts. High-density energy and human ingenuity have removed the natural constraints on our behaviour of distance, time, oceans and mountains. For now, these phenomena are largely confined to developed nations—people living in a hut in Botswana or a yurt in Mongolia cannot as easily be exposed to the 'hijacking stimuli' of an average westerner, especially one living in a big city in the West, like London or Los Angeles.

According to *Time Magazine,* July 2007,

- **2 million+** pathological gamblers
- **4 Million+** addicted to food
- **15 million+** compulsive shoppers
- **16 million+** addicted to sex or pornography
- **19 million+** addicted to alcohol (7.7%)
- **3.6 million** addicted to illegal drugs
- **71.5 million** addicted to nicotine
- **80–90% of adults** routinely ingest caffeine
- **But USA only has 300 million people!!**

Many activities in an energy-rich society unintentionally target the difference between expected and unexpected reward. Take sportfishing for example. If my brother and I are on a lake fishing and we get a bite, it sends a surge of excitement through our bodies—what kind of fish is it? How big is it? etc. We land an 8-inch perch! Great! A minute later we catch another 8 inch perch—wow, there must be a school! After 45 minutes of catching nothing but 8-inch perch, our brain comes to expect this outcome, and we need something bigger, or a

different species, to generate the same level of excitement, so we will likely move to a different part of the lake in search of 'bigger' and/or 'different' fish. (Though my brother claims he would never tire of catching 8-inch perch I think he's exaggerating). Recreational fishing is benign (if not to the fish), but one can visualise other more resource-intensive pastimes activating similar circuitry. New shoes, new cars, new vacations, new home improvements, new girlfriends are all present on the modern unexpected reward smorgasbord.

The habituation process explains how some initially benign activities can morph into things more destructive. Weekly church bingo escalates to $50 blackjack tables; the *Sports Illustrated* swimsuit edition results, several years down the road, in the monthly delivery (in unmarked brown packaging) of *Jugs* magazine or webcams locked in on a bedroom in Eastern Europe; youthful rides on a rollercoaster evolve into annual heli-skiing trips, etc. The World Wide Web is especially capable of hijacking our neural reward pathways. The 24/7 ubiquity and nearly unlimited options for distraction on the internet almost seem to be perfectly designed to hone in on our brains' g-spot. Shopping, pornography, gambling, social networking, information searches, etc. easily out-compete the non-virtual, more mundane (and necessary) activities of yesteryear. Repetitive internet use can be highly addictive, though psychiatrists in different countries are debating whether it is a true addiction. For better or worse, the first things I do in the morning is a) check what time it is, b) start the coffee machine then c) check my e-mail, to see what 'novelty' might be in my inbox. Bills to pay, and e-mails from people who are not important or interesting, wait until later in the day, or are forgotten altogether.

There are few healthy men on the planet today who do not respond in social settings to the attention of a high-status, attractive 20- to 30-something woman. This is *salient* stimuli, irrespective of the man's marital status. But here is one example of where nature and nurture mesh. Despite the fact that 99+% of our history was polygynous, modern culture precludes men from running around pell-mell chasing women; we have rules, laws, and institutions such as marriage. However, habituation to various matrimonial aspects combined with exposure to dozens or even hundreds of alternatives annually in the jet age may at least partially explain the 60%+ divorce rate in modern society.

The entire brain and behaviour story is far more complex than just one neurotransmitter but the pursuit of this particular 'substance' is clearly correlated with anxiety, obesity, and the general increasing of conspicuous consumption in our society. That dopamine is directly involved is pretty clear. Parkinson's Disease is a condition where dopamine is lacking in an area of the brain necessary for motor coordination. The drug, Mirapex, increases dopamine levels in that area of the brain, but since pills are not lasers, it also increases dopamine in other areas of the body, including (surprise) the reward pathways. There are numerous lawsuits currently pending by Parkinson's patients who after taking the drug, developed sex, gambling, shopping and overeating compulsions.[12]

Our brain can also be tricked by the food choices prevalent in an abundant-energy society. We evolved in situations where salt and sugar were rare and lacking and signaled nutrition. So now, when we taste Doritos or Ben and Jerry's Chocolate Fudge Brownie ice cream, our reward pathways say 'yes yes—this is good for you!!' Our 'rational' brain attempts to remind us of the science showing obesity comes from eating too much of the wrong type of foods, but often loses out to the desire of the moment. Fully 30% of Americans are now categorised as obese. And, since we are exporting our culture (via the global market system) to developing countries, it is no surprise that China is following in our footsteps. From 1991 to 2004 the percentage of adults who are overweight or obese in China increased from 12.9% to 27.3%.[13] Furthermore, we can become habituated to repeated presentation of the same food type; we quickly get tired of it and crave something different.[14] We like variety—in food and in other things. Finally, when we overstimulate the brain pleasure centres with highly palatable food, these systems adapt by decreasing their own activity. Many of us now require constant stimulation from palatable (fatty) food to avoid entering a persistent state of negative reward. It is this dynamic that has led scientists to recently declare that fatty foods such as cheesecake and bacon are addictive in the same manner as cocaine.[15] And as we shall see, both what we eat and experience not only alters our own health, but also makes it more difficult to act in environmentally benign ways.

Impulsivity, Discount Rates and Preparing for the Future

Overconsumption fueled by increasing neural high water marks is a problem enough in itself, but such widespread neural habituation also diminishes our ability to think and act about the coming societal transition away from fossil fuels. Economists measure how much **we prefer the present over the future** via something called a 'discount rate'. . . . A discount rate of 100% means we prefer the present completely and put no value on the future. A discount rate of 0% means we treat the future 1000 years from now equally the same as 5 minutes from now.

Certain types of people have steeper discount rates than others; in general, gamblers, drinkers, drug users, men (vs. women), low IQ scorers, risk-takers, those exhibiting cognitive load, etc. all tend to show more preference for small short-term rewards rather than waiting for larger, long-term ones.[16] On average, heroin addicts' discount rates are over double those of control groups. Furthermore, in tests measuring discount rates and preferences among opium addicts, opioid-dependent participants discounted delayed monetary rewards significantly more than did non-drug using controls. Also, the opioid-dependent participants discounted delayed opium significantly more than delayed money, more evidence that brain chemicals are central to an organism's behaviour and that money and other abstractions are secondary.[17] Research has also shown that subjects deprived of addictive substances have an even greater preference for immediate consumption over delayed gratification.[18]

Even if we are not snorting cocaine or binge drinking on a Tuesday night, in a world with so much choice and so many stimulating options vying for our attention, more and more of our time is taken up feeding neural compulsions. In any case,

facing large long-term risks like peak oil and climate change requires dedicated long-term thinking—so having neural wiring that, due to cultural stimuli, focuses more and more on the present instead, is a big problem.

The Fallacy of Reversibility A.K.A. "The Ratchet Effect"

Though our natural tendency is to want more of culturally condoned pursuits, many such desires do have negative feedbacks. For instance, I can only eat about three cheeseburgers before my stomach sends a signal to my brain that I am full—and at 4 or 5 my stomach and esophagus would fill to the level I couldn't physically eat another. However, this is not so with virtual wealth, or many of the "wanting" stimuli promoted in our economic 'more equals better' culture. Professor Juliet Schor of Boston University has demonstrated that irrespective of their baseline salary, Americans always say they'd like to make a little more the following year.[19] Similar research by UCLA economist Richard Easterlin (whose "Easterlin Paradox" points out that average happiness has remained constant over time despite sharp rises in GDP per capita.) followed a cohort of people over a 16-year period. The participants were asked at the onset to list 10 items that they desired (e.g. sports car, snowmobile, house, private jet, etc.) During the 16-year study, all age groups tested did acquire some/many of the things they originally desired. But in each case, their *desires increased more than their acquisitions.*[20] This phenomenon is termed the "Hedonic Treadmill". I believe this behaviour is at the heart of the Limits to Growth problem, and gives me less confidence that we are just going to collectively 'tighten our belts' when the events accompanying resource depletion get a little tougher. That is, unless we somehow change what it is that we want more of.

The Ratchet Effect is a term for a situation in which, once a certain level is reached, there is no going back, at least not all the way. In evolution the effect means once a suite of genes become ubiquitous in a population, there is no easy way to 'unevolve' it. A modern example of this is obesity—as we get fatter the body creates more lipocytes (cells composing adipose tissue). But this system doesn't work in reverse; even though we can lose some of the weight gain, the body can't eliminate these new cells—they are there to stay.

After peak oil/peak credit, the ratchet effect is likely to mean that any rules requiring a more equitable distribution of wealth will not be well received by those who amassed wealth and status when oil was abundant. In biology, we see that animals will expend more energy defending freshly gained territory than they would to gain it if it was unclaimed. In humans, the pain from losing money is greater than the pleasure of gaining it. Economists describe and quantify this phenomenon as the endowment effect and loss aversion. And, as an interesting but disturbing aside, recent research suggests that the dopamine that males receive during acts of aggression rivals that of food or sex.[20,21] All these different dynamics of 'what we have' and 'what we are used to' will come into play in a world with less resources available per head.

Old Brain, New Choices

Humans have always lived in the moment but our gradual habituation to substances and activities that hijack our reward system may be forcing us, in aggregate, to live so much for the present that we are ignoring the necessity for urgent societal change. Unwinding this cultural behaviour may prove difficult. The sensations we seek in the modern world are not only available and cheap, but most are legal, and the vast majority are actually condoned and promoted by our culture. If the rush we get from an accomplishment is tied to something that society rewards we call it ambition, if it is attached to something a little scary, then we label the individual a 'risk taker' and if it is tied to something illegal—only then have we become an 'addict' or substance abuser. So it seems culture has voted on which ways of engaging our evolutionarily derived neurotransmitter cocktails are 'good' to pursue.

Drug addiction is defined as "*the compulsive seeking and taking of a drug despite adverse consequences*". If we substitute the word 'resource' for 'drug', have we meaningfully violated or changed this definition? That depends on the definition of 'drug'. "*A substance that a person chemically comes to rely upon*" is the standard definition but ultimately it is any activity or substance that generates brain chemicals that we come to require/need. Thus, it is not crude oil's intrinsic qualities we crave but the biochemical sensations to which we have become accustomed arising from the use of its embodied energy.

Take stock trading for example. Neuroscience scans show that stock trading lights up the same brain areas as picking nuts and berries do in other primates.

I think people trade for

1. money/profit (to compete/move up the mating ladder),
2. the feeling of being 'right' (whether they ever spend the money or not) and
3. the excitement/dopamine they get from the unexpected nature of the market puzzle.

While these three are not mutually exclusive, it is not clear to me which objective dominates, especially among people who have already attained infinite wealth. (Technically, infinite wealth is their annual expenses divided by the interest rate on Treasury bills. This gives the sum of money that would provide them with an income to buy all they want forever). When I worked for Lehman Brothers, my billionaire clients seemed less 'happy' on average than the $30k-a-year clerks processing their trades. They had more exciting lives perhaps, but they were not happier; that is, their reward baseline reset to zero each morning irrespective of the financial wealth they had amassed in previous days or years,. They wanted 'more' because they were habituated to getting more—it was how they kept score. Clearly, unless you inherit, you don't get to be a billionaire if you are easily satisfied.

MRI scans show that objects associated with wealth and social dominance activate reward-related brain areas. In one study, people's anterior cingulate (a brain region linked to reward) had more blood and oxygen response to visual cues of sports cars than to limousines or small cars.[22]

If compulsive shopping was a rational process, and our choices were influenced only by need, then brand-name t-shirts would sell no better than less expensive shirts of equal quality. The truth is that many shopping decisions are biased by corporate advertising campaigns or distorted by a desire to satisfy some competitive urge or emotional need. For most of us, the peak 'neurotransmitter cocktail' is the moment we decide to buy that new 'item'. After a brief euphoria and a short respite, the clock starts ticking on the next craving/purchase.

Adaptation Executors

There is a shared mythology in America that we can each enjoy fame and opulence at the top of the social pyramid. 78% of Americans still believe that anybody in America can become rich and live the good life.[23] Although in our economic system, not everyone can be a Warren Buffet or Richard Branson—there are not enough resources—it is the carrot of potential reward that keeps people working 50 hours a week until they retire at 65. All cannot be first. All cannot be wealthy, which makes our current version of capitalism, given the finite resources of the planet, not dissimilar from a Ponzi scheme.

Envy for status is a strong motivator. Increasing evidence in the fields of psychology and economics shows that above a minimum threshold of income/wealth, it's one's relative wealth that matters, not absolute. In an analysis of more than 80,000 observations, the relative rank of an individual's income predicted the individual's general life satisfaction whereas absolute income and reference income had little to no effect.[24] The "aspiration gap" is economic-speak for the relative fitness/status drive towards who/what is at the top of the cultural status hierarchy. For decades (centuries?), China has had a moderate aspiration gap, but since the turbo-capitalist global cues have spread across Asia, hundreds of millions of Chinese have raised their pecuniary wealth targets.

Economist Robert Frank asked people in the US if they would prefer living in a 4,000-square-foot house where all the neighboring houses were 6,000 square feet or a 3,000-square-foot house where the surrounding houses were 2,000 square feet. The majority of people chose the latter—*smaller in absolute terms but bigger in relative size*. A friend of mine says that when he last visited Madagascar, the 5th poorest nation on earth, the villagers huddled around the one TV in the village watching the nation's most popular TV show *Melrose Place,* giving them a window of desire into Hollywood glitz and glamour, and a beacon to dream about and strive for. Recently, a prince in the royal family of U.A.E. paid $14 million for a licence plate with the single numeral "1". "I bought it because I want to be the best in the world", Saeed Abdul Ghafour Khouri explained. What environmental cues do the kids watching TV in the U.A.E. or the U.S. receive?

As a species, we are both cooperative and competitive depending on the circumstances, but it's very important to understand that our neurophysiological scaffolding was assembled during long mundane periods of privation in the ancestral environment. This is still not integrated into the Standard Social Science Model that forms the basis of most liberal arts

8

educations (and economic theory). A new academic study on relative income as a primary driver of life satisfaction had over 50 references, *none of which* linked to the biological literature on status, sexual selection or relative fitness. Furthermore, increasing cognitive neuroscience and evolutionary psychology research illustrates that we are not the self-interested 'utility maximisers' that economists claim, but are highly 'other regarding'—we care about other people's welfare as well as our own. Though high-perceived relative fitness is a powerful behavioural carrot, inequality has pernicious effects on societies; it erodes trust, increases anxiety and illness, and leads to excessive consumption.[25] Health steadily worsens as one descends the social ladder, even within the upper and middle classes.[26]

When a child is born, he has all the genetic material he will ever have. All his ancestors until that moment had their neural wiring shaped for fitness maximisation—but when he is born, his genes will interact with environment cues showing those ways to compete for status, respect, mating prospects, and resources etc. which are socially acceptable. From this point forward, the genes are 'fixed' and the infant goes through life as an *'adaptation executor'* NOT a fitness maximiser. What will a child born in the 21st century 'learn' to compete for? Historically, we have always pursued social status, though status has been measured in dramatically different ways throughout history. Currently, most people pursue money as a short-cut fitness marker, though some compete in other ways—politics, knowledge, etc. Thus, a large looming problem is that the Chinese and other rapidly developing nations don't just aspire to the wealth of average Americans—they want to go the whole hog to be millionaires.

Conclusions

We are a clever, ambitious species that evolved to live almost entirely off of solar flows. Eventually we worked out how to access stored sunlight in the form of fossil fuels which required very little energy/natural resource input to extract. The population and growth trajectory that ensued eventually oversatisfied the "more is better" mantra of evolution and we've now developed a habit of requiring more fossil fuels and more clever ways to use them every year. There also exists a pervasive belief that human ingenuity will create unlimited substitutes for finite natural resources like oil and water. Put simply, it is likely that our abundant natural resources are not only required, but will be taken for granted until they are gone.

This essay has explored some of the underlying drivers of resource depletion and planetary consumption: more humans competing for more stuff that has more novelty. The self-ambition and curiosity that Adam Smith hailed as the twin engines of economic growth have been quite effective over the past 200 years. But Adam Smith did caution in *Moral Sentiments* that human envy and a tendency toward compulsions, if left unchecked, would undermine the empathic social relationships that would be essential to the successful long-term operation of free markets. Amidst so much novel choice and pressure to create wealth, we are discovering some uncomfortable facts, backed up by modern neurobiology, that confirm his concerns.

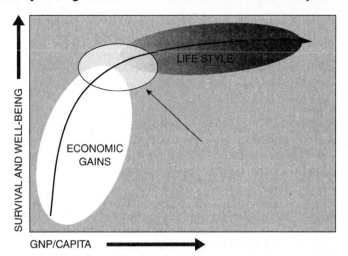

Meeting in the middle? The arrowed circle on this Inglehart Curve represents the highest level of well-being/survival consistent with a low level of resource use. It is therefore a target at which a society should aim.

Source: N. Hagens and R. Inglehart 1997

In an era of material affluence, when wants have not yet been fully constrained by limited resources, the evidence from our ongoing American experiment conclusively shows that humans have trouble setting limits on our instinctual cravings. What's more, our rational brains have quite a hard time acknowledging this uncomfortable but glaring fact.

This essay undoubtedly raises more questions than it answers. If we can be neurally hijacked, what does it suggest about television, advertising, media, etc? The majority of the neuro-economic sources I used in writing this were a *byproduct* of studies funded by neuromarketing research! How does 'rational utility' function in a society where we are being expertly marketed to pull our evolutionary triggers to funnel the money upwards? How does Pareto optimality—the assumption that all parties to an exchange will be made better off—hold up when considering neuro-economic findings? Recent studies show that American young people (between ages of 8–18) use 7.5 hours of electronic media (internet, Ipod, Wii, etc) per day and, thanks to multi-tasking, had a total of 11 hours 'gadget' exposure per day![27] The children with the highest hours of use had markedly poorer grades and more behavioural problems. How will these stimuli-habituated children adapt to a world of fewer resources?

Not all people pursue money, but our cultural system does. An unbridled pursuit of profits has created huge disparities in digitally amassed monetary wealth both within and between nations, thus holding a perpetually unattainable carrot in front of most of the world's population. So it is not just the amount we consume that is unsustainable, but also the message we send to others, internationally, nationally and in our neighbourhoods.

At the same time, traditional land, labour and capital inputs have been subsidised by the ubiquity of cheap energy inputs, and more recently by a large increase in both government and private debt, a spatial and temporal reallocator of resources. These cheap energy/cheap credit drivers will soon be a thing of the past, and this will curtail future global growth aspirations. When this happens, and we face the possibility of currency reform and what it might mean to start afresh with the same resources but a new basket of claims and assumptions, we will need to remember the neural backdrops of competition for relative status, and how people become habituated to high neural stimuli. Perhaps, given the supply-side limits and neural aspirations, some new goals can be attempted at lower absolute levels of consumption by at least partially lowering the amplitude of social rank.

We cannot easily change our penchant to want more. We can only change cultural cues on how we define the 'more' and thereby reduce resource use. In the cross-cultural study referenced in the diagram above, we can see that well-being increases only slightly as GNP increases above some minimum threshold. The arrowed circle would be a logical place for international policymakers concerned about planetary resource and sink capacity to aim to reach via taxes, disincentives to conspicuous consumption and subsidies. However, I fear that nations and governments will do little to slow their consumption and will get increasingly locked into defending the status quo instead.

In a society with significant overall surpluses, people who actively lower their own economic and ecological footprint might get by very well because their relative status—which is typically above average—allows them to make such reductions without reaching limits that compromise their well-being. As these people allocate time and resources away from financial marker capital and towards social, human, built and natural capital, they have an opportunity to redefine what sort of 'wealth' we compete for and thus potentially lead by example. However, personal experience with people in the lifestyle section of the chart leads me to believe that they will probably continue to pursue more resources and status even if it doesn't improve their well-being.

Put aside peak oil and climate change for the moment. Though it is difficult, we have it in us as individuals and as a culture to make small changes to the way our brains get 'hijacked' and, as a result, achieve more benign consequences. For example, we can choose to go for a jog/hike instead of sending ten emails and websurfing, we can choose to have a salad instead of a cheeseburger, we can choose to play a game or read a story with our children instead of making business phone calls. But most of these types of choices require both prior planning and discipline if our brains are not to fall into the neural grooves that modern culture has created. It takes conscious plans to change these behaviours, and for some this will be harder than for others But in choosing to do so, besides slowing and eventually reversing the societal stimulation feedback loop, we are likely to make ourselves healthier and happier. In neuro-speak, many of the answers facing a resource-constrained global society involve the rational neo-cortex suppressing and overriding the primitive and stronger limbic impulses.

So, ultimately, we must start to address new questions. In addition to asking source/sink questions like 'how much do we have' we should begin asking questions like 'how much is enough?' Reducing our addictive behaviours collectively will make it easier to face the situations likely to arise during an energy descent. Changing the environmental cues on what we compete for, via taxes or new social values, will slow down resource throughput and give future policymakers time to forge a new economic system consistent with our evolutionary heritage and natural resource balance sheet. We will always seek status and have hierarchies in human society but unless we first understand and then integrate our various demand-side constraints into our policies, culture and institutions, sustainability will be another receding horizon. Though there is probably no blanket policy to solve our resource crisis that would both work and gain social approval, an understanding of the main points of this essay might be a springboard to improve one's own happiness and well-being. Which would be a start. . . .

Critical Thinking

1. What role does status play in the human species? Is the achievement of status different for different people in different parts of the world?

2. How can the habitation process lead to destructive environmental behaviors?

3. How does an energy-abundant society contribute to over-consumption in their quest for status?

4. Summarize the basic psychological roots of resource over-consumption.

Endnotes

1. news.nationalgeographic.com/news/2003/02/0214_030214_genghis.html.

2. Darwin, C. (1871) *The Descent of Man and Selection in Relation to Sex* John Murray, London.

3. Miller, G. F. (1999). "Sexual selection for cultural displays" in R. Dunbar, C. Knight, & C. Power (Eds.), *The evolution of culture.* Edinburgh U. Press, pp. 71–91.

4. Zahavi, A. and Zahavi, A. (1997). *The handicap principle: a missing piece of Darwin's puzzle.* Oxford University Press.

5. Dulawa et al, "Dopamine D4 Receptor-Knock-Out Mice Exhibit Reduced Exploration of Novel Stimuli", *Journal of NeuroScience,* 19:9550–9556, 1999.

6. Gerald, M. S. & Higley, J. D. (2002) "Evolutionary Underpinnings of Excessive Alcohol Consumption". *Addiction,* 97, 415–425.

7. Whybrow, Peter, "American Mania".

8. Waelti, P., Dickinson, A. and Schultz, W.: "Dopamine responses comply with basic assumptions of formal learning theory". *Nature* 412: 43–48, 2001.

9. Rescorla R.A., Wagner A.R., "A theory of Pavlovian conditioning: Variations in the effectiveness of reinforcement and nonreinforcement" in: *Classical Conditioning II: Current Research and Theory* (Eds Black A.H., Prokasy W.F.) New York: Appleton Century Crofts, pp. 64–99, 1972.

10. Schultz, W., et al., "A Neural Substrate of Prediction and Reward", *Science,* 275:1593–1599.

11. Dudley, R. (2002) "Fermenting Fruit and the Historical Ecology of Ethanol Ingestion: Is Alcoholism in Modern Humans an Evolutionary Hangover?" *Addiction,* 97, 381–388.

12. Dodd et al., "Pathological Gambling Caused by Drugs Used to Treat Parkinson Disease", *Arch Neurol.* 2005;62:1377–1381.

13. Popkin, Barry. "The World Is Fat", *Scientific American,* September, 2007, pp. 94. ISSN 0036-8733.

14. Ernst, M., Epstein, L. "Habituation of Responding for Food in Humans", *Appetite* Volume 38, Issue 3, June 2002, Pages 224–234.

15. Johnson, P., Kenny, P., "Addiction-Like Reward Dysfunction and Compulsive Eating in Obese Rats: Role for Dopamine D2 Receptors", *Nature: Neuroscience* 3/28/2010.

16. Chablis et al, "Intertemporal Choice"—*The New Palgrave Dictionary of Economics,* 2007.

17. Madden et al., "Impulsive and Self-Control Choices in Opioid-Dependent Patients and Non-Drug Using Control Participants: Drug and Monetary Rewards", *Environmental and Clinical Psychopharmacology* (1997), vol 5 no 3 256–262.

18. Giorodano, L et al, "Mild opioid deprivation increases the degree that opioid-dependent outpatients discount delayed heroin and money", *Psychopharmacology* (2002) 163: 174–182.

19. Schor, Juliet, *The Overspent American: Why We Want What We Don't Need,* Harper Perennial 1999.

20. Easterlin, Richard "Explaining Happiness" September 4, 2003, 10.1073/pnas.1633144100 (Especially Table 3).

21. Couppis, M., Kennedy C., "The rewarding effect of aggression", *Psychopharmacology,* Volume 197, Number 3 / April, 2008.

22. Erk, S, M Spitzer, A Wunderlich, L Galley, H Walter "Cultural objects modulate reward circuitry." *Neuroreport.* 2002 Dec 20;13 (18):2499–503 12499856.

23. Samuelson, Robert, "Ambition and it Enemies" *Newsweek* Aug 23, 1999.

24. Boyce, C., et al, "**Money and Happiness—Rank of Income, Not Income, Affects Life Satisfaction**", *Psychological Science* Feb 2010.

25. Wilkinson, Richard; Pickett, Kate "*The Spirit Level—Why Greater Equality Makes Societies Stronger*", Bloomsbury Press 2010.

26. Marmot, Michael, "*The Status Syndrome: How Social Standing Affects Our Health and Longevity*", Holt Publishing 2005.

27. **Generation M2 – Media in the Lives of 8–18 Year Olds,** Kaiser Family Foundation 2010According to *Time Magazine,* July 2007,From Ms., Summer 2007, pp. 41–45. Copyright © 2007 by Rebecca Clarren. Reprinted by permission of Ms. Magazine.

Why Do We Over-Consume?

Darek Gondor

Jared Diamond famously stated that "the biggest problems facing the world today are not at all beyond our control, rather they are all of our own making, and entirely in our power to deal with" when talking about his book Collapse: How Societies Choose to Fail or Succeed.

But why have human ingenuity, technology, knowledge, and wealth grown step in step with unsustainability? If you compare the Human Development Index with resource use, we can see that as soon as countries meet the development standard of "high human development" they inevitably cross the line of unsustainability.

Opponents of this view will say that human well-being has on average increased in the world. However, while this is true, the indicators for species extinctions, habitat loss, greenhouse gas emissions and resource depletion have all been negative for a prolonged period of time.

Personal consumption data is even more telling. When the richest 10% account for 60% of all private consumption, we have to ask ourselves if these top-tier consumers could possibly improve their well-being any further through material gains?

Back to Our Pre-Modern Roots

Researchers like E.O. Wilson explain this paradox with a theory rarely incorporated into decisions—evolution.

The characteristics of human behaviour that became fixed in our population through natural selection occurred over the 95% of our pre-modern existence where we lived in sparsely populated hunter-gatherer bands with local community connections. Then the resource problem was one of local access.

Early human societies had primitive and inefficient ways of collecting resources, so those that thrived were ones that developed high rates of consumption and new innovations for resource gathering. They also had built up strong identity with their own community and competitiveness with others, and short-term thinking (discounting the future).

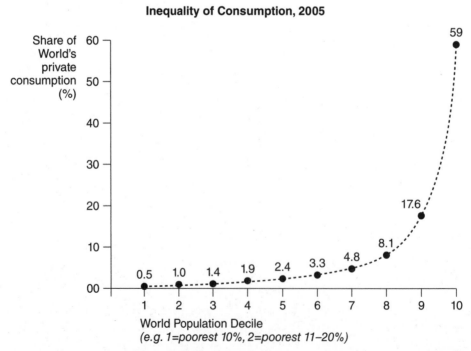

Inequality of Consumption, 2005

Share of World's private consumption (%)

World Population Decile
(e.g. 1=poorest 10%, 2=poorest 11–20%)

Source: World Bank Development Indicators 2008

Why Do We Always Need More Stuff?

Those characteristics endure today, in concrete but perhaps increasingly extraneous ways.

One of the basic human needs is food: the accumulation of which, along with other resources, is directly linked to the ability to reproduce and provide for a family. In pre-agricultural times, it was unlikely that a single family or tribe could gather enough food to make any further consumption undesirable, so there was little need for the evolution of a trait to limit consumption.

The second greatest human need was to secure a partner for reproduction. Unsurprisingly, it seems that those that could secure more resources through their hunting skills or status also had the best choice of mates. Research shows that women in all cultures, more than men, prefer partners with higher social status and that flaunting what you've got helps to seal the deal, so to speak.

Acquiring enough resources is not the end of it. Status is a comparative mark (dependent on one's immediate peers), and relates to competition. Often cited research by the likes of J.F. Helliwell shows that happiness levels peak at an income level of $10,000 per year in the US, after which happiness is determined by one's relative affluence.

That most of us want to earn more is therefore very well explained by sexual selection: the process of choosing a sexual partner. Some individual male birds—a species whose mating relationships most resemble humans—will spend a great deal of energy building elaborate, colourful and useless displays on the forest floor to attract females. But in doing so they signal to the female counterpart that they get along just fine nonetheless: a sign of a healthy, capable individual.

Researchers think that people buy yachts, numerous cars and expensive jewelry in the same way. This over-consumption pattern just gets more intense as we move up the social ladder, and seems to have little to do with satisfying living needs. That is, when we become successful enough to own yachts and expensive cars, the absolute amount of possessions does not dull the drive to consume—because we tend to hang out with other people who own yachts and expensive cars and they put a damper on our relative status.

Today, advertising and marketing professionals exploit this drive, as they do many traits of human nature, to keep the consumption train going. This may in part explain the continued wealth disparities between individuals.

Why Can't We All Just Get Along?

Competition is closely linked with consumption, as it produces social hierarchies among members of a group depending on their ability to secure resources. The idea of "us" and "them" was a very important one when humans lived in territorial bands and formed allegiances against common enemies.

Membership in a group provided security against aggression from other groups and means to cooperate. Internally, there was still a hierarchy that enabled the strong to control relationships and resources.

From the perspective of evolutionary fitness, the strongest individual had the opportunity to pass along the most genes, while receiving the protection of the group. Since evolutionary pressures act on individuals, competition and consumption do not have a shut-off point when the survival of the species is at stake, and there are many examples of human societies (think Easter Island) that likely competed themselves to extinction.

Today, we can look at political divisions to see how competing loyalties and different identities stall our efforts at cooperation. One reason why the United Nations organisation could not unanimously interfere in acts deplored by members (like genocide in Sudan or Rwanda) is that it is a collection of leaders whose allegiances lie elsewhere, such as with their in-group that provides security and shares commonalities like language, religion or culture.

In federal states like Canada, the national government has little power to enforce national policies in Alberta, where oil sand development is provincial jurisdiction, even though it may impact aboriginal communities falling under national protection. There inter-provincial and intergovernmental competition is a defining feature of that country's politics.

Is it any wonder we are just now beginning to attempt to halt carbon dioxide emissions, nearly 20 years after the need was demonstrated? How will any deal reached at COP15 in Copenhagen be implemented in countries with competing subnational identities?

With new global problems like poverty, climate change and biodiversity loss, we are now being asked to be global citizens, and care about those we have never met, and areas we will never visit. This runs counter to our evolutionary past.

Evolution of Culture and Ideas

But are we slaves to our genes? No serious biologist believes that is the case regarding behaviour, we simply have genetic predispositions to do some things and not others. So the question is: how can we put our ingrained traits to benefit, or even overcome them?

There are certainly ways that human characteristics can be considered and utilised in working towards sustainable future paths.

The melting Antarctic ice sheet—no matter how bleak the images on TV—does not seem able to provoke wide enough behaviour change, because most of us can all go back to our daily lives, unaffected. Our individual interests have to be tapped to create the political support for implementing progressive ideas, and one way to do this is with money. The recent call for rich nations to put up at least $10 billion a year to entice developing countries into an agreement at COP15 is a starter because dollars can easily be mentally translated into benefits.

Regulation can also be swallowed more easily with the aid of self-interest—like the very recent EPA announcement that GHGs are health dangers, clearing the way for laws restricting their release.

Other ways of playing to individual interests is through reputation—rewarding and shaming. Or by setting an extreme baseline for policy and then intermittently moving it back. For example, like closing the tuna fishery and then opening it back up slowly. Expectations for improvement from an undesirable baseline can be more acceptable than an unsustainable benefit with dire future predictions.

Another manner of influencing behaviour is perhaps the most obvious. Environmental education for the world's children, that builds on human-nature relationships, is indispensible to nurturing an identity that recognises the intrinsic value of nature and equality of cultures.

Such education is bound to pay off. We often point to that which makes humans unique—our language, intelligence, art and culture—as the root of cultural evolution. In other words, the development and passing down of ideas and values by societies that have lots of spare time after they have met their livelihood needs.

There is evidence that shows we increasingly live for ourselves, forego reproduction, enjoy life past reproductive age (thanks to the evolution of menopause), turn to cooperation over conflict, and choose partners based on humour and personality—traits that may not be indicators of reproductive success and survival.

Cultural evolution is quicker and can be more powerful than our ingrained instincts. Our modern environment has changed from locally centered to global, and biologically we have not caught up. Our ideas have to make up the difference.

Critical Thinking

1. How can the richest 10 percent possibly improve their well-being any further through material gain?

2. Why do we always need so much stuff? Why do you personally? Do all humans all over the world have the same desire to accumulate material wealth?

3. How long can we maintain this level of overconsumption as more and more countries become "middle class"?

4. How can education change a culture of overconsumption?

The Gospel of Consumption

And the Better Future We Left behind

JEFFREY KAPLAN

PRIVATE CARS WERE RELATIVELY SCARCE in 1919 and horse-drawn conveyances were still common. In residential districts, electric streetlights had not yet replaced many of the old gaslights. And within the home, electricity remained largely a luxury item for the wealthy.

Just ten years later things looked very different. Cars dominated the streets and most urban homes had electric lights, electric flat irons, and vacuum cleaners. In upper-middle-class houses, washing machines, refrigerators, toasters, curling irons, percolators, heating pads, and popcorn poppers were becoming commonplace. And although the first commercial radio station didn't begin broadcasting until 1920, the American public, with an adult population of about 122 million people, bought 4,438,000 radios in the year 1929 alone.

But despite the apparent tidal wave of new consumer goods and what appeared to be a healthy appetite for their consumption among the well-to-do, industrialists were worried. They feared that the frugal habits maintained by most American families would be difficult to break. Perhaps even more threatening was the fact that the industrial capacity for turning out goods seemed to be increasing at a pace greater than people's sense that they needed them.

It was this latter concern that led Charles Kettering, director of General Motors Research, to write a 1929 magazine article called "Keep the Consumer Dissatisfied." He wasn't suggesting that manufacturers produce shoddy products. Along with many of his corporate cohorts, he was defining a strategic shift for American industry—from fulfilling basic human needs to creating new ones.

In a 1927 interview with the magazine *Nation's Business,* Secretary of Labor James J. Davis provided some numbers to illustrate a problem that the *New York Times* called "need saturation." Davis noted that "the textile mills of this country can produce all the cloth needed in six months' operation each year" and that 14 percent of the American shoe factories could produce a year's supply of footwear. The magazine went on to suggest, "It may be that the world's needs ultimately will be produced by three days' work a week."

Business leaders were less than enthusiastic about the prospect of a society no longer centered on the production of goods. For them, the new "labor-saving" machinery presented not a vision of liberation but a threat to their position at the center of power. John E. Edgerton, president of the National Association of Manufacturers, typified their response when he declared: "I am for everything that will make work happier but against everything that will further subordinate its importance. The emphasis should be put on work—more work and better work." "Nothing," he claimed, "breeds radicalism more than unhappiness unless it is leisure."

By the late 1920s, America's business and political elite had found a way to defuse the dual threat of stagnating economic growth and a radicalized working class in what one industrial consultant called "the gospel of consumption"—the notion that people could be convinced that however much they have, it isn't enough. President Herbert Hoover's 1929 Committee on Recent Economic Changes observed in glowing terms the results: "By advertising and other promotional devices . . . a measurable pull on production has been created which releases capital otherwise tied up." They celebrated the conceptual breakthrough: "Economically we have a boundless field before us; that there are new wants which will make way endlessly for newer wants, as fast as they are satisfied."

Today "work and more work" is the accepted way of doing things. If anything, improvements to the labor-saving machinery since the 1920s have intensified the trend. Machines *can* save labor, but only if they go idle when we possess enough of what they can produce. In other words, the machinery offers us an opportunity to work less, an opportunity that as a society we have chosen not to take. Instead, we have allowed the owners of those machines to define their purpose: not reduction of labor, but "higher productivity"—and with it the imperative to consume virtually everything that the machinery can possibly produce.

FROM THE EARLIEST DAYS of the Age of Consumerism there were critics. One of the most influential was Arthur Dahlberg, whose 1932 book *Jobs, Machines, and Capitalism* was well known to policymakers and elected officials in Washington. Dahlberg declared that "failure to shorten the length of the working day . . . is the primary cause of our rationing of opportunity, our excess industrial plant, our enormous wastes of competition, our high pressure advertising, [and] our economic imperialism." Since much of what industry produced was no longer aimed at satisfying human physical needs, a four-hour workday, he claimed, was necessary to prevent society from becoming disastrously materialistic. "By not shortening the working day when all the wood is in," he suggested, the profit motive becomes "both the creator and satisfier of spiritual needs." For when the profit motive can turn nowhere else, "it wraps our soap in pretty boxes and tries to convince us that that is solace to our souls."

There was, for a time, a visionary alternative. In 1930 Kellogg Company, the world's leading producer of ready-to-eat cereal, announced that all of its nearly fifteen hundred workers would move from an eight-hour to a six-hour workday. Company president Lewis Brown and owner W. K. Kellogg noted that if the company ran "four six-hour shifts . . . instead of three eight-hour shifts, this will give work and paychecks to the heads of three hundred more families in Battle Creek."

This was welcome news to workers at a time when the country was rapidly descending into the Great Depression. But as Benjamin Hunnicutt explains in his book *Kellogg's Six-Hour Day,* Brown and Kellogg wanted to do more than save jobs. They hoped to show that the "free exchange of goods, services, and labor in the free market would not have to mean mindless consumerism or eternal exploitation of people and natural resources." Instead "workers would be liberated by increasingly higher wages and shorter hours for the final freedom promised by the Declaration of Independence—the pursuit of happiness."

To be sure, Kellogg did not intend to stop making a profit. But the company leaders argued that men and women would work more efficiently on shorter shifts, and with more people employed, the overall purchasing power of the community would increase, thus allowing for more purchases of goods, including cereals.

A shorter workday did entail a cut in overall pay for workers. But Kellogg raised the hourly rate to partially offset the loss and provided for production bonuses to encourage people to work hard. The company eliminated time off for lunch, assuming that workers would rather work their shorter shift and leave as soon as possible. In a "personal letter" to employees, Brown pointed to the "mental income" of "the enjoyment of the surroundings of your home, the place you work, your neighbors, the other pleasures you have [that are] harder to translate into dollars and cents." Greater leisure, he hoped, would lead to "higher standards in school and civic . . . life" that would benefit the company by allowing it to "draw its workers from a community where good homes predominate."

It was an attractive vision, and it worked. Not only did Kellogg prosper, but journalists from magazines such as *Forbes* and *BusinessWeek* reported that the great majority of company employees embraced the shorter workday. One reporter described "a lot of gardening and community beautification, athletics and hobbies . . . libraries well patronized and the mental background of these fortunate workers . . . becoming richer."

A U.S. Department of Labor survey taken at the time, as well as interviews Hunnicutt conducted with former workers, confirm this picture. The government interviewers noted that "little dissatisfaction with lower earnings resulting from the decrease in hours was expressed, although in the majority of cases very real decreases had resulted." One man spoke of "more time at home with the family." Another remembered: "I could go home and have time to work in my garden." A woman noted that the six-hour shift allowed her husband to "be with 4 boys at ages it was important."

Those extra hours away from work also enabled some people to accomplish things that they might never have been able to do otherwise. Hunnicutt describes how at the end of her interview an eighty-year-old woman began talking about ping-pong. "We'd get together. We had a ping-pong table and all my relatives would come for dinner and things and we'd all play ping-pong by the hour." Eventually she went on to win the state championship.

Many women used the extra time for housework. But even then, they often chose work that drew in the entire family, such as canning. One recalled how canning food at home became "a family project" that "we all enjoyed," including her sons, who "opened up to talk freely." As Hunnicutt puts it, canning became the "medium for something more important than preserving food. Stories, jokes, teasing, quarreling, practical instruction, songs, griefs, and problems were shared. The modern discipline of alienated work was left behind for an older . . . more convivial kind of working together."

This was the stuff of a human ecology in which thousands of small, almost invisible, interactions between family members, friends, and neighbors create an intricate structure that supports social life in much the same way as topsoil supports our biological existence. When we allow either one to become impoverished, whether out of greed or intemperance, we put our long-term survival at risk.

Our modern predicament is a case in point. By 2005 per capita household spending (in inflation-adjusted dollars) was twelve times what it had been in 1929, while per capita spending for durable goods—the big stuff such as cars and appliances—was thirty-two times higher. Meanwhile, by 2000 the average married couple with children was working almost five hundred hours a year more than in 1979. And according to reports by the Federal Reserve Bank in 2004 and 2005, over 40 percent of American families spend more than they earn. The average household carries $18,654 in debt, not including home-mortgage debt, and the ratio of household debt to income is at record levels, having roughly doubled over the last two decades. We are quite literally working ourselves into a frenzy just so we can consume all that our machines can produce.

Yet we could work and spend a lot less and still live quite comfortably. By 1991 the amount of goods and services produced for each hour of labor was double what it had been in 1948. By 2006 that figure had risen another 30 percent. In other words, if as a society we made a collective decision to get by on the amount we produced and consumed seventeen years ago, we could cut back from the standard forty-hour week to 5.3 hours per day—or 2.7 hours if we were willing to return to the 1948 level. We were already the richest country on the planet in 1948 and most of the world has not yet caught up to where we were then.

Rather than realizing the enriched social life that Kellogg's vision offered us, we have impoverished our human communities with a form of materialism that leaves us in relative isolation from family, friends, and neighbors. We simply don't have time for them. Unlike our great-grandparents who passed the time, we spend it. An outside observer might conclude that we are in the grip of some strange curse, like a modern-day King Midas whose touch turns everything into a product built around a microchip.

Of course not everybody has been able to take part in the buying spree on equal terms. Millions of Americans work long hours at poverty wages while many others can find no work at all. However, as advertisers well know, poverty does not render one immune to the gospel of consumption.

Meanwhile, the influence of the gospel has spread far beyond the land of its origin. Most of the clothes, video players, furniture, toys, and other goods Americans buy today are made in distant countries, often by underpaid people working in sweatshop conditions. The raw material for many of those products comes from clearcutting or strip mining or other disastrous means of extraction. Here at home, business activity is centered on designing those products, financing their manufacture, marketing them—and counting the profits.

KELLOGG'S VISION, DESPITE ITS POPULARITY with his employees, had little support among his fellow business leaders. But Dahlberg's book had a major influence on Senator (and future Supreme Court justice) Hugo Black who, in 1933, introduced legislation requiring a thirty-hour workweek. Although Roosevelt at first appeared to support Black's bill, he soon sided with the majority of businessmen who opposed it. Instead, Roosevelt went on to launch a series of policy initiatives that led to the forty-hour standard that we more or less observe today.

By the time the Black bill came before Congress, the prophets of the gospel of consumption had been developing their tactics and techniques for at least a decade. However, as the Great Depression deepened, the public mood was uncertain, at best, about the proper role of the large corporation. Labor unions were gaining in both public support and legal legitimacy, and the Roosevelt administration, under its New Deal program, was implementing government regulation of industry on an unprecedented scale. Many corporate leaders saw the New Deal as a serious threat. James A. Emery, general counsel for the National Association of Manufacturers (NAM), issued a "call to arms" against the "shackles of irrational regulation" and the "back-breaking burdens of taxation," characterizing the New Deal doctrines as "alien invaders of our national thought."

In response, the industrial elite represented by NAM, including General Motors, the big steel companies, General Foods, DuPont, and others, decided to create their own propaganda. An internal NAM memo called for "re-selling all of the individual Joe Doakes on the advantages and benefits he enjoys under a competitive economy." NAM launched a massive public relations campaign it called the "American Way." As the minutes of a NAM meeting described it, the purpose of the campaign was to link "free enterprise in the public consciousness with free speech, free press and free religion as integral parts of democracy."

Consumption was not only the linchpin of the campaign; it was also recast in political terms. A campaign booklet put out by the J. Walter Thompson advertising agency told readers that under "private capitalism, the *Consumer,* the *Citizen* is boss," and "he doesn't have to wait for election day to vote or for the Court to convene before handing down his verdict. The consumer 'votes' each time he buys one article and rejects another."

According to Edward Bernays, one of the founders of the field of public relations and a principal architect of the American Way, the choices available in the polling booth are akin to those at the department store; both should consist of a limited set of offerings that are carefully determined by what Bernays called an "invisible government" of public-relations experts and advertisers working on behalf of business leaders. Bernays claimed that in a "democratic society" we are and should be "governed, our minds . . . molded, our tastes formed, our ideas suggested, largely by men we have never heard of."

NAM formed a national network of groups to ensure that the booklet from J. Walter Thompson and similar material appeared in libraries and school curricula across the country. The campaign also placed favorable articles in newspapers (often citing "independent" scholars who were paid secretly) and created popular magazines and film shorts directed to children and adults with such titles as "Building Better Americans," "The Business of America's People Is Selling," and "America Marching On."

Perhaps the biggest public relations success for the American Way campaign was the 1939 New York World's Fair. The fair's director of public relations called it "the greatest public relations program in industrial history," one that would battle what he called the "New Deal propaganda." The fair's motto was "Building the World of Tomorrow," and it was indeed a forum in which American corporations literally modeled the future they were determined to create. The most famous of the exhibits was General Motors' 35,000-square-foot Futurama, where visitors toured Democracity, a metropolis of multilane highways that took its citizens from their countryside homes to their jobs in the skyscraper-packed central city.

For all of its intensity and spectacle, the campaign for the American Way did not create immediate, widespread, enthusiastic support for American corporations or the corporate vision of the future. But it did lay the ideological groundwork for changes that came after the Second World War, changes that established what is still commonly called our post-war society.

The war had put people back to work in numbers that the New Deal had never approached, and there was considerable fear that unemployment would return when the war ended. Kellogg workers had been working forty-eight-hour weeks during the war and the majority of them were ready to return to a six-hour day and thirty-hour week. Most of them were able to do so, for a while. But W. K. Kellogg and Lewis Brown had turned the company over to new managers in 1937.

The new managers saw only costs and no benefits to the six-hour day, and almost immediately after the end of the war they began a campaign to undermine shorter hours. Management offered workers a tempting set of financial incentives if they would accept an eight-hour day. Yet in a vote taken in 1946, 77 percent of the men and 87 percent of the women wanted to return to a thirty-hour week rather than a forty-hour one. In making that choice, they also chose a fairly dramatic drop in earnings from artificially high wartime levels.

The company responded with a strategy of attrition, offering special deals on a department-by-department basis where eight hours had pockets of support, typically among highly skilled male workers. In the culture of a post-war, post-Depression U.S., that strategy was largely successful. But not everyone went along. Within Kellogg there was a substantial, albeit slowly dwindling group of people Hunnicutt calls the "mavericks," who resisted longer work hours. They clustered in a few departments that had managed to preserve the six-hour day until the company eliminated it once and for all in 1985.

The mavericks rejected the claims made by the company, the union, and many of their co-workers that the extra money they could earn on an eight-hour shift was worth it. Despite the enormous difference in societal wealth between the 1930s and the 1980s, the language the mavericks used to explain their preference for a six-hour workday was almost identical to that used by Kellogg workers fifty years earlier. One woman, worried about the long hours worked by her son, said, "He has no time to live, to visit and spend time with his family, and to do the other things he really loves to do."

Several people commented on the link between longer work hours and consumerism. One man said, "I was getting along real good, so there was no use in me working any more time than I had to." He added, "Everybody thought they were going to get rich when they got that eight-hour deal and it really didn't make a big difference. . . . Some went out and bought automobiles right quick and they didn't gain much on that because the car took the extra money they had."

The mavericks, well aware that longer work hours meant fewer jobs, called those who wanted eight-hour shifts plus overtime "work hogs." "Kellogg's was laying off people," one woman commented, "while some of the men were working really fantastic amounts of overtime—that's just not fair." Another quoted the historian Arnold Toynbee, who said, "We will either share the work, or take care of people who don't have work."

PEOPLE IN THE DEPRESSION-WRACKED 1930s, with what seems to us today to be a very low level of material goods, readily chose fewer work hours for the same reasons as some of their children and grandchildren did in the 1980s: to have more time for themselves and their families. We could, as a society, make a similar choice today.

But we cannot do it as individuals. The mavericks at Kellogg held out against company and social pressure for years, but in the end the marketplace didn't offer them a choice to work less and consume less. The reason is simple: that choice is at odds with the foundations of the marketplace itself—at least as it is currently constructed. The men and women who masterminded the creation of the consumerist society understood that theirs was a political undertaking, and it will take a powerful political movement to change course today.

Bernays's version of a "democratic society," in which political decisions are marketed to consumers, has many modern proponents. Consider a comment by Andrew Card, George W. Bush's former chief of staff. When asked why the administration waited several months before making its case for war against Iraq, Card replied, "You don't roll out a new product in August." And in 2004, one of the leading legal theorists in the United States, federal judge Richard Posner, declared that "representative democracy . . . involves a division between rulers and ruled," with the former being "a governing class," and the rest of us exercising a form of "consumer sovereignty" in the political sphere with "the power not to buy a particular product, a power to choose though not to create."

Sometimes an even more blatant antidemocratic stance appears in the working papers of elite think tanks. One such example is the prominent Harvard political scientist Samuel Huntington's 1975 contribution to a Trilateral Commission report on "The Crisis of Democracy." Huntington warns against an "excess of democracy," declaring that "a democratic political system usually requires some measure of apathy and non-involvement on the part of some individuals and groups." Huntington notes that "marginal social groups, as in the case of the blacks, are now becoming full participants in the political system" and thus present the "danger of overloading the political system" and undermining its authority.

According to this elite view, the people are too unstable and ignorant for self-rule. "Commoners," who are viewed as factors of production at work and as consumers at home, must adhere to their proper roles in order to maintain social stability. Posner, for example, disparaged a proposal for a national day of deliberation as "a small but not trivial reduction in the amount of productive work." Thus he appears to be an ideological descendant of the business leader who warned that relaxing the imperative for "more work and better work" breeds "radicalism."

As far back as 1835, Boston workingmen striking for shorter hours declared that they needed time away from work to be good citizens: "We have rights, and we have duties to perform as American citizens and members of society." As those workers well understood, any meaningful democracy requires citizens who are empowered to create and re-create their government, rather than a mass of marginalized voters who merely choose from what is offered by an "invisible" government. Citizenship requires a commitment of time and attention, a commitment people cannot make if they are lost to themselves in an ever-accelerating cycle of work and consumption.

We can break that cycle by turning off our machines when they have created enough of what we need. Doing so will give us an opportunity to re-create the kind of healthy communities that were beginning to emerge with Kellogg's six-hour day, communities in which human welfare is the overriding concern rather than subservience to machines and those who own them. We can create a society where people have time to play together as well as work together, time to act politically in their common interests, and time even to argue over what those common interests might be. That fertile mix of human relationships is necessary for healthy human societies, which in turn are necessary for sustaining a healthy planet.

If we want to save the Earth, we must also save ourselves from ourselves. We can start by sharing the work *and* the wealth. We may just find that there is plenty of both to go around.

Critical Thinking

1. Outline briefly how the American work ethic has contributed to the "gospel of consumption."

2. How was Kellogg's "six-hour day" idea intended originally to avoid the mindless consumerism and exploitation of people and resources that he feared our work obsession would create?

3. Why is working too much not good for societies or for the environment?

4. Do you think different cultures (rural, urban, agricultural, etc) have different "work ethics" and different work goals? Explain.

5. Can the industrialized world's obsession with work and production be compatible with the ideals of sustainability? Explain.

Kaplan, Jeffrey. From *Orion*, May/June 2008, pp. 1–8. Copyright © 2008 by Jeffrey Kaplan. Reprinted by permission of the author.

Do We Consume Too Much?

MARK SAGOFF

In 1994, when delegates from around the world gathered in Cairo for the International Conference on Population and Development, representatives from developing countries protested that a baby born in the United States will consume during its lifetime twenty times as much of the world's resources as an African or an Indian baby. The problem for the world's environment, they argued, is overconsumption in the North, not overpopulation in the South.

Discussions of the future of the planet are dominated by those who believe that an expanding world economy will use up natural resources and those who see no reasons, environmental or otherwise, to limit economic growth. Neither side has it right

---

Consumption in industrialized nations "has led to overexploitation of the resources of developing countries," a speaker from Kenya declared. A delegate from Antigua reproached the wealthiest 20 percent of the world's population for consuming 80 percent of the goods and services produced from the earth's resources.

Do we consume too much? To some, the answer is self-evident. If there is only so much food, timber, petroleum, and other material to go around, the more we consume, the less must be available for others. The global economy cannot grow indefinitely on a finite planet. As populations increase and economies expand, natural resources must be depleted; prices will rise, and humanity—especially the poor and future generations at all income levels—will suffer as a result.

Other reasons to suppose we consume too much are less often stated though also widely believed. Of these the simplest—a lesson we learn from our parents and from literature since the Old Testament—may be the best: although we must satisfy basic needs, a good life is not one devoted to amassing material possessions; what we own comes to own us, keeping us from fulfilling commitments that give meaning to life, such as those to family,

friends, and faith. The appreciation of nature also deepens our lives. As we consume more, however, we are more likely to transform the natural world, so that less of it will remain for us to appreciate.

The reasons for protecting nature are often religious or moral. As the philosopher Ronald Dworkin points out, many Americans believe that we have an obligation to protect species which goes beyond our own well-being; we "think we should admire and protect them because they are important in themselves, and not just if or because we or others want or enjoy them." In a recent survey Americans from various walks of life agreed by large majorities with the statement "Because God created the natural world, it is wrong to abuse it." The anthropologists who conducted this survey concluded that "divine creation is the closest concept American culture provides to express the sacredness of nature."

During the nineteenth century preservationists forthrightly gave ethical and spiritual reasons for protecting the natural world. John Muir condemned the "temple destroyers, devotees of ravaging commercialism" who "instead of lifting their eyes to the God of the mountains, lift them to the Almighty dollar." This was not a call for better cost-benefit analysis: Muir described nature not as a commodity but as a companion. Nature is sacred, Muir held, whether or not resources are scarce.

Philosophers such as Emerson and Thoreau thought of nature as full of divinity. Walt Whitman celebrated a leaf of grass as no less than the journeywork of the stars: "After you have exhausted what there is in business, politics, conviviality, love, and so on," he wrote in *Specimen Days,* and "found that none of these finally satisfy, or permanently wear—what remains? Nature remains." These philosophers thought of nature as a refuge from economic activity, not as a resource for it.

Today those who wish to protect the natural environment rarely offer ethical or spiritual reasons for the policies they favor. Instead they say we are running out of resources or causing the collapse of ecosystems on which we depend. Predictions of resource scarcity appear objective and scientific, whereas pronouncements that nature is sacred or that greed is bad appear judgmental or even embarrassing in a secular society. Prudential and economic arguments, moreover, have succeeded better than moral or spiritual ones in swaying public policy.

These prudential and economic arguments are not likely to succeed much longer. It is simply wrong to believe that nature sets physical limits to economic growth—that is, to prosperity and the production and consumption of goods and services on which it is based. The idea that increasing consumption will inevitably lead to depletion and scarcity, as plausible as it may seem, is mistaken both in principle and in fact. It is based on four misconceptions.

Misconception No. 1: We Are Running Out of Raw Materials

In the 1970s Paul Ehrlich, a biologist at Stanford University, predicted that global shortages would soon send prices for food, fresh water, energy, metals, paper, and other materials sharply higher. "It seems certain," Paul and Anne Ehrlich wrote in *The End of Affluence* (1974), "that energy shortages will be with us for the rest of the century, and that before 1985 mankind will enter a genuine age of scarcity in which many things besides energy will be in short supply." Crucial materials would near depletion during the 1980s, Ehrlich predicted, pushing prices out of reach. "Starvation among people will be accompanied by starvation of industries for the materials they require."

Things have not turned out as Ehrlich expected. In the early 1990s real prices for food overall fell. Raw materials—including energy resources—are generally more abundant and less expensive today than they were twenty years ago. When Ehrlich wrote, economically recoverable world reserves of petroleum stood at 640 billion barrels. Since that time reserves have *increased* by more than 50 percent, reaching more than 1,000 billion barrels in 1989. They have held steady in spite of rising consumption. The pre-tax real price of gasoline was lower during this decade than at any other time since 1947. The World Energy Council announced in 1992 that "fears of imminent [resource] exhaustion that were widely held 20 years ago are now considered to have been unfounded."

The World Resources Institute, in a 1994–1995 report, referred to "the frequently expressed concern that high levels of consumption will lead to resource depletion and to physical shortages that might limit growth or development opportunity." Examining the evidence, however, the institute said that "the world is not yet running out of most nonrenewable resources and is not likely to, at least in the next few decades." A 1988 report from the Office of Technology Assessment concluded, "The nation's future has probably never been less constrained by the cost of natural resources."

It is reasonable to expect that as raw materials become less expensive, they will be more rapidly depleted. This expectation is also mistaken. From 1980 to 1990, for example, while the prices of resource-based commodities declined (the price of rubber by 40 percent, cement by 40 percent, and coal by almost 50 percent), reserves of most raw materials increased. Economists offer three explanations.

First, with regard to subsoil resources, the world becomes ever more adept at discovering new reserves and exploiting old ones. Exploring for oil, for example, used to be a hit-or-miss proposition, resulting in a lot of dry holes. Today oil companies can use seismic waves to help them create precise computer images of the earth. New methods of extraction—for example, using bacteria to leach metals from low-grade ores—greatly increase resource recovery. Reserves of resources "are actually functions of technology," one analyst has written. "The more advanced the technology, the more reserves become known and recoverable."

Second, plentiful resources can be used in place of those that become scarce. Analysts speak of an Age of Substitutability and point, for example, to nanotubes, tiny cylinders of carbon whose molecular structure forms fibers a hundred times as strong as steel, at one sixth the weight. As technologies that use more-abundant resources substitute for those needing less-abundant ones—for example, ceramics in place of tungsten, fiber optics in place of copper wire, aluminum cans in place of tin ones—the demand for and the price of the less-abundant resources decline.

One can easily find earlier instances of substitution. During the early nineteenth century whale oil was the preferred fuel for household illumination. A dwindling supply prompted innovations in the lighting industry, including the invention of gas and kerosene lamps and Edison's carbon-filament electric bulb. Whale oil has substitutes, such as electricity and petroleum-based lubricants. Whales are irreplaceable.

Third, the more we learn about materials, the more efficiently we use them. The progress from candles to carbon-filament to tungsten incandescent lamps, for example, decreased the energy required for and the cost of a unit of household lighting by many times. Compact fluorescent lights are four times as efficient as today's incandescent bulbs and last ten to twenty times as long. Comparable energy savings are available in other appliances: for example, refrigerators sold in 1993 were 23 percent more efficient than those sold in 1990 and 65 percent more efficient than those sold in 1980, saving consumers billions in electric bills.

Amory Lovins, the director of the Rocky Mountain Institute, has described a new generation of ultralight automobiles that could deliver the safety and muscle of today's cars but with far better mileage—four times as much in prototypes and ten times as much in projected models (see "Reinventing the Wheels," January, 1995, *Atlantic*). Since in today's cars only 15 to 20 percent of the fuel's energy reaches the wheels (the rest is lost in the engine and the transmission), and since materials lighter and stronger than steel are available or on the way, no expert questions the feasibility of the high-mileage vehicles Lovins describes.

Computers and cameras are examples of consumer goods getting lighter and smaller as they get better. The game-maker Sega is marketing a hand-held children's game, called Saturn, that has more computing power than the 1976 Cray supercomputer, which the United States tried to keep out of the hands of the Soviets. Improvements that extend the useful life of objects also save resources. Platinum spark plugs in today's cars last for 100,000 miles, as do "fill-for-life" transmission fluids. On average, cars bought in 1993 have a useful life more than 40 percent longer than those bought in 1970.

As lighter materials replace heavier ones, the U.S. economy continues to shed weight. Our per capita consumption of raw materials such as forestry products and metals has, measured by weight, declined steadily over the past twenty years. A recent World Resources Institute study measured the "materials intensity" of our economy—that is, "the total material input and the hidden or indirect material flows, including deliberate landscape alterations" required for each dollar's worth of economic output. "The result shows a clearly declining pattern of materials intensity, supporting the conclusion that economic activity is growing somewhat more rapidly than natural resource use." Of course, we should do better. The Organization for Economic Cooperation and Development, an association of the world's industrialized nations, has proposed that its members strive as a long-range goal to decrease their materials intensity by a factor of ten.

Communications also illustrates the trend toward lighter, smaller, less materials-intensive technology. Just as telegraph cables replaced frigates in transmitting messages across the Atlantic and carried more information faster, glass fibers and microwaves have replaced cables—each new technology using less materials but providing greater capacity for sending and receiving information. Areas not yet wired for telephones (in the former Soviet Union, for example) are expected to leapfrog directly into cellular communications. Robert Solow, a Nobel laureate in economics, says that if the future is like the past, "there will be prolonged and substantial reductions in natural-resource requirements per unit of real output." He asks, "Why shouldn't the productivity of most natural resources rise more or less steadily through time, like the productivity of labor?"

Misconception No. 2: We Are Running Out of Food and Timber

The United Nations projects that the global population, currently 5.7 billion, will peak at about 10 billion in the next century and then stabilize or even decline. Can the earth feed that many people? Even if food crops increase sufficiently, other renewable resources, including many

fisheries and forests, are already under pressure. Should we expect fish stocks to collapse or forests to disappear?

The world already produces enough cereals and oilseeds to feed 10 billion people a vegetarian diet adequate in protein and calories. If, however, the idea is to feed 10 billion people not healthful vegetarian diets but the kind of meat-laden meals that Americans eat, the production of grains and oilseeds may have to triple—primarily to feed livestock. Is anything like this kind of productivity in the cards?

Maybe. From 1961 to 1994 global production of food doubled. Global output of grain rose from about 630 million tons in 1950 to about 1.8 billion tons in 1992, largely as a result of greater yields. Developing countries from 1974 to 1994 increased wheat yields per acre by almost 100 percent, corn yields by 72 percent, and rice yields by 52 percent. "The generation of farmers on the land in 1950 was the first in history to double the production of food," the Worldwatch Institute has reported. "By 1984, they had outstripped population growth enough to raise per capita grain output an unprecedented 40 percent." From a two-year period ending in 1981 to a two-year period ending in 1990 the real prices of basic foods fell 38 percent on world markets, according to a 1992 United Nations report. Prices for food have continually decreased since the end of the eighteenth century, when Thomas Malthus argued that rapid population growth must lead to mass starvation by exceeding the carrying capacity of the earth.

Farmers worldwide could double the acreage in production, but this should not be necessary. Better seeds, more irrigation, multi-cropping, and additional use of fertilizer could greatly increase agricultural yields in the developing world, which are now generally only half those in the industrialized countries. It is biologically possible to raise yields of rice to about seven tons per acre—about four times the current average in the developing world. Super strains of cassava, a potato-like root crop eaten by millions of Africans, promise to increase yields tenfold. American farmers can also do better. In a good year, such as 1994, Iowa corn growers average about 3.5 tons per acre, but farmers more than double that yield in National Corn Growers Association competitions.

In drier parts of the world the scarcity of fresh water presents the greatest challenge to agriculture. But the problem is regional, not global. Fortunately, as Lester Brown, of the Worldwatch Institute, points out, "there are vast opportunities for increasing water efficiency" in arid regions, ranging from installing better water-delivery systems to planting drought-resistant crops. He adds, "Scientists can help push back the physical frontiers of cropping by developing varieties that are more drought resistant, salt tolerant, and early maturing. The payoff on the first two could be particularly high."

As if in response, Novartis Seeds has announced a program to develop water-efficient and salt-tolerant crops, including genetically engineered varieties of wheat. Researchers in Mexico have announced the development of drought-resistant corn that can boost yields by a third. Biotechnologists are converting annual crops into perennial ones, eliminating the need for yearly planting. They also hope to enable cereal crops to fix their own nitrogen, as legumes do, minimizing the need for fertilizer (genetically engineered nitrogen-fixing bacteria have already been test-marketed to farmers). Commercial varieties of crops such as corn, tomatoes, and potatoes which have been genetically engineered to be resistant to pests and diseases have been approved for field testing in the United States; several are now being sold and planted. A new breed of rice, 25 percent more productive than any currently in use, suggests that the Gene Revolution can take over where the Green Revolution left off. Biotechnology, as the historian Paul Kennedy has written, introduces "an entirely new stage in humankind's attempts to produce more crops and plants."

Biotechnology cannot, however, address the major causes of famine: poverty, trade barriers, corruption, mismanagement, ethnic antagonism, anarchy, war, and male-dominated societies that deprive women of food. Local land depletion, itself a consequence of poverty and institutional failure, is also a factor. Those who are too poor to use sound farming practices are compelled to overexploit the resources on which they depend. As the economist Partha Dasgupta has written, "Population growth, poverty and degradation of local resources often fuel one another." The amount of food in world trade is constrained less by the resource base than by the maldistribution of wealth.

Analysts who believe that the world is running out of resources often argue that famines occur not as a result of political or economic conditions but because there are "too many people." Unfortunately, as the economist Amartya Sen has pointed out, public officials who think in Malthusian terms assume that when absolute levels of food supplies are adequate, famine will not occur. This conviction diverts attention from the actual causes of famine, which has occurred in places where food output kept pace with population growth but people were too destitute to buy it.

We would have run out of food long ago had we tried to supply ourselves entirely by hunting and gathering. Likewise, if we depend on nature's gifts, we will exhaust many of the world's important fisheries. Fortunately, we are learning to cultivate fish as we do other crops. Genetic engineers have designed fish for better flavor and color as well as for faster growth, improved disease resistance, and other traits. Two farmed species—silver carp and grass carp—already rank among the ten most-consumed fish worldwide. A specially bred tilapia, known as the "aquatic chicken," takes six months to grow to a harvestable size of about one and a half pounds.

Aquaculture produced more than 16 million tons of fish in 1993; capacity has expanded over the past decade at an annual rate of 10 percent by quantity and 14 percent by value. In 1993 fish farms produced 22 percent of all food fish consumed in the world and 90 percent of all oysters sold. The World Bank reports that aquaculture could provide 40 percent of all fish consumed and more than half the value of fish harvested within the next fifteen years.

Salmon ranching and farming provide examples of the growing efficiency of aquacultural production. Norwegian salmon farms alone produce 400 million pounds a year. A biotech firm in Waltham, Massachusetts, has applied for government approval to commercialize salmon genetically engineered to grow four to six times as fast as their naturally occurring cousins. As a 1994 article in *Sierra* magazine noted, "There is so much salmon currently available that the supply exceeds demand, and prices to fishermen have fallen dramatically."

For those who lament the decline of natural fisheries and the human communities that grew up with them, the successes of aquaculture may offer no consolation. In the Pacific Northwest, for example, overfishing in combination with dams and habitat destruction has reduced the wild salmon population by 80 percent. Wild salmon—but not their bio-engineered aquacultural cousins—contribute to the cultural identity and sense of place of the Northwest. When wild salmon disappear, so will some of the region's history, character, and pride. What is true of wild salmon is also true of whales, dolphins, and other magnificent creatures—as they lose their economic importance, their aesthetic and moral worth becomes all the more evident. Economic considerations pull in one direction, moral considerations in the other. This conflict colors all our battles over the environment.

The transition from hunting and gathering to farming, which is changing the fishing industry, has taken place more slowly in forestry. Still there is no sign of a timber famine. In the United States forests now provide the largest harvests in history, and there is more forested U.S. area today than there was in 1920. Bill McKibben has observed ... that the eastern United States, which loggers and farmers in the eighteenth and nineteenth centuries nearly denuded of trees, has become reforested during this century (see "An Explosion of Green," April, 1995, *Atlantic*). One reason is that farms reverted to woods. Another is that machinery replaced animals; each draft animal required two or three cleared acres for pasture.

Natural reforestation is likely to continue as biotechnology makes areas used for logging more productive. According to Roger Sedjo, a respected forestry expert, advances in tree farming, if implemented widely, would permit the world to meet its entire demand for industrial wood using just 200 million acres of plantations—an area equal to only five percent of current forest land. As less land is required for commercial tree production, more natural forests may be protected—as they should be, for aesthetic, ethical, and spiritual reasons.

Often natural resources are so plentiful and therefore inexpensive that they undercut the necessary transition to technological alternatives. If the U.S. government did not protect wild forests from commercial exploitation, the timber industry would have little incentive to invest in tree plantations, where it can multiply yields by a factor of ten and take advantage of the results of genetic research. Only by investing in plantation silviculture can North American forestry fend off price competition from rapidly developing tree plantations in the Southern Hemisphere. Biotechnology-based silviculture can in the near future be expected to underprice "extractive" forestry worldwide. In this decade China will plant about 150 million acres of trees; India now plants four times the area it harvests commercially.

The expansion of fish and tree farming confirms the belief held by Peter Drucker and other management experts that our economy depends far more on the progress of technology than on the exploitation of nature. Although raw materials will always be necessary, knowledge has become the essential factor in the production of goods and services. "Where there is effective management," Drucker has written, "that is, application of knowledge to knowledge, we can always obtain the other resources." If we assume, along with Drucker and others, that resource scarcities do not exist or are easily averted, it is hard to see how economic theory, which after all concerns scarcity, provides the conceptual basis for valuing the environment. The reasons to preserve nature are ethical more often than they are economic.

Misconception No. 3: We Are Running Out of Energy

Probably the most persistent worries about resource scarcity concern energy. "The supply of fuels and other natural resources is becoming the limiting factor constraining the rate of economic growth," a group of experts proclaimed in 1986. They predicted the exhaustion of domestic oil and gas supplies by 2020 and, within a few decades, "major energy shortages as well as food shortages in the world."

Contrary to these expectations, no global shortages of hydrocarbon fuels are in sight. "One sees no immediate danger of 'running out' of energy in a global sense," writes John P. Holdren, a professor of environmental policy at Harvard University. According to Holdren, reserves of oil and natural gas will last seventy to a hundred years if exploited at 1990 rates. (This does not take into account huge deposits of oil shale, heavy oils, and gas from unconventional sources.) He concludes that "running out of energy resources in any global sense is not what the energy problem is all about."

The global energy problem has less to do with depleting resources than with controlling pollutants. Scientists generally agree that gases, principally carbon dioxide, emitted in the combustion of hydrocarbon fuels can build up in and warm the atmosphere by trapping sunlight. Since carbon dioxide enhances photosynthetic activity, plants to some extent absorb the carbon dioxide we produce. In 1995 researchers reported in *Science* that

vegetation in the Northern Hemisphere in 1992 and 1993 converted into trees and other plant tissue 3.5 billion tons of carbon—more than half the carbon produced by the burning of hydrocarbon fuels worldwide.

However successful this and other feedback mechanisms may be in slowing the processes of global warming, a broad scientific consensus, reflected in a 1992 international treaty, has emerged for stabilizing and then decreasing emissions of carbon dioxide and other "greenhouse" gases. This goal is well within the technological reach of the United States and other industrialized countries. Amory Lovins, among others, has described commercially available technologies that can "support present or greatly expanded worldwide economic activity while stabilizing global climate—and saving money." He observes that "even very large expansions in population and industrial activity need not be energy-constrained."

Lovins and other environmentalists contend that pollution-free energy from largely untapped sources is available in amounts exceeding our needs. Geothermal energy—which makes use of heat from the earth's core—is theoretically accessible through drilling technology in the United States in amounts thousands of times as great as the amount of energy contained in domestic coal reserves. Tidal energy is also promising. Analysts who study solar power generally agree with Lester Brown, of the *Worldwatch Institute,* that "technologies are ready to begin building a world energy system largely powered by solar resources." In the future these and other renewable energy sources may be harnessed to the nation's system of storing and delivering electricity.

Last year Joseph Romm and Charles Curtis described ... advances in photovoltaic cells (which convert sunlight into electricity), fuel cells (which convert the hydrogen in fuels directly to electricity and heat, producing virtually no pollution), and wind power ("Mideast Oil Forever?" April, 1996, *Atlantic*). According to these authors, genetically engineered organisms used to ferment organic matter could, with further research and development, bring down the costs of ethanol and other environmentally friendly "biofuels" to make them competitive with gasoline.

Environmentalists who, like Amory Lovins, believe that our economy can grow and still reduce greenhouse gases emphasize not only that we should be able to move to renewable forms of energy but also that we can use fossil fuels more efficiently. Some improvements are already evident. In developed countries the energy intensity of production—the amount of fuel burned per dollar of economic output—has been decreasing by about two percent a year.

From 1973 to 1986, for example, energy consumption in the United States remained virtually flat while economic production grew by almost 40 percent. Compared with Germany or Japan, this is a poor showing. The Japanese, who tax fuel more heavily than we do, use only half as much energy as the United States per unit of economic output. (Japanese environmental regulations are also generally stricter than ours; if anything, this has improved the competitiveness of Japanese industry.) The United States still wastes hundreds of billions of dollars annually in energy inefficiency. By becoming as energy-efficient as Japan, the United States could expand its economy and become more competitive internationally.

If so many opportunities exist for saving energy and curtailing pollution, why have we not seized them? One reason is that low fossil-fuel prices remove incentives for fuel efficiency and for converting to other energy sources. Another reason is that government subsidies for fossil fuels and nuclear energy amounted to many billions of dollars a year during the 1980s, whereas support for renewables dwindled to $114 million in 1989, a time when it had been proposed for near elimination. "Lemon socialism," a vast array of subsidies and barriers to trade, protects politically favored technologies, however inefficient, dangerous, filthy, or obsolete. "At heart, the major obstacles standing in the way [of a renewable-energy economy] are not technical in nature," the energy consultant Michael Brower has written, "but concern the laws, regulations, incentives, public attitudes, and other factors that make up the energy market."

In response to problems of climate change, the World Bank and other international organizations have recognized the importance of transferring advanced energy technologies to the developing world. Plainly, this will take a large investment of capital, particularly in education. Yet the "alternative for developing countries," according to José Goldemberg, a former Environment Minister of Brazil, "would be to remain at a dismally low level of development which . . . would aggravate the problems of sustainability."

Technology transfer can hasten sound economic development worldwide. Many environmentalists, however, argue that economies cannot expand without exceeding the physical limits nature sets—for example, with respect to energy. These environmentalists, who regard increasing affluence as a principal cause of environmental degradation, call for economic retrenchment and retraction—a small economy for a small earth. With Paul Ehrlich, they reject "the hope that development can greatly increase the size of the economic pie and pull many more people out of poverty." This hope is "basically a humane idea," Ehrlich has written, "made insane by the constraints nature places on human activity."

In developing countries, however, a no-growth economy "will deprive entire populations of access to better living conditions and lead to even more deforestation and land degradation," as Goldemberg warns. Moreover, citizens of developed countries are likely to resist an energy policy that they associate with poverty, discomfort, sacrifice, and pain. Technological pessimism, then, may not be the best option for environmentalists. It is certainly not the only one.

Misconception No. 4: The North Exploits the South

William Reilly, when he served as administrator of the Environmental Protection Agency in the Bush Administration, encountered a persistent criticism at international meetings on the environment. "The problem for the world's environment is your consumption, not our population," delegates from the developing world told him. Some of these delegates later took Reilly aside. "The North buys too little from the South," they confided. "The real problem is too little demand for our exports."

The delegates who told Reilly that the North consumes too little of what the South produces have a point. "With a few exceptions (notably petroleum)," a report from the World Resources Institute observes, "most of the natural resources consumed in the United States are from domestic sources." Throughout the 1980s the United States and Canada were the world's leading exporters of raw materials. The United States consistently leads the world in farm exports, running huge agricultural trade surpluses. The share of raw materials used in the North that it buys from the South stands at a thirty-year low and continues to decline; industrialized nations trade largely among themselves. The World Resources Institute recently reported that "the United States is largely self-sufficient in natural resources." Again, excepting petroleum, bauxite (from which aluminum is made), "and a few other industrial minerals, its material flows are almost entirely internal."

Sugar provides an instructive example of how the North excludes—rather than exploits—the resources of the South. Since 1796 the United States has protected domestic sugar against imports. American sugar growers, in part as a reward for large contributions to political campaigns, have long enjoyed a system of quotas and prohibitive tariffs against foreign competition. American consumers paid about three times world prices for sugar in the 1980s, enriching a small cartel of U.S. growers. *Forbes* magazine has estimated that a single family, the Fanjuls, of Palm Beach, reaps more than $65 million a year as a result of quotas for sugar.

The sugar industry in Florida, which is larger than that in any other state, makes even less sense environmentally than economically. It depends on a publicly built system of canals, levees, and pumping stations. Fertilizer from the sugarcane fields chokes the Everglades. Sugar growers, under a special exemption from labor laws, import Caribbean laborers to do the grueling and poorly paid work of cutting cane.

As the United States tightened sugar quotas (imports fell from 6.2 to 1.5 million tons annually from 1977 to 1987), the Dominican Republic and other nations with climates ideal for growing cane experienced political turmoil and economic collapse. Many farmers in Latin America, however, did well by switching from sugar to coca, which is processed into cocaine—perhaps the only high-value imported crop for which the United States is not developing a domestic substitute.

Before the Second World War the United States bought 40 percent of its vegetable oils from developing countries. After the war the United States protected its oilseed markets—for example, by establishing price supports for soybeans. Today the United States is one of the world's leading exporters of oil and oilseeds, although it still imports palm and coconut oils to obtain laurate, an ingredient in soap, shampoo, and detergents. Even this form of "exploitation" will soon cease. In 1994 farmers in Georgia planted the first commercial acreage of a high-laurate canola, genetically engineered by Calgene, a biotechnology firm.

About 100,000 Kenyans make a living on small plots of land growing pyrethrum flowers, the source of a comparatively environmentally safe insecticide of which the United States has been the largest importer. The U.S. Department of Commerce, however, awarded $1.2 million to a biotechnology firm to engineer pyrethrum genetically. Industrial countries will soon be able to synthesize all the pyrethrum they need and undersell Kenyan farmers.

An article in *Foreign Policy* in December of 1995 observed that the biotechnological innovations that create "substitutes for everything from vanilla to cocoa and coffee threaten to eliminate the livelihood of millions of Third World agricultural workers." Vanilla cultured in laboratories costs a fifth as much as vanilla extracted from beans, and thus jeopardizes the livelihood of tens of thousands of vanilla farmers in Madagascar. In the past, farms produced agricultural commodities and factories processed them. In the future, factories may "grow" as well as process many of the most valuable commodities—or the two functions will become one. As one plant scientist has said, "We have to stop thinking of these things as plant cells, and start thinking of them as new microorganisms, with all the potential that implies"—meaning, for instance, that the cells could be made to grow in commercially feasible quantities in laboratories, not fields.

The North not only balks at buying sugar and other crops from developing countries; it also dumps its excess agricultural commodities, especially grain, on them. After the Second World War, American farmers, using price supports left over from the New Deal, produced vast wheat surpluses, which the United States exported at concessionary prices to Europe and then the Third World. These enormous transfers of cereals to the South, institutionalized during the 1950s and 1960s by U.S. food aid, continued during the 1970s and 1980s, as the United States and the European Community vied for markets, each outdoing the other in subsidizing agricultural exports.

Grain imports from the United States "created food dependence within two decades in countries which had been mostly self-sufficient in food at the end of World War II," the sociologist Harriet Friedmann has written.

Tropical countries soon matched the grain gluts of the North with their own surpluses of cocoa, coffee, tea, bananas, and other export commodities. Accordingly, prices for these commodities collapsed as early as 1970, catching developing nations in a scissors. As Friedmann describes it, "One blade was food import dependency. The other blade was declining revenues for traditional exports of tropical crops."

It might be better for the environment if the North exchanged the crops for which it is ecologically suited—wheat, for example—for crops easily grown in the South, such as coffee, cocoa, palm oil, and tea. Contrary to common belief, these tropical export crops—which grow on trees and bushes, providing canopy and continuous root structures to protect the soil—are less damaging to the soil than are traditional staples such as cereals and root crops. Better markets for tropical crops could help developing nations to employ their rural populations and to protect their natural resources. Allen Hammond, of the World Resources Institute, points out that "if poor nations cannot export anything else, they will export their misery—in the form of drugs, diseases, terrorism, migration, and environmental degradation."

Peasants in less-developed nations often confront intractable poverty, an entrenched land-tenure system, and a lack of infrastructure; they have little access to markets, education, or employment. Many of the rural poor, according to the environmental consultant Norman Myers, "have no option but to over-exploit environmental resource stocks in order to survive"—for example, by "increasingly encroaching onto tropical forests among other low-potential lands." These poorest of the poor "are causing as much natural-resource depletion as the other three billion developing-world people put together."

Myers observes that traditional indigenous farmers in tropical forests moved from place to place without seriously damaging the ecosystem. The principal agents of tropical deforestation are refugees from civil war and rural poverty, who are forced to eke out a living on marginal lands. Activities such as road building, logging, and commercial agriculture have barely increased in tropical forests since the early 1980s, according to Myers; slash-and-burn farming by displaced peasants accounts for far more deforestation—roughly three fifths of the total. Its impact is fast expanding. Most of the wood from trees harvested in tropical forests—that is, those not cleared for farms—is used locally for fuel. The likeliest path to protecting the rain forest is through economic development that enables peasants to farm efficiently, on land better suited to farming than to forest.

Many have argued that economic activity, affluence, and growth automatically lead to resource depletion, environmental deterioration, and ecological collapse. Yet greater productivity and prosperity—which is what economists mean by growth—have become prerequisite for controlling urban pollution and protecting sensitive ecological systems such as rain forests. Otherwise, destitute people who are unable to acquire food and fuel will create pollution and destroy forests. Without economic growth, which also correlates with lower fertility, the environmental and population problems of the South will only get worse. For impoverished countries facing environmental disaster, economic growth may be the one thing that is sustainable.

What Is Wrong With Consumption?

Many of us who attended college in the 1960s and 1970s took pride in how little we owned. We celebrated our freedom when we could fit all our possessions—mostly a stereo—into the back of a Beetle. Decades later, middle-aged and middle-class, many of us have accumulated an appalling amount of stuff. Piled high with gas grills, lawn mowers, excess furniture, bicycles, children's toys, garden implements, lumber, cinder blocks, ladders, lawn and leaf bags stuffed with memorabilia, and boxes yet to be unpacked from the last move, the two-car garages beside our suburban homes are too full to accommodate the family minivan. The quantity of resources, particularly energy, we waste and the quantity of trash we throw away (recycling somewhat eases our conscience) add to our consternation.

Even if predictions of resource depletion and ecological collapse are mistaken, it seems that they *should* be true, to punish us for our sins. We are distressed by the suffering of others, the erosion of the ties of community, family, and friendship, and the loss of the beauty and spontaneity of the natural world. These concerns reflect the most traditional and fundamental of American religious and cultural values.

Simple compassion instructs us to give to relieve the misery of others. There is a lot of misery worldwide to relieve. But as bad as the situation is, it is improving. In 1960 nearly 70 percent of the people in the world lived at or below the subsistence level. Today less than a third do, and the number enjoying fairly satisfactory conditions (as measured by the United Nations Human Development Index) rose from 25 percent in 1960 to 60 percent in 1992. Over the twenty-five years before 1992 average per capita consumption in developing countries increased 75 percent in real terms. The pace of improvements is also increasing. In developing countries in that period, for example, power generation and the number of telephone lines per capita doubled, while the number of households with access to clean water grew by half.

What is worsening is the discrepancy in income between the wealthy and the poor. Although world income measured in real terms has increased by 700 percent since the Second World War, the wealthiest people have absorbed most of the gains. Since 1960 the richest fifth of the world's people have seen their share of the world's income increase from 70 to 85 percent. Thus one fifth of the world's population possesses much more than four fifths of the world's wealth, while the share held by all others has correspondingly fallen; that of the world's poorest 20 percent has declined from 2.3 to 1.4 percent.

Benjamin Barber ("Jihad vs. McWorld," March, 1992, *Atlantic*) described market forces that "mesmerize the world with fast music, fast computers, and fast food—with MTV, Macintosh, and McDonald's, pressing nations into one commercially homogeneous global network: one McWorld tied together by technology, ecology, communications, and commerce." Affluent citizens of South Korea, Thailand, India, Brazil, Mexico, and many other rapidly developing nations have joined with Americans, Europeans, Japanese, and others to form an urban and cosmopolitan international society. Those who participate in this global network are less and less beholden to local customs and traditions. Meanwhile, ethnic, tribal, and other cultural groups that do not dissolve into McWorld often define themselves in opposition to it—fiercely asserting their ethnic, religious, and territorial identities.

The imposition of a market economy on traditional cultures in the name of development—for example, the insistence that everyone produce and consume more—can dissolve the ties to family, land, community, and place on which indigenous peoples traditionally rely for their security. Thus development projects intended to relieve the poverty of indigenous peoples may, by causing the loss of cultural identity, engender the very powerlessness they aim to remedy. Pope Paul VI, in the encyclical *Populorum Progressio* (1967), described the tragic dilemma confronting indigenous peoples: "either to preserve traditional beliefs and structures and reject social progress; or to embrace foreign technology and foreign culture, and reject ancestral traditions with their wealth of humanism."

The idea that everything is for sale and nothing is sacred—that all values are subjective—undercuts our own moral and cultural commitments, not just those of tribal and traditional communities. No one has written a better critique of the assault that commerce makes on the quality of our lives than Thoreau provides in *Walden*. The cost of a thing, according to Thoreau, is not what the market will bear but what the individual must bear because of it: it is "the amount of what I will call life which is required to be exchanged for it, immediately or in the long run."

Many observers point out that as we work harder and consume more, we seem to enjoy our lives less. We are always in a rush—a "Saint Vitus' dance," as Thoreau called it. Idleness is suspect. Americans today spend less time with their families, neighbors, and friends than they did in the 1950s. Juliet B. Schor, an economist at Harvard University, argues that "Americans are literally working themselves to death." A fancy car, video equipment, or a complex computer program can exact a painful cost in the form of maintenance, upgrading, and repair. We are possessed by our possessions; they are often harder to get rid of than to acquire.

That money does not make us happier, once our basic needs are met, is a commonplace overwhelmingly confirmed by sociological evidence. Paul Wachtel, who teaches social psychology at the City University of New York, has concluded that bigger incomes "do not yield an increase in feelings of satisfaction or well-being, at least for populations who are above a poverty or subsistence level." This cannot be explained simply by the fact that people have to work harder to earn more money: even those who hit jackpots in lotteries often report that their lives are not substantially happier as a result. Well-being depends upon health, membership in a community in which one feels secure, friends, faith, family, love, and virtues that money cannot buy. Robert Lane, a political scientist at Yale University, using the concepts of economics, has written, "If 'utility' has anything to do with happiness, above the poverty line the long-term marginal utility of money is almost zero."

Economists in earlier times predicted that wealth would not matter to people once they attained a comfortable standard of living. "In ease of body and peace of mind, all the different ranks of life are nearly upon a level," wrote Adam Smith, the eighteenth-century English advocate of the free market. In the 1930s the British economist John Maynard Keynes argued that after a period of great expansion further accumulation of wealth would no longer improve personal well-being. Subsequent economists, however, found that even after much of the industrial world had attained the levels of wealth Keynes thought were sufficient, people still wanted more. From this they inferred that wants are insatiable.

Perhaps this is true. But the insatiability of wants and desires poses a difficulty for standard economic theory, which posits that humanity's single goal is to increase or maximize wealth. If wants increase as fast as income grows, what purpose can wealth serve?

Critics often attack standard economic theory on the ground that economic growth is "unsustainable." We are running out of resources, they say; we court ecological disaster. Whether or not growth is sustainable, there is little reason to think that once people attain a decent standard of living, continued growth is desirable. The economist Robert H. Nelson recently wrote in the journal *Ecological Economics* that it is no longer possible for most people to believe that economic progress will "solve all the problems of mankind, spiritual as well as material." As long as the debate over sustainability is framed in terms of the physical limits to growth rather than the moral purpose of it, mainstream economic theory will have the better of the argument. If the debate were framed in moral or social terms, the result might well be otherwise.

Making a Place for Nature

According to Thoreau, "a man's relation to Nature must come very near to a personal one." For environmentalists in the tradition of Thoreau and John Muir, stewardship is a form of fellowship; although we must use nature, we do not value it primarily for the economic purposes it serves. We take our bearings from the natural world—our sense of time from its days and seasons, our sense of place from the character of a landscape and the particular plants and animals native to it. An intimacy with nature ends our isolation in the world. We know where we belong, and we can find the way home.

In defending old-growth forests, wetlands, or species we make our best arguments when we think of nature chiefly in aesthetic and moral terms. Rather than having the courage of our moral and cultural convictions, however, we too often rely on economic arguments for protecting nature, in the process attributing to natural objects more instrumental value than they have. By claiming that a threatened species may harbor lifesaving drugs, for example, we impute to that species an economic value or a price much greater than it fetches in a market. When we make the prices come out right, we rescue economic theory but not necessarily the environment.

There is no credible argument, moreover, that all or even most of the species we are concerned to protect are essential to the functioning of the ecological systems on which we depend. (If whales went extinct, for example, the seas would not fill up with krill.) David Ehrenfeld, a biologist at Rutgers University, makes this point in relation to the vast ecological changes we have already survived. "Even a mighty dominant like the American chestnut," Ehrenfeld has written, "extending over half a continent, all but disappeared without bringing the eastern deciduous forest down with it." Ehrenfeld points out that the species most likely to be endangered are those the biosphere is least likely to miss. "Many of these species were never common or ecologically influential; by no stretch of the imagination can we make them out to be vital cogs in the ecological machine."

Species may be profoundly important for cultural and spiritual reasons, however. Consider again the example of the wild salmon, whose habitat is being destroyed by hydroelectric dams along the Columbia River. Although this loss is unimportant to the economy overall (there is no shortage of salmon), it is of the greatest significance to the Amerindian tribes that have traditionally subsisted on wild salmon, and to the region as a whole. By viewing local flora and fauna as a sacred heritage—by recognizing their intrinsic value—we discover who we are rather than what we want. On moral and cultural grounds society might be justified in making great economic sacrifices—removing hydroelectric dams, for example—to protect remnant populations of the Snake River sockeye, even if, as critics complain, hundreds or thousands of dollars are spent for every fish that is saved.

Even those plants and animals that do not define places possess enormous intrinsic value and are worth preserving for their own sake. What gives these creatures value lies in their histories, wonderful in themselves, rather than in any use to which they can be put. The biologist E. O. Wilson elegantly takes up this theme: "Every kind of organism has reached this moment in time by threading one needle after another, throwing up brilliant artifices to survive and reproduce against nearly impossible odds." Every plant or animal evokes not just sympathy but also reverence and wonder in those who know it.

In *Earth in the Balance* (1992) Al Gore, then a senator, wrote, "We have become so successful at controlling nature that we have lost our connection to it." It is all too easy, Gore wrote, "to regard the earth as a collection of 'resources' having an intrinsic value no larger than their usefulness at the moment." The question before us is not whether we are going to run out of resources. It is whether economics is the appropriate context for thinking about environmental policy.

Even John Stuart Mill, one of the principal authors of utilitarian philosophy, recognized that the natural world has great intrinsic and not just instrumental value. More than a century ago, as England lost its last truly wild places, Mill condemned a world

> *with nothing left to the spontaneous activity of nature; with every rood of land brought into cultivation, which is capable of growing food for human beings; every flowery waste or natural pasture ploughed up; all quadrupeds or birds which are not domesticated for man's use exterminated as his rivals for food, every hedgerow or superfluous tree rooted out, and scarcely a place left where a wild shrub or flower could grow without being*

> *eradicated as a weed in the name of improved agriculture.*

The world has the wealth and the resources to provide everyone the opportunity to live a decent life. We consume too much when market relationships displace the bonds of community, compassion, culture, and place. We consume too much when consumption becomes an end in itself and makes us lose affection and reverence for the natural world.

Critical Thinking

1. It has been said that "the global economy cannot grow infinitely on a finite planet." If this is true, then why do many humans behave as though it were not? And if this is not *true,* then why do we have colleges and universities offering classes and degrees in "environmental science"?

2. List the author's "four misconceptions" and consider how the observations made regarding those misconceptions are "place dependent" (e.g., might vary for countries, regions, cultures, geopolitical economies).

3. According to the author, how is the North *not* exploiting the South? Think of some examples of how the industrialized world *is* exploiting developing nations.

4. Aside from wealth, what are some geopolitical-economy factors that make it easy for Americans to amass so many material possessions?

5. The author believes we consume too much when . . .? Do you agree or disagree? Explain.

How Much Should a Person Consume?

RAMACHANDRA GUHA

This paper takes as its point of departure an old essay by John Kenneth Galbraith, an essay so ancient and obscure that it might very well have been forgotten even by its prolific author. The essay was written in 1958, the same year that Galbraith published *The Affluent Society,* a book that wryly anatomised the social consequences of the mass-consumption age.

In his book, Galbraith highlighted the "preoccupation with productivity and production" in postwar America and western Europe. The population in these societies had for the most part been adequately housed, clothed and fed; now they expressed a desire for "more elegant cars, more exotic food, more erotic clothing, more elaborate entertainment".[1] When Galbraith termed 1950s America the "affluent society", he meant not only that this was a society most of whose members were hugely prosperous when reckoned against other societies and other times, but also that this was a society so dedicated to affluence that the possession and consumption of material goods became the exclusive standard of individual and collective achievement. He quoted the anthropologist Geoffrey Gorer, who remarked that this was a culture in which "any device or regulation which interferes, or can be conceived as interfering, with [the] supply of more and better things is resisted with unreasoning horror, as the religious resist blasphemy, or the warlike pacifism".[2]

The Unasked Question

The essay I speak of was written months after the book, which made Galbraith's name and reputation. "How Much Should a Country Consume?" is its provocative title, and it can be read as a reflective footnote to *The Affluent Society.* In the book itself, Galbraith had noted the disjunction between "private affluence and public squalor", of how this single-minded pursuit of wealth had diverted attention and resources from the nurturing of true democracy, which he defined as the provision of public infrastructure, the creation of decent schools, parks and hospitals. Now the economist turned his attention, all too fleetingly, to

the long-term and global consequences of this collective promotion of consumption, of the "gargantuan and growing appetite" for resources in contemporary America. The American conservation movement, he remarked, had certainly noted the massive exploitation of resources and materials in the postwar period. However, its response was to look for more efficient methods of extraction, or the substitution of one material for another through technological innovation. There was, wrote Galbraith, a marked "selectivity in the conservationist's approach to materials consumption". For

> if we are concerned about our great appetite for materials, it is plausible to seek to increase the supply, or decrease waste, to make better use of the stocks that are available, and to develop substitutes. But what of the appetite itself? Surely this is the ultimate source of the problem. If it continues its geometric course, will it not one day have to be restrained? Yet in the literature of the resource problem this is the forbidden question. Over it hangs a nearly total silence. It is as though, in the discussion of the chance for avoiding automobile accidents, we agree not to make any mention of speed![3]

A cultural explanation for this silence had been previously provided by the great Berkeley geographer Carl Sauer. Writing in 1938, Sauer remarked that "the doctrine of a passing frontier of nature replaced by a permanent and sufficiently expanding frontier of technology is a contemporary and characteristic expression of occidental culture, itself a historical–geographical product". This frontier attitude, he went on, "has the recklessness of an optimism that has become habitual, but which is residual from the brave days when north European freebooters overran the world and put it under tribute". Warning that the surge of growth at the expense of nature would not last indefinitely, Sauer—speaking for his fellow Americans—noted wistfully that "we have not yet learned the difference between yield and loot. We do not like to be economic realists".[4]

Galbraith himself identified two major reasons for the silence as regards consumption. One was ideological, the worship of the great god Growth. The principle of Growth (always with a capital G) was a cardinal belief of the American people, which necessarily implied a continuous increase in the production of consumer goods. The second reason was political, the widespread scepticism about the state. For the America of the fifties had witnessed the "resurgence of a notably over-simplified view of economic life which [ascribed] a magical automatism to the price system". Now, Galbraith was himself an unreconstructed New Dealer, who would tackle the problem of overconsumption as he would tackle the problem of underemployment, that is, through purposive state intervention. At the time he wrote, however, free-market economics ruled, and "since consumption could not be discussed without raising the question of an increased role for the state, it was not discussed".[5]

Four years later, Rachel Carson published *Silent Spring,* and the modern American environmental movement gathered pace. One might have expected this new voice of civil society to undertake what the market could not. As it happened, consumption continued to be the great unasked question of the conservation movement. The movement principally focused on two things: the threats to human health posed by pollution, and the threats to wild species and wild habitats posed by economic expansion. The latter concern became, in fact, the defining motif of the movement. The dominance of wilderness protection in American environmentalism has promoted an essentially negativist agenda—the protection of the parks and their animals by freeing them of human habitation and productive activities. As the historian Samuel Hays points out, "natural environments, which formerly had been looked upon as 'useless', waiting only to be developed, now came to be thought of as 'useful' for filling human wants and needs. They played no less a significant role in the advanced consumer society than did such material goods as hi-fi sets or indoor gardens".[6] While saving these islands of biodiversity, environmentalists paid scant attention to what was happening outside them. In the American economy as a whole, the consumption of energy and materials continued to rise.

The growing popular interest in the wild and the beautiful thus not merely accepted the parameters of the affluent society but was wont to see nature itself as merely one more good to be consumed. The uncertain commitment of most nature lovers to a more comprehensive environmental ideology is illustrated by the paradox that they were willing to drive thousands of miles—consuming scarce oil and polluting the atmosphere—to visit national parks and sanctuaries, thus using anti-ecological means to marvel at the beauty of forests, swamps or mountains protected as specimens of a "pristine" and "untouched" nature.[7]

The Real Population Problem

The selectivity of the conservationist approach to consumption was underlined in the works of biologists obsessed with the "population problem". Influential American scientists such as Paul Ehrlich and Garret Hardin identified human population growth as the single most important reason for environmental degradation. This is how Ehrlich began the first chapter of his best-selling book, *The Population Bomb:*

> I have understood the population explosion intellectually for a long time. I came to understand it emotionally one stinking hot night in Delhi a couple of years ago. My wife and daughter and I were returning to our hotel in an ancient taxi. The seats were hopping with fleas. The only functional gear was third. As we crawled through the city, we entered a crowded slum area. The temperature was well over 100, and the air was a haze of dust and smoke. The streets seemed alive with people. People eating, people washing, people sleeping. People visiting, people arguing and screaming. People thrusting their hands through the taxi window, begging. People defecating and urinating. People clinging to buses. People herding animals. People, people, people, people.[8]

Here exploding numbers are blamed for increasing pollution, stinking hot air and even technological obsolescence (that ancient taxi!). During the 1970s and 80s, neo-Malthusian interpretations gained wide currency. Countries such as India and Bangladesh were commonly blamed for causing an environmental crisis. Not surprisingly, activists in these countries have been quick to take offence, pointing out that the West, and especially the United States, consumes, per capita as well as in the aggregate, a far greater proportion of the world's resources. Table 1 gives partial evidence of this. For apart from its overuse of nature's stock (which the table documents), the Western world has also placed an unbearable burden on nature's sink (which the table ignores). Thus, the atmosphere and the oceans can absorb about 13 billion metric tons of carbon dioxide annually. This absorptive capacity, if distributed fairly among all the people of the world, would give each human being the right to emit about 2.3 tons of carbon dioxide per year. At present, an American discharges in excess of 20 tons annually, a German 12 tons, a Japanese 9 tons, an Indian less than one ton. If one looks at the process historically the charges mount, for it is the industrialised countries, led by the United States,

Table 1 US Share of World Consumption of Key Materials, 1995
(figures in millions of metric tons)

Material	World Production	US Consumption	US Consumption as Percentage of World Production
Minerals	7,641	2,410	31.54
Wood products	724	170	23.48
Metals	1,196	132	11.03
Synthetics	252	131	51.98
All materials	9,813	2,843	28.97

Source: Computed from *State of the World 1999* (New York: Worldwatch Institute and W. W. Norton, 1999).

Note: US population is approximately 4.42 percent of total world population.

which have principally been responsible for the build-up of greenhouse gases over the past hundred years.

These figures explain why Third World scholars and activists like to argue that the real "population problem" is in America, since the birth of a child there has an impact on the global environment equivalent to the birth of (say) seventy Indonesian or Indian children. There was a Bangladeshi diplomat who made this case whenever he could, in the United Nations and elsewhere. But after a visit to an American supermarket he was obliged to modify his argument—to state instead that the birth of an American dog (or cat) was the equivalent, ecologically speaking, of the birth of a dozen Bangladeshi children.[9]

As a long-time admirer of American scholarship, I might add my own words of complaint here. Consider the rich and growing academic field of environmental history, which is most highly developed in the United States. Scholars in other parts of the world have taken much inspiration from the works of American exemplars, from their methodological subtlety and fruitful criss-crossing of disciplinary boundaries. For all this, there is a studied insularity among the historians of North America. There were, at last count, more than three hundred professional environmental historians in the United States, and yet not one has seriously studied the global consequences of the consumer society, the impact on land, soil, forests, climate, etc., of the American Way of Life.

One striking example of this territorial blindness is the Gulf War. In that prescient essay of 1958, John Kenneth Galbraith remarked that "it remains a canon of modern diplomacy that any preoccupation with oil should be concealed by calling on our still ample reserves of sanctimony".[10] To be sure, there were Americans who tore the veil of this sanctimonious hypocrisy, who pointed out that it was the US government that had carefully armed and consolidated the dictator it now wished to overthrow. Yet the essentially material imperatives of the war remained unexamined. It was the left-wing British newspaper, the *Guardian,* which claimed that the Gulf War was carried out to safeguard the American Way of Driving. No American historian, however, has taken to heart the wisdom in that throwaway remark, to reveal in all its starkness the ecological imperialism of the world's sole superpower.

Germany's Greens

I would now like to contrast the American case with the German one. Environmentalists in Germany have been more forthright in their criticisms of the consumer society. "The key to a sustainable development model worldwide," writes Helmut Lippelt, "is the question of whether West European societies really are able to reconstruct their industrial systems in order to permit an ecologically and socially viable way of production and consumption." That Lippelt does not here include the United States or Japan is noteworthy, an expression of his (and his movement's) willingness to take the burden upon themselves. West Europeans should reform themselves, rather than transfer their existing "patterns of high production and high consumption to eastern Europe and the 'Third World' [and thus] destroy the earth".[11]

For the German greens, economic growth in Europe and North America has been made possible only through the economic and ecological exploitation of the Third World. Rudolf Bahro is characteristically blunt: "The present way of life of the most industrially advanced nations," he remarked in 1984, "stands in a global and antagonistic contradiction to the natural conditions of human existence. We are eating up what other nations and future generations need to live on." From this perspective, indeed,

The working class here [in the North] is the richest lower class in the world. And if I look at the problem from the point of view of the whole of humanity, not just from that of Europe, then I must say that the metropolitan working class is the worst exploiting class in history.... What made poverty bearable in eighteenth- or nineteenth-century Europe was the prospect of escaping it through exploitation of the periphery. But this is no longer a possibility, and continued industrialism in the Third World will mean poverty for whole generations and hunger for millions.[12]

Bahro was a famous "Fundi", a leader of that section of the German greens which stood in the most uncompromising antagonism to modern society. But even the most hardheaded members of the other, or "Realo", faction acknowledge the unsustainability, on the global plane, of industrial society. The parliamentarian (and now foreign minister) Joschka Fischer, asked by a reporter where he planned to spend his old age, replied: "In the Frankfurt cemetery, although by that time we may pose an environmental hazard with all the poisons, heavy metals and dioxin that we carry around in our bodies." Or as a party document more matter-of-factly put it: "The global spread of industrial economic policies and lifestyles is exhausting the basic ecological health of our planet faster than it can be replenished." This global view, coupled with the stress on accountability, calls for "far-reaching voluntary commitments to restraint by wealthy nations". The industrialised countries, which consume three-quarters of the world's energy and resources, and which contribute the lion's share of "climate-threatening gaseous emissions", must curb their voracious appetite while allowing Southern nations to grow out of poverty. The greens urge the cancellation of all international debt, the banning of trade in products that destroy vulnerable ecosystems, and most radical of all, the freer migration of peoples from poor countries to rich ones.[13]

These elements in the green programme were, of course, forged as an alternative to the policies promoted by the two dominant political parties in Germany, themselves committed to the great god Growth. Since October 1998, the greens find themselves sharing power at the federal level, junior partners, but partners nevertheless, in a coalition dominated by the Social Democrats. Being in power will certainly tame them. They will work only for incremental change, instead of the wholesale restructuring of the consumption and production system some of them previously advocated.

Gandhi

Fifty years before the founding of the German green party, and thirty years before the article by Galbraith alluded to above, an Indian politician pointed to the unsustainability,

at the global level, of the Western model of economic development. "God forbid," he wrote, "that India should ever take to industrialization after the manner of the West. The economic imperialism of a single tiny island kingdom (England) is today keeping the world in chains. If an entire nation of 300 million took to similar economic exploitation, it would strip the world bare like locusts."[14]

The man was Mahatma Gandhi, writing in the weekly journal *Young India* in December 1928. Two years earlier, Gandhi had claimed that to "make India like England and America is to find some other races and places of the earth for exploitation". As it appeared that the Western nations had already "divided all the known races outside Europe for exploitation and there are no new worlds to discover", he pointedly asked: "What can be the fate of India trying to ape the West?"[15]

Gandhi's critique of Western industrialisation has, of course, profound implications for the way we live and relate to the environment today. For him, "the distinguishing characteristic of modern civilisation is an indefinite multiplicity of wants," whereas ancient civilisations were marked by an "imperative restriction upon, and a strict regulating of, these wants".[16] In uncharacteristically intemperate tones, he spoke of his "wholeheartedly detest[ing] this mad desire to destroy distance and time, to increase animal appetites, and [to] go to the ends of the earth in search of their satisfaction. If modern civilization stands for all this, and I have understood it to do so, I call it satanic".[17]

At the level of the individual, Gandhi's code of voluntary simplicity also offered a sustainable alternative to modern lifestyles. One of his best-known aphorisms, that the "world has enough for everybody's need, but not enough for everybody's greed", is in effect an exquisitely phrased one-line environmental ethic. This was an ethic he himself practised, for resource recycling and the minimisation of wants were integral to his life.

Gandhi's arguments have been revived and elaborated by the present generation of Indian environmentalists. Their country is veritably an ecological disaster zone, marked by excessively high rates of deforestation, species loss, land degradation, and air and water pollution. The consequences of this wholesale abuse of nature have chiefly been borne by the poor in the countryside—the peasants, tribespeople, fisherfolk and pastoralists who have seen their resources snatched away or depleted by more powerful economic interests. For in the last few decades, the men who rule India have attempted precisely to "make India like England and America". Without the access to resources and markets enjoyed by those two nations when they began to industrialise, India has had perforce to rely on the exploitation of its own people and environment. The natural resources of the country-side have been increasingly channelled to meet the needs of the urban–industrial sector, the diversion of forests, water, etc., to the elite having accelerated environmental degradation even as it has deprived rural and tribal communities of their traditional rights of access and use. Meanwhile, the modern sector has moved aggressively into the remaining resource frontiers of India—the northeast and the Andaman and Nicobar islands. This bias towards urban–industrial development has resulted only in a one-sided exploitation of the hinterland, thus proving Gandhi's contention that "the blood of the villages is the cement with which the edifice of the cities is built".[18]

The preceding paragraph brutally summarises arguments and evidence provided in a whole array of Indian environmentalist tracts.[19] Simplifying still further, one might say that the key contribution of the Indian environmental movement has been to point to inequalities of consumption *within* a society (or nation). In this respect they have complemented the work of their German counterparts, who have most effectively documented and criticised the inequalities of consumption *between* societies and nations.

Omnivores and Others

The criticisms of these environmentalists are strongly flavoured by morality, by the sheer injustice of one group or country consuming more than its fair share of the earth's resources, by the political imperative of restoring some sense of equality in global or national consumption. I now present an analytical framework that might more dispassionately explain these asymmetries in patterns of consumption.[20] Derived in the first instance from the Indian experience, this model rests on a fundamental opposition between two groups, *omnivores* and *ecosystem people*. These are distinguished above all by the size of their "resource catchment". Thus, omnivores, who include industrialists, rich farmers, state officials and the growing middle class based in the cities (estimated at in excess of one hundred million people), are able to draw upon the natural resources of the whole of India to maintain their lifestyles. Ecosystem people, on the other hand—who would include roughly two-thirds of the rural population, say about four hundred million people—rely for the most part on the resources of their own vicinity, from a catchment of a few dozen square miles at best. Such are the small and marginal farmers in rain-fed tracts, the landless labourers and also the heavily resource-dependent communities of hunter–gatherers, swidden agriculturists, animal herders and woodworking artisans, all stubborn pre-modern survivals in an increasingly postmodern landscape.

The process of development in independent India has been characterised by a basic asymmetry between omnivores and ecosystem people. A one-sentence definition

of development, as it has unfolded over the last fifty years, would be: "Development is the channelling of an ever-increasing volume of natural resources, through the intervention of the state apparatus and at the cost of the state exchequer, to serve the interests of the rural and urban omnivores." Some central features of this process have been:

1. The concentration of political power/decision-making in the hands of omnivores.[21]
2. Hence the use of the state machinery to divert natural resources to islands of omnivore prosperity, especially through subsidies. Wood for paper mills, fertilisers for rich farmers, water and power for urban dwellers are all supplied by the state to omnivores at well below market prices.
3. The culture of subsidies has fostered an indifference among omnivores to the environmental degradation they cause, aided by their ability to pass on its costs to ecosystem people or to society at large.
4. Projects based on the capture of wood, water or minerals—such as eucalyptus plantations, large dams or opencast mining—have tended to dispossess the ecosystem people who previously enjoyed ready access to those resources. This has led to a rising tide of protests by the victims of development—Chipko, Narmada and dozens of other protests that we know collectively as the "Indian environmental movement".
5. But development has also *permanently* displaced large numbers of ecosytem people from their homes. Some twenty million Indians have been uprooted by steel mills, dams and the like; countless others have been forced to move to the cities in search of a legitimate livelihood denied to them in the countryside (sometimes as a direct consequence of environmental degradation).[22] Thus has been created a third class, of *ecological refugees,* living in slums and temporary shelters in the towns and cities of India.

This framework, which divides the Indian population into the three socio-ecological classes of omnivores, ecosystem people and ecological refugees, can help us understand why economic development since 1947 has destroyed nature while failing to remove poverty. The framework synthesises the insights of ecology with sociology, in that it distinguishes social classes by their respective resource catchments, by their cultures and styles of consumption, and also by their widely varying powers to influence state policy.

The framework is analytical as well as value-laden, descriptive and prescriptive. It helps us understand and interpret nature-based conflicts at various spatial levels: from the village community upwards through the district and region and on to the nation. Stemming from the study of the history of modern India, it might also throw light on the dynamics of socio-ecological change in other large developing "Third World" countries such as Brazil and Malaysia, where conflicts between omnivores and ecosystem people have also erupted and whose cities are likewise marked by a growing population of "ecological refugees". At a pinch, it might explain asymmetries and inequalities at the global level, too. More than a hundred years ago a famous German radical proclaimed, "Workers of the World, Unite!" But as another German radical[23] recently reminded this writer, the reality of our times is very nearly the reverse: the process of globalisation, whose motto might well be, "Omnivores of the World, Unite!"

Conflicting Fallacies

What, then, is the prospect for the future? Consider two well-known alternatives already prominent in the market place of ideas:

1. *The Fallacy of the Romantic Economist,* which states that everyone can become an omnivore if only we allow the market full play. That is the hope, and the illusion, of globalisation, which promises a universalisation of American styles of consumption. But this is nonsense, for although businessmen and economists resolutely refuse to recognise it, there are clear ecological limits to a global consumer society, to all Indians or Mexicans attaining the lifestyle of an average middle-class North American. Can there be a world with one billion cars, an India with two hundred million cars?

2. *The Fallacy of the Romantic Environmentalist,* which claims that ecosystem people want to remain ecosystem people. This is the anti-modern, anti-Western, anti-science position of some of India's best-known, neo-Gandhian environmentalists.[24] This position is also gaining currency among some sections of Western academia. Anthropologists in particular are almost falling over themselves in writing epitaphs for development, in works that seemingly dismiss the very prospect of directed social change in much of the Third World. It is implied that development is a nasty imposition on the innocent peasant and tribesperson who, left to themselves, would not willingly partake of Enlightenment rationality, modern technology or modern consumer goods.[25] This literature has become so abundant and so influential that it has even been anthologised, in a volume called (what else?) *The Post-Development Reader.*[26]

The editor of this volume is a retired Iranian diplomat now living in the south of France. The authors of those other demolitions of the development project mentioned in footnote 25 are, without exception, tenured professors at well-established American universities. I rather suspect that the objects of their sympathy would cheerfully exchange their own social position for that of their chroniclers. For it is equally a fallacy that ecosystem people want to remain as they are, that they do not want to enhance their own resource consumption, to get some of the benefits of science, development and modernity.

This point can be made more effectively by way of anecdote. Some years ago, a group of Indian scholars and activists gathered in the southern town of Manipal for a national meeting in commemoration of the 125th anniversary of the birth of Mahatma Gandhi. They spoke against the backdrop of a life-size portrait of Gandhi, depicting him clad in the loincloth he wore for the last thirty-three years of his life. Speaker after speaker invoked his mode of dress as symbolising the Mahatma's message. Why did we all not follow his example and give up everything, thus to mingle more definitively with the masses?

Then, on the last evening of the conference, the Dalit (low-caste) poet Devanur Mahadeva got up to speak. He read out a short poem in the Kannada language of southwest India, written not by him but by a Dalit woman of his acquaintance. The poem spoke reverentially of the great Untouchable leader B. R. Ambedkar (1889–1956) and especially of the dark blue suit that Ambedkar invariably wore in the last three decades of *his* life. Why did the Dalit lady focus on Ambedkar's suit, asked Mahadeva? Why, indeed, did the countless statues of Ambedkar put up in Dalit hamlets always have him clad in suit and tie? His answer was deceptively and eloquently simple: if Gandhi wears a loincloth, we all marvel at his *tyaga,* his sacrifice. The scantiness of dress is in this case a marker of what the man has given up. A high-caste, well-born, English-educated lawyer had voluntarily chosen to renounce power and position and live the life of an Indian peasant. That is why we memorialise that loincloth.

However, if Ambedkar had worn a loincloth, that would not occasion wonder or surprise. He is a Dalit, we would say—what else should he wear? Millions of his caste fellows wear nothing else. It is the fact that he has escaped this fate, that his extraordinary personal achievements (a law degree from Lincoln's Inn, a PhD from Columbia University, the drafting of the Constitution of India) have allowed him to escape the fate that society and history had

Table 2 Hierarchies of Resource Consumption

Fuel Used	Mode of Housing	Mode of Transport
Grass	Cave	Feet
Wood	Thatched hut	Bullock cart
Coal	Wooden house	Bicycle
Gas	Stone house	Motor scooter
Electricity	Cement house	Car

allotted him, that is so effectively symbolised in that blue suit. Modernity, not tradition, development, not stagnation, are responsible for this inversion, for this successful and all-too infrequent storming of the upper-caste citadel.

A Blueprint for India

Let me now attempt to represent the story of Dr B. R. Ambedkar's suit in more material terms. Consider the simple hierarchies of fuel, housing and transportation set out in Table 2.

To move down any level of this table is to move towards a more reliable, more efficient, longer-lasting and generally safer mode of consumption. Why, then, would one abjure cheap and safe cooking fuel, for example, or quick and reliable transport, or stable houses that can outlive one monsoon? To prefer gas to dung for your stove, a car to a bullock-cart for your mobility, a wooden home to a straw hut for your family's shelter, is to choose greater comfort, wellbeing and freedom. These are choices that, despite specious talk of cultural difference, must be made available to all humans.

At the same time, to move down these levels is generally to move towards a more intensive and possibly unsustainable use of resources. Unsustainable at the global level, that is, for while a car admittedly expands freedom, there is no possibility whatsoever of every human on Earth being able to possess a car. As things stand, some people consume too much, while others consume far too little. It is these asymmetries that a responsible politics would seek to address. Confining ourselves to India, for instance, one would work to enhance the social power of ecological refugees and ecosystem people, their ability to govern their lives and to gain from the transformation of nature into artefact. This policy would simultaneously force omnivores to internalise the costs of their profligate behaviour. A new, "left–green" development strategy would feature the following five central elements:

1. A move towards a genuinely participatory democracy, with a strengthening of the institutions of local governance (at village, town or district levels) mandated by the Constitution of India but aborted by successive central governments in New Delhi. The experience of the odd state, such as West Bengal and Karnataka, which has experimented successfully with the *panchayat* or self-government system suggests that local control is conducive to the successful management of forests, water, etc.

2. Creation of a process of natural resource use which is open, accessible and accountable. This would include a Freedom of Information Act, so that citizens are fully informed about the intentions of the state and better able to challenge or welcome them, thus making officials more responsive to their public.

3. The use of decentralisation to stop the widespread undervaluing of natural resources. The removal of subsidies and the putting of proper price tags will make resource use more efficient and less destructive of the environment.

4. The encouragement of a shift to private enterprise for producing goods and services, while ensuring that there are no hidden subsidies and that firms properly internalise externalities. There is at present an unfortunate distaste for the market among Indian radicals, whether Gandhian or Marxist. But one cannot turn one's back on the market; the task rather is to tame it. The people and environment of India have already paid an enormous price for allowing state monopolies in sectors such as steel, energy, transport and communications.

5. This kind of development can, however, succeed only if India is a far more equitable society than is the case at present. Three key ways of enhancing the social power of ecological refugees and ecosystem people (in all of which India has conspicuously failed) are land reform, literacy (especially female literacy) and proper health care. These measures would also help bring population growth under control. In the provision of health and education the state might be aided by the voluntary sector, paid for by communities out of public funds.

Remedying Global Inequalities

The charter of sustainable development outlined here[27] applies, of course, only to one country, albeit a large and representative one. Its *raison d'etre* is the persistent and grave inequalities of consumption within the nation. What, then, of inequalities of consumption between nations? This question has been authoritatively addressed in a recent study of the prospects for a "sustainable Germany" sponsored by the Wüppertal Institute for Climate and Ecology.[28] Its fundamental premise is that the North lays excessive claim to the "environmental space" of the South. The way the global economy is currently structured,

> The North gains access to cheap raw materials and hinders access to markets for processed products from those countries; it imposes a system (World Trade Organisation) that favours the strong; it makes use of large areas of land in the South, tolerating soil degradation, damage to regional eco-systems, and disruption of local self-reliance; it exports toxic waste; it claims patent rights to utilisation of biodiversity in tropical regions, etc.[29]

Seen "against the backdrop of a divided world," says the report, "the excessive use of nature and its resources in the North is a principal block to greater justice in the world. . . . A retreat of the rich from overconsumption is thus a necessary first step towards allowing space for improvement of the lives of an increasing number of people." The problem thus identified, the report goes on to itemise, in meticulous detail, how Germany can take the lead in reorienting its economy and society towards a more sustainable path. It begins with an extended treatment of overconsumption, of the excessive use of the global commons by the West over the past two hundred years, of the terrestrial consequences of profligate lifestyles—soil erosion, forest depletion, biodiversity loss, air and water pollution. It then outlines a long-term plan for reducing the "throughput" of nature in the economy and cutting down on emissions. The report sets targets for substantial cuts by the year 2010 in the consumption of energy (at least 30 percent) and non-renewable raw materials (25 percent), and in the release of substances such as carbon dioxide (35 percent), sulphur dioxide (80–90 percent), synthetic nitrogen fertilisers (100 percent) and agricultural biocides (100 percent).

The policy and technical changes necessary to achieve these targets are identified as including the elimination of subsidies for chemical farming, the levying of ecological taxes (on gasoline, for example), the adoption of slower and fuel-efficient cars and the movement of goods by rail instead of road. Some examples of resource conservation in practice are given, such as the replacement of concrete girders by those made with steel, water conservation and recycling within the city, and a novel contract between the Munich municipal authorities and organic farmers in the countryside. By adopting such measures, Germany would transform itself from a nature-abusing to a nature-saving country.

The Wüppertal Institute study is notable for its mix of moral ends with material means, as well as its judicious blending of economic and technical options. More striking still has been its reception. The original German book sold forty thousand copies, an abbreviated version selling an additional hundred thousand copies. It was made into an award-winning television film and discussed by trade unions, political parties, consumer groups, scholars, church congregations and countless lay citizens. In several German towns and regions attempts have begun to put some of its proposals into practice.

Inequalities of consumption thus need to be addressed at both national and international levels. Indeed, the two are interconnected. The Spanish economist Juan Martinez-Alier provides one telling example. In the poorer countries of Asia and Africa, firewood and animal dung are often the only sources of cooking fuel. These are inefficient and polluting, and their collection involves mwuch drudgery. The provision of oil or liquefied petroleum gas for the cooking stoves of Somali or Nepalese peasant women would greatly improve the quality of their lives. This could easily be done, says Martinez-Alier, if the rich were very moderately taxed. He calculates that to replace the fuel used by the world's three thousand million poor people would require about two hundred million tons of oil a year. Now, this is only a quarter of the United States' annual consumption. But the bitter irony is that "oil at $15 a barrel is so cheap that it can be wasted by rich countries, but [is] too expensive to be used as domestic fuel by the poor". The solution is simple: oil consumption in the rich countries should be taxed, while the use of liquefied petroleum gas or kerosene for fuel in the poor countries should be subsidised.[30]

Allowing the poor to ascend just one rung up the hierarchies of resource consumption requires a very moderate sacrifice by the rich. In the present climate, however, any proposal with even the slightest hint of redistribution would be shot down as smacking of "socialism". But this might change, as conflicts over consumption begin to sharpen, as they assuredly shall. Within countries, access to water, land, and forest and mineral resources will be fiercely fought over by contending groups. Between countries, there will be bitter arguments about the "environmental space" occupied by the richer nations.[31] As these divisions become more manifest, the global replicability of Western styles of living will be more directly and persistently challenged. Sometime in the middle decades of the twenty-first century, Galbraith's great unasked question, "How Much Should a Country Consume?", with its corollary, "How Much Should a Person Consume?", will come, finally, to dominate intellectual and political debate.

Notes

1. John Kenneth Galbraith, *The Affluent Society* (London: Hamish Hamilton, 1958), pp. 109–10.

2. Ibid., p. 96.

3. John Kenneth Galbraith, "How Much Should a Country Consume?", in *Perspectives on Conservation*, ed. Henry Jarret (Baltimore: Johns Hopkins University Press, 1958), pp. 91–2.

4. Carl Sauer, "Theme of Plant and Animal Destruction in Economic History" (1938), in his *Land and Life* (Berkeley: University of California Press, 1963), p. 154.

5. Galbraith, "How Much Should a Country Consume?", p. 97.

6. Samuel Hays, "From Conservation to Environment: Environmental Politics in the United States since World War Two", *Environmental Review* 6, no. 1 (1982), p. 21.

7. For details, see my essays, "Radical American Environmentalism and Wilderness Preservation: A Third World Critique", *Environmental Ethics* 11, no. 1 (spring 1989), and "The Two Phases of American Environmentalism: A Critical History", in *Decolonizing Knowledge,* ed. Stephen A. Marglin and Frederique Apffel-Marglin (Oxford: Clarendon Press, 1996).

8. Paul R. Ehrlich, *The Population Bomb* (New York: Ballantine Books, 1968), p. 15.

9. See Satyajit Singh, "Environment, Class and State in India: A Perspective on Sustainable Irrigation" (PhD dissertation, Delhi University, 1994).

10. Galbraith, "How Much Should a Country Consume?", p. 90.

11. Helmut Lippelt, "Green Politics in Progress: Germany", *International Journal of Sociology and Social Policy* 12, nos. 4–7 (1992), p. 197.

12. Rudolf Bahro, *From Red to Green: Interviews with New Left Review* (London: Verso, 1984), p. 184.

13. This paragraph is based on Werner Hülsberg, *The German Greens: A Social and Political Profile* (London: Verso, 1988); but see also Margit Mayer and John Ely, eds., *Between Movement and Party: The Paradox of the German Greens* (Philadelphia: Temple University Press, 1997), and Saral Sarkar, "The Green Movement in West Germany", *Alternatives* 11, no. 2 (1986).

14. Mahatma Gandhi, "Discussion with a Capitalist", *Young India,* 20 December 1928, in the *Collected Works of Mahatma Gandhi* (hereafter *CWMG*) [New Delhi: Publications Division, n.d.], vol. 38, p. 243.

15. "The Same Old Argument", *Young India,* 7 October 1926 (*CWMG*, vol. 31, p. 478).

16. "Choice before Us", *Young India,* 2 June 1927 (*CWMG*, vol. 33, pp. 417–8).

17. "No and Yes", *Young India,* 17 March 1927 (*CWMG*, vol. 33, p. 163).

18. See *Harijan,* 23 June 1946.

19. See especially the two *Citizens Reports on the Indian Environment,* published in 1982 and 1985 by the New Delhi–based Centre for Science and Environment. See also the magisterial essay by the centre's director, Anil Agarwal, "Human–Nature Interactions in a Third World Country", *Environmentalist* 6, no. 3 (1986).

20. The following paragraphs expand and elaborate on some ideas first presented in Madhav Gadgil and Ramachandra Guha, *Ecology and Equity: The Use and Abuse of Nature in Contemporary India* (London: Routledge, 1995).

21. See Pranab Bardhan, *The Political Economy of India's Development* (Oxford: Clarendon Press, 1984).

22. See Eenakshi Ganguly-Thukral, ed., *Big Dams, Displaced People* (New Delhi: Sage Publishers, 1992).

23. The environmentalist and social critic Wolfgang Sachs.

24. See, for example, Ashis Nandy, ed., *Science, Hegemony and Violence: A Requiem for Modernity* (New Delhi: Oxford University Press, 1989); Vandana Shiva, *Staying Alive: Women, Ecology and Development* (London: Zed Books, 1989).

25. See, for example, Arturo Escobar, *Encountering Development: The Making and Unmaking of the Third World* (Princeton, N.J.: Princeton University Press, 1995); James Scott, *Seeing Like a State: How Certain Schemes to Improve the Human Condition Have Failed* (New Haven: Yale University Press, 1998); and especially Wolfgang Sachs, ed., *The Development Dictionary: A Guide to Knowledge as Power* (London: Zed Books, 1992).

26. Majid Rahnema, ed., *The Post-Development Reader* (London: Zed Books, 1998).

27. And elaborated in more detail in Gadgil and Guha, *Ecology and Equity.*

28. Wolfgang Sachs et al., *Greening the North: A Post-Industrial Blueprint for Ecology and Equity* (London: Zed Books, 1998), on which the rest of this section is based. See also F. Schmidt-Beek, ed., *Carnoules Declaration: Factor 10 Club* (Wüppertal: WIKUE, 1994), which sets the target of a 90 percent reduction in material use by the industrialised countries.

29. Sachs et al., *Greening the North,* p. 159.

30. See Martinez-Alier's essay, "Poverty and the Environment", in Ramachandra Guha and Juan Martinez-Alier, *Varieties of Environmentalism: Essays North and South* (London: Earthscan, 1997). See also Juan Martinez-Alier, *Ecological Economics: Energy, Environment, Society,* rev. ed. (London: Basil Blackwell, 1991).

31. In this connection, see Anil Agarwal and Sunita Narain, *Global Warming in an Unequal World: A Case of Environmental Colonialism?* (New Delhi: Centre for Science and Environment, 1992).

Critical Thinking

1. Based on your readings in this text, describe some of the potential global consequences of the "consumer society."

2. What would be the global environmental consequences (potential) of both India and China "developing" the way the United States has?

3. With regard to patterns of consumption, what is the difference between "omnivores" and "ecosystem people?"

4. Explain the idea of "inequalities of consumption between nations."

5. What are some implications this article has for global development? For feeding humanity? For a thirsty planet? For our quest for energy? Pick one and discuss.

6. Identify in this article three key terms, concepts, or principles that are used in your textbook (environmental science, economics, sociology, history, geography, etc,) or employed in the discipline you are currently studying. (Note: The terms, concepts. or principles may be implicit, explicit, implied, or inferred.)

RAMACHANDRA GUHA is a historian and anthropologist who lives in Bangalore. His paper is the product of a research and writing grant from the John D. and Catherine T. MacArthur Foundation.

Consumption, Not Population Is Our Main Environmental Threat

FRED PEARCE

It's the great taboo, I hear many environmentalists say. Population growth is the driving force behind our wrecking of the planet, but we are afraid to discuss it.

It sounds like a no-brainer. More people must inevitably be bad for the environment, taking more resources and causing more pollution, driving the planet ever farther beyond its carrying capacity. But hold on. This is a terribly convenient argument—"over-consumers" in rich countries can blame "over-breeders" in distant lands for the state of the planet. But what are the facts?

The world's population quadrupled to six billion people during the 20th century. It is still rising and may reach 9 billion by 2050. Yet for at least the past century, rising per-capita incomes have outstripped the rising head count several times over. And while incomes don't translate precisely into increased resource use and pollution, the correlation is distressingly strong.

Moreover, most of the extra consumption has been in rich countries that have long since given up adding substantial numbers to their population.

By almost any measure, a small proportion of the world's people take the majority of the world's resources and produce the majority of its pollution.

Take carbon dioxide emissions—a measure of our impact on climate but also a surrogate for fossil fuel consumption. Stephen Pacala, director of the Princeton Environment Institute, calculates that the world's richest half-billion people—that's about 7 percent of the global population—are responsible for 50 percent of the world's carbon dioxide emissions. Meanwhile the poorest 50 percent are responsible for just 7 percent of emissions.

Although overconsumption has a profound effect on greenhouse gas emissions, the impacts of our high standard of living extend beyond turning up the temperature of the planet. For a wider perspective of humanity's effects on the planet's life support systems, the best available measure is the "ecological footprint," which estimates the area of land required to provide each of us with food, clothing, and other resources, as well as to soak up our pollution. This analysis has its methodological problems, but its comparisons between nations are firm enough to be useful.

They show that sustaining the lifestyle of the average American takes 9.5 hectares, while Australians and Canadians require 7.8 and 7.1 hectares respectively; Britons, 5.3 hectares; Germans, 4.2; and the Japanese, 4.9. The world average is 2.7 hectares. China is still below that figure at 2.1, while India and most of Africa (where the majority of future world population growth will take place) are at or below 1.0.

The United States always gets singled out. But for good reason: It is the world's largest consumer. Americans take the greatest share of most of the world's major commodities: corn, coffee, copper, lead, zinc, aluminum, rubber, oil seeds, oil, and natural gas. For many others, Americans are the largest per-capita consumers. In "super-size-me" land, Americans gobble up more than 120 kilograms of meat a year per person, compared to just 6 kilos in India, for instance.

I do not deny that fast-rising populations can create serious local environmental crises through overgrazing, destructive farming and fishing, and deforestation. My argument here is that viewed at the global scale, it is overconsumption that has been driving humanity's impacts on the planet's vital life-support systems during at least the past century. But what of the future?

We cannot be sure how the global economic downturn will play out. But let us assume that Jeffrey Sachs, in his book *Common Wealth,* is right to predict a 600 percent increase in global economic output by 2050. Most projections put world population then at no more than 40 percent above today's level, so its contribution to future growth in economic activity will be small.

Of course, economic activity is not the same as ecological impact. So let's go back to carbon dioxide emissions. Virtually all of the extra 2 billion or so people expected on this planet in the coming 40 years will be in the poor half of the world. They will raise the population of the poor world from approaching 3.5 billion to about 5.5 billion, making them the poor two-thirds.

Sounds nasty, but based on Pacala's calculations—and if we assume for the purposes of the argument that per-capita emissions in every country stay roughly the same as today—those extra two billion people would raise the share of emissions contributed by the poor world from 7 percent to 11 percent.

Look at it another way. Just five countries are likely to produce most of the emissions and UN fertility projections. He found that an extra child in the United States today will, down the generations, produce an eventual carbon footprint seven times that of an extra Chinese child, 46 times that of a Pakistan child, 55 times that of an Indian child, and 86 times that of a Nigerian child.

Of course those assumptions may not pan out. I have some confidence in the population projections, but per-capita emissions of carbon dioxide will likely rise in poor countries for some time yet, even in optimistic scenarios. But that is an issue of consumption, not population.

In any event, it strikes me as the height of hubris to downgrade the culpability of the rich world's environmental footprint because generations of poor people not yet born might one day get to be as rich and destructive as us. Overpopulation is not driving environmental destruction at the global level; overconsumption is. Every time we talk about too many babies in Africa or India, we are denying that simple fact.

At root this is an ethical issue. Back in 1974, the famous environmental scientist Garret Hardin proposed something he called "lifeboat ethics." In the modern, resource-constrained world, he said, "each rich nation can be seen as a lifeboat full of comparatively rich people. In the ocean outside each lifeboat swim the poor of the world, who would like to get in." But there were, he said, not enough places to go around. If any were let on board, there would be chaos and all would drown. The people in the lifeboat had a duty to their species to be selfish—to keep the poor out.

Hardin's metaphor had a certain ruthless logic. What he omitted to mention was that each of the people in the lifeboat was occupying ten places, whereas the people in the water only wanted one each. I think that changes the argument somewhat.

Critical Thinking

1. Articulate in one sentence the primary argument employed by the author to support his thesis.

2. What implications does the author's argument have with regard to any one of the unit topics presented in this text? Pick one topic and describe some connections/implications.

3. How does the author connect scientist Garret Hardin's "lifeboat ethics" to the consumption argument?

4. Do you agree or disagree with Question 3? Why or why not?

5. Why does such a small portion of the earth's population consume the majority of the earth's resources?

FRED PEARCE is a freelance author and journalist based in the UK. He is an environment consultant for New Scientist magazine and author of recent books *When The Rivers Run Dry* and *With Speed and Violence* (Beacon Press).

The Issue: Natural Resources, What Are They?

WORLD RESOURCE FORUM

Global Resource Use—Worldwide Patterns of Resource Extraction

Economic and thus human development have always been closely linked to the control and production of materials. Due to continued growth of the global economy, the demand for natural resources, such as fossil fuels, metals and minerals, and biomass from agriculture (crops), forestry, fishery, etc, provided by Planet Earth is rapidly increasing, and they are being exploited without metres and bounds. This results in serious environmental damages through the extraction process itself, but also due to the ever longer transport distances between extraction, processing and final consumption.

Global resource extraction grew more or less steadily over the past 25 years, from 40 billion tons in 1980 to 58 billion tons in 2005, representing an aggregated growth rate of 45%.

Figure 1 illustrates the overall material basis and the growing resource extraction of the global economy between 1980 and 2005. However, growth rates were unevenly distributed among the main material categories. Particularly the extraction of metal ores increased (by more than 65%), indicating the continued importance of this resource category for industrial development. Increases in biomass extraction were below average. The share of renewable resources in total resource extraction thus is decreasing on the global level.

Taking into account all the materials that are extracted but not actually used to create value in economic processes (i.e. overburden in mining or "ecological rucksacks"), the resource extraction more than doubled in the last 25 years. Due to simultaneously increasing world population numbers, the average resource extraction per capita remained almost stable, today amounting for nearly nine tons.

Figure 1   Global used extraction of natural resources in categories.

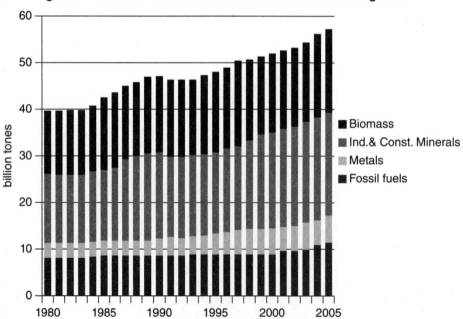

Regarding material intensity, i.e. economic output per unit of domestic natural resource extraction, Europe is the most 'eco-efficient' region, while Africa produces the smallest economic output per domestic extraction. Nonetheless, Europe's share in worldwide resource extraction is 1.5 times higher than the share of the African continent and Europe is increasingly importing natural resources from other world regions.

Disaggregating the extraction data by world regions, it can be seen that, as a consequence of rapid industrialisation in countries such as China and India, Asia's share in global resource extraction has increased steadily, especially since the early 1990s. From 1980 to 2005 the extraction of fossil fuels in China, for instance, tripled; the total increase in used extraction amounted for 150%.

Europe's resource extraction grew only by 3%, but studies show that these raw materials are increasingly being substituted by imports from other world regions. Latin America, for example, is specialising noticeably in the extraction of resource-intensive export products, such as metal ores or biomass for biofuels. In 2005, Chile extracted fivefold the amount of copper of 1980, Brazil threefold the amount of sugar cane—being the raw material for ethanol fuel.

North America brings in the highest net regional metal imports, receiving 82% of all regional net metal imports. The two territories importing the most metals worldwide (US$ net) are the United States and Mexico. . . . Net imports are imports minus exports. . . . Mineral depletion is the loss of potential future income at current prices due to current quantities of minerals extracted. Included here are gold, lead, zinc, iron, copper, nickel, silver, bauxite, and phosphate. Territories with the highest mineral depletion are Australia, Brazil, Chile and China. Australia is the largest producer of bauxite, Brazil of industrial diamonds, China of tungsten, and South Africa of platinum and gold. Mineral extraction often causes environmental damage, itself a form of depletion. . . .

Global Material Extraction and Resource Efficiency

The above explained exchange of domestic extraction by imported materials does also affect the countries' eco-efficiency, expressed by the material intensity indicator . . . , reflecting economic output per unit of domestic natural resource extraction. One observes that industrialised economies are characterised by the lowest material intensities (or highest eco-efficiency), with Europe being world-leader with around 1.25 tons per 1000 US $ GDP in the 1980s and improving to 0.75 tons at the beginning of this decade. Without a doubt this development is partly the result of the use of new technologies with improved material and energy performance and structural change of economies towards service sectors characterised by less material input per economic output.

Nonetheless, the picture generated by this indicator is distorted, as this leading position is gained, to a certain degree, at the expense of the exporting countries. Figure 2 clearly illustrates that, on a world-wide level, it has been possible to decouple economic growth (GDP) and resource use (extraction); nonetheless, absolute numbers of resource extraction are still increasing, mirroring the fact that efficiency gains through structural or technological effects are overcompensated by scale effects brought about by economic growth. The finiteness of important resources as well as constricted regeneration capacities make a reversal of this trend indispensable.

Another effect of an extraction for export is that the added value remaining in the exporting country is very low, which also affects the material intensity. Thus, while steadily increasing raw material prices result in enormous revenue growth for resource rich countries (e.g. through taxes) as well as for the extracting companies, the trickle-down effect to the countries' population is limited, depending on the political strategies of

Figure 2   Trends of worldwide growth of GDP and resource use

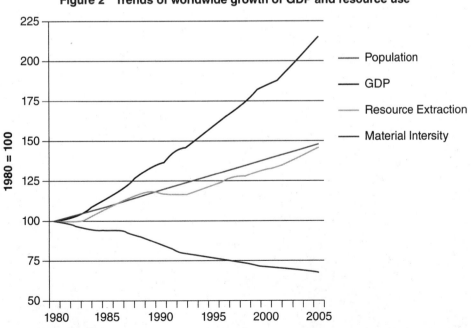

the local government. Additionally, the dependence on resource exports brings about a high grade of vulnerability, as price changes, or even slumps, have especially severe consequences on the local economy.

Scenarios of Future Resource Extraction

Scenarios on future natural resource extraction, applying integrated economic-environmental models show that in a baseline scenario without additional policies to limit resource use, used domestic extraction within the EU remains roughly constant until 2020, while unused domestic extraction decreases (particularly overburden from mining activities). . . .

The stabilisation of domestic extraction, however, is accompanied by growing imports of material intensive products. This indicates that the material requirements of the European economy will increasingly be met through imports from other world regions, causing shifts of environmental pressures related to material extraction and processing away from Europe towards resource-rich countries.

In order to quantify the use of resources as presented above, the Sustainable Europe Research Institute (SERI) in Vienna built up and maintains the only worldwide comprehensive data base on resource extraction, which comprises data for almost 200 countries, 270 types of resources, and currently a time series of 26 years (1980–2005). The complete aggregated data set is freely accessible on the website www.materialflows.net.

Critical Thinking

1. Using information provided in Articles 1–6, make some predictions about how the chart in Figure 1 will look in 2050.

2. According to the data in this article, what nations are consuming the Earth?

3. How will the growing material requirements of the consuming nations cause new shifts of environmental pressures?

Consumption and Consumerism

ANUP SHAH

Global inequality in consumption, while reducing, is still high.

Using latest figures available, in 2005, the wealthiest 20% of the world accounted for 76.6% of total private consumption. The poorest fifth just 1.5%.

Breaking that down slightly further, the poorest 10% accounted for just 0.5% and the wealthiest 10% accounted for 59% of all the consumption (Figure 1).

In 1995, the inequality in consumption was wider, but the United Nations also provided some eye-opening statistics (which do not appear available, yet, for the later years) worth noting here:

Today's consumption is undermining the environmental resource base. It is exacerbating inequalities. And the dynamics of the consumption-poverty-inequality-environment nexus are accelerating. If the trends continue without change—not redistributing from high-income to low-income consumers, not shifting from polluting to cleaner goods and production technologies, not promoting goods that empower poor producers, not shifting priority from consumption for conspicuous display to meeting basic needs—today's problems of consumption and human development will worsen.

. . . The real issue is not consumption itself but its patterns and effects.

. . . Inequalities in consumption are stark. Globally, the 20% of the world's people in the highest-income countries account for 86% of total private consumption expenditures—the poorest 20% a minuscule 1.3%. More specifically, the richest fifth:

- Consume 45% of all meat and fish, the poorest fifth 5%
- Consume 58% of total energy, the poorest fifth less than 4%
- Have 74% of all telephone lines, the poorest fifth 1.5%
- Consume 84% of all paper, the poorest fifth 1.1%
- Own 87% of the world's vehicle fleet, the poorest fifth less than 1%

Runaway growth in consumption in the past 50 years is putting strains on the environment never before seen.

— *Human Development Report 1998 Overview,*[1]
United Nations Development Programme (UNDP)

— *Emphasis Added. Figures quoted use data from 1995*

If they were available, it would likely be that the breakdowns shown for the 1995 figures will not be as wide in 2005. However, they are likely to still show wide inequalities in consumption. Furthermore, as a few developing countries continue to develop and help make the numbers show a narrowing gap, there are at least two further issues:

- Generalized figures hide extreme poverty and inequality of consumption on the whole (for example, between 1995 and 2005, the inequality in consumption for the poorest fifth of humanity has hardly changed)
- If emerging nations follow the same path as today's rich countries, their consumption patterns will also be damaging to the environment

And consider the following, reflecting world priorities:

Global Priority	$U.S. Billions
Cosmetics in the United States	8
Ice cream in Europe	11
Perfumes in Europe and the United States	12
Pet foods in Europe and the United States	17
Business entertainment in Japan	35
Cigarettes in Europe	50
Alcoholic drinks in Europe	105
Narcotics drugs in the world	400
Military spending in the world	780

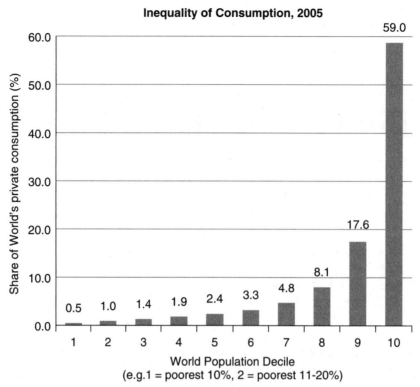

Inequality of Consumption, 2005

Source: World Bank Development Indicators 2008

And compare that to what was estimated as *additional* costs to achieve universal access to basic social services in all developing countries:

Global Priority	$U.S. Billions
Basic education for all	6
Water and sanitation for all	9
Reproductive health for all women	12

"Over" population is usually blamed as the major cause of environmental degradation, but the above statistics strongly suggests otherwise. As we will see, consumption patterns today are not to meet everyone's needs. The system that drives these consumption patterns also contributes to inequality of consumption patterns too.

This section of the globalissues.org web site will attempt to provide an introductory look at various aspects of what we consume and how.

- We will see possible "hidden" costs of convenient items to society, the environment and individuals, as well as the relationship with various sociopolitical and economic effects on those who do consume, and those who are unable to consume as much (due to poverty and so on).
- We will look at how some luxuries were turned into necessities in order to increase profits.
- This section goes beyond the "don't buy this product" type of conclusion to the deeper issues and ramifications.

- We will see just a hint at how wasteful all this is on resources, society and capital. The roots of such disparities in consumption are inextricably linked to the roots of poverty. There is such enormous waste in the way we consume that an incredible amount of resources is wasted as well. Furthermore, the processes that lead to such disparities in unequal consumption are themselves wasteful and is structured deep into the system itself. Economic efficiency is for making profits, not necessarily for social good (which is treated as a side effect). The waste in the economic system is, as a result, deep. Eliminating the causes of this type of waste are related to the elimination of poverty and bringing rights to all. Eliminating the waste also allows for further equitable consumption for all, as well as a decent standard of consumption.
- So these issues go beyond just consumption, and this section only begins to highlight the enormous waste in our economy which is not measured as such.
- A further bold conclusion is also made that elimination of so much wasted capital would actually require a reduction of people's workweek. This is because the elimination of such waste means entire industries are halved in size in some cases. So much labor redundancy cannot be tolerated, and hence the answer is therefore to share the remaining productive jobs, which means reducing the workweek!

- We will see therefore, that political causes of poverty are very much related to political issues and roots of consumerism. Hence solutions to things like hunger, environmental degradation, poverty and other problems have many commonalities that would need to be addressed.

Critical Thinking

1. The report states: "The real issue is not consumption itself but its patterns and effects." Refer to the text world map, and sketch on the map what you see are the current "patterns" of consumption.

2. Explain the statement: "Today's consumption is undermining the environmental resource base."

3. What does this report imply about the current use of our fossil fuels? Will future energy alternatives be used as well to maintain consumption?

4. The report argues "There are important issues around consumerism that need to be understood." Select 3 issues and provide a brief answer for each. Try to center your answers around the concept of "environment."

5. Identify in this article three key terms, concepts, or principles that are used in your textbook (environmental science, economics, sociology, history, geography, etc,) or employed in the discipline you are currently studying. (Note: The terms, concepts, or principles may be implicit, explicit, implied, or inferred.)

From *Global Issues*, March 6, 2011. Copyright © 2011 by Anup Shah. Reprinted by permission. www.globalissues.org/issue/235/consumption-and-consumerism

UNIT 2

The Human Factor: Environmental Virus or Symbiosis?

Unit Selections

Learning Outcomes

After reading this unit, you should be able to:

• Identify linkages between the environmental issue of Human Population Growth and global development patterns, demographic changes, consumer trends, and the environment.

• Explain why loss of biodiversity and habitat may be the most difficult environmental problem to "fix."

• Outline the reasons that population growth and its subsequent consumption demands will vary among peoples and places.

• Offer insights on what the impacts of changing population demographics might have on global development and natural resources.

• Identify underlying connections between current consumption trends and possible future environmental issues.

• Elaborate on the four demographic megatrends that are predicted to change the world and the potential environmental impacts they may imply.

• Provide examples of how human population growth over the last century might appear more "bacterial" than "primate."

Student Website

www.mhhe.com/cls

Internet References

Can Cities Save the Future?
www.huduser.org/publications/econdev/habitat/prep2.html

Earth Renewal
www.earthrenewal.org/global_economics.htm

Geography and Socioeconomic Development
www.ksg.harvard.edu/cid/andes/Documents/Background%20
Papers/Geography Socioeconomic%20Development.pdf

Global Footprint Network
http://footprints@footprintnetwork.org

Global Trends Project
www.globaltrendsproject.org

Graphs Comparing Countries
http://humandevelopment.bu.edu/use_existing_index/start_comp_
graph.cfm

IISDnet
www.iisd.org

Linkages on Environmental Issues and Development
www.iisd.ca/linkages

People and Planet
www.peopleandplanet.org

Population Action International
www.populationaction.org

Population Reference Bureau
www.prb.org

SocioSite: Sociological Subject Areas
www.pscw.uva.nl/sociosite/TOPICS

Man will survive as a species for one reason: He can adapt to the destructive effects of our power-intoxicated technology and of our ungoverned population growth, to the dirt, pollution and noise of a New York or Tokyo. And that is the tragedy. It is not man the ecological crisis threatens to destroy but the quality of human life.

—René Dubos, quoted in Life, 28, July 1970

© Tom Mareschal/Alamy

There can be no true discourse on the planetary consumption issue without including the primary driving agent of that issue: humans. But it's not simply the nature of our consumption that threatens the planet's ability to sustain our species. It's our population growth *patterns* and the *demographic characteristics* of those patterns that are creating new challenges to the future health of our environment and our own survival as a species. Many scholars have suggested that the earth can support a human population of 9, 10, even 11 billion people. However, the earth's ability to support *quantity* of life is not the issue. It's *quality* of life. How do we want our billions to live? To begin to answer this, we first need to recognize—and admit—that humans must consume the Earth: transform raw resources into a form we can use to ensure our well-being as a species as well as our survival. Second, we need to realize that our current human consumption patterns are proving to be unsustainable and damaging to the ecosystems upon which all living things must depend. Third, and most important, we need to better understand what possible future human population growth patterns—geographic, demographic, social, and economic—may unfold and how those patterns may impact our ability to ensure *quality of life* for all humans and maintain a *sustainable relationship* with our planet.

Science is about being able to make accurate predictions. If we can predict human–environment interaction outcomes better, we can make better decisions, and we can often change the inevitability of unpleasant surprises and even fatal mistakes. Unit 2 seeks to stimulate the reader's thinking about the characteristics, patterns, and subsequent challenges to consider regarding the future of human population growth and our need to consume resources.

What will our future relationship with the planet evolve to—environmental symbiosis or environmental virus? Unit 2 opens the examination of this question with Article 9, "People and the Planet: Executive Summary," an overview report by the Royal Society Policy Center, which highlights the people–environment–consumption nexus. The report states that emerging global demographic changes and consumption patterns are leading to three pressing challenges: reducing poverty and thus global inequality; urgently reducing consumption in developed and emerging economies; and slowing and stabilizing global population growth.

Article 10, "The Human Factor," continues to consider the Unit 2 question in an interview with renowned environmental scientist E. O. Wilson who sees the pattern of population growth of the past 20th century "more bacterial than primate." He believes our greatest threat to survival is not war, energy, economy, or political instability but the destruction of natural habitats that our resource consumption creates and the subsequent loss of genetic and species diversity.

Although Article 11, "Global Aging and Crisis of the 2020s" by Neal Howe and Richard Jackson, does not directly address the environmental consumption issue, it is included because it provides the reader an excellent opportunity to uncover hidden linkages to and potential consequences for the future of our environment and resource consumption patterns. As global population demographics change, so to do humans' relationship to the environment in terms of what is needed to sustain their well-being. Will an aging population place fewer material demands upon the environment? Will such populations require and demand *more* clean air and water, *more* sustainable food production? Or will they consume *less* energy than its 20th century predecessors?

Unit 2 concludes with writer Jack Goldstone offering predictions (in 2010) regarding what he believes to be the next "population bomb" and the global megatrends it will set in motion in Article 12, "The New Population Bomb: Four Megatrends That Will Change the World." The trends of global population growth in the world's poorest countries, continued urbanization of our population, changing demographics of the industrialized nations' populations, and subsequent political economy consequences are all linked to consumption patterns. However, it is also imperative that we expand our thinking on the issues presented in this article and assess the potential future scenarios in light of both place—local and regional impacts within nations—and the potential environmental impacts (which will vary considerably by place) resulting from these changing human earth consumption patterns.

To conclude with regard to the human factor focused on in Unit 2, we must constantly be cognizant of the question: *Who* is currently consuming the Earth and at what rate? And *who* will be consuming the Earth in the future, at what rate, and sustainably or unsustainably?

People and the Planet: Executive Summary

ROYAL SOCIETY POLICY CENTER

Summary

The 21st century is a critical period for people and the planet. The global population reached 7 billion during 2011 and the United Nations projections indicate that it will reach between 8 and 11 billion by 2050. Human impact on the Earth raises serious concerns, and in the richest parts of the world per capita material consumption is far above the level that can be sustained for everyone in a population of 7 billion or more. This is in stark contrast to the world's 1.3 billion poorest people, who need to consume more in order to be raised out of extreme poverty.

The highest fertility rates are now seen primarily in the least developed countries while the lowest fertility rates are seen in the more developed countries, and increasingly in Asia and Latin America. Despite a decline in fertility almost everywhere, global population is still growing at about 80 million per year, because of the demographic momentum inherent in a large cohort of young people. The global rate of population growth is already declining, but the poorest countries are neither experiencing, nor benefiting from, this decline.

Population and consumption are both important: the combination of increasing global population and increasing overall material consumption has implications for a finite planet. As both continue to rise, signs of unwanted impacts and feedback (eg climate change reducing crop yields in some areas) and of irreversible changes (eg the increased rate of species extinction) are growing alarmingly. The relationship between population, consumption and the environment is not straightforward, as the natural environment and human socioeconomic systems are complex in their own right. The Earth's capacity to meet human needs is finite, but how the limits are approached depends on lifestyle choices and associated consumption; these depend on what is used, and how, and what is regarded as essential for human wellbeing.

Demographic change is driven by economic development, social and cultural factors as well as environmental change. A transition from high to low birth and death rates has occurred in various cultures, in widely different socio-economic settings, and at different rates. Countries such as Iran and South Korea have moved through the phases of this transition much more rapidly than Europe or North America. This has brought with it challenges different from those that were experienced by the more developed countries as they reached the late stages of the transition.

Population is not only about the growing numbers of people: changes in age structure, migration, urbanisation and population decline present both opportunities and challenges to human health,
wellbeing and the environment. Migrants often provide benefits to their countries of origin, through remittances, and to their host countries by helping to offset a workforce gap in ageing populations. Current and future migration will be affected by environmental change, although lack of resources may mean that the most vulnerable to these changes are the least able to migrate. Policy makers should prepare for international migration and its consequences, for integration of migrants and for protection of their human rights.

Developing countries will be building the equivalent of a city of a million people every five days from now to 2050. The continuing and rapid growth of the urban population is having a marked bearing on lifestyle and behaviour: how and what they consume, how many children they have, the type of employment they undertake. Urban planning is essential to avoid the spread of slums, which are highly deleterious to the welfare of individuals and societies.

The demographic changes and consumption patterns described above lead to three pressing challenges.

First, the world's 1.3 billion poorest people need to be raised out of extreme poverty. This is critical to reducing global inequality, and to ensuring the wellbeing of all people. It will require increased per capita consumption for this group, allowing improved nutrition and healthcare, and reduction in family size in countries with high fertility rates.

Second, in the most developed and the emerging economies unsustainable consumption must be urgently reduced. This will entail scaling back or radical transformation of damaging material consumption and emissions and the adoption of sustainable technologies, and is critical to ensuring a sustainable future for all. At present, consumption is closely linked to economic models based on growth. Improving the wellbeing of individuals so that humanity flourishes rather than survives requires moving from current economic measures to fully valuing natural capital. Decoupling economic activity from material and environmental throughputs is needed urgently for example by reusing equipment and recycling materials, reducing waste, obtaining energy from renewable sources, and by consumers paying for the wider costs of their consumption. Changes to the current socio-economic model and institutions are needed to allow both people and the planet to flourish by collaboration as well as competition during this and subsequent centuries. This requires farsighted political leadership concentrating on long term goals.

Third, global population growth needs to be slowed and stabilised, but this should by no means be coercive. A large unmet need for contraception remains in both developing and developed countries. Voluntary family planning is a key part of continuing the downward trajectory in fertility rates, which brings benefits to the individual wellbeing of men and women around the world. In the long term a stabilised population is an essential prerequisite for individuals to flourish. Education will play an important role: well educated people tend to live longer healthier lives, are more able to choose the number of children they have and are more resilient to, and capable of, change. Education goals have been repeatedly agreed by the international community, but implementation is poor.

Science and technology have a crucial role to play in meeting these three challenges by improving the understanding of causes and effects (such as stratospheric ozone depletion), and in developing ways to limit the most damaging trends (such as enhancing agricultural production with reduced environmental impact). However, attention must be paid to the socio-economic dimensions of technological deployment, as barriers will not be overcome solely by technology but in combination with changes in usage and governance.

Demographic changes and their associated environmental impacts will vary across the globe, meaning that regional and national policy makers will need to adopt their own range of solutions to deal with their specific issues. At an international level, this year's Rio+20 Conference on Sustainable Development, the discussions at the UN General Assembly revisiting the International Conference on Population and Development (ICPD+20) scheduled for 2014/2015 and the review of the Millennium Development Goals in 2015 present opportunities to reframe the relationship between people and the planet. Successfully reframing this relationship will open up a prosperous and flourishing future, for present and future generations.

Recommendations

Recommendation 1

The international community must bring the 1.3 billion people living on less than $1.25 per day out of absolute poverty, and reduce the inequality that persists in the world today. This will require focused efforts in key policy areas including economic development, education, family planning and health.

Recommendation 2

The most developed and the emerging economies must stabilise and then reduce material consumption levels through: dramatic improvements in resource use efficiency, including: reducing waste; investment in sustainable resources, technologies and infrastructures; and systematically decoupling economic activity from environmental impact.

Recommendation 3

Reproductive health and voluntary family planning programmes urgently require political leadership and financial commitment, both nationally and internationally. This is needed to continue the downward trajectory of fertility rates, especially in countries where the unmet need for contraception is high.

Recommendation 4

Population and the environment should not be considered as two separate issues. Demographic changes, and the influences on them, should be factored into economic and environmental

debate and planning at international meetings, such as the Rio+20 Conference on Sustainable Development and subsequent meetings.

Recommendation 5

Governments should realise the potential of urbanisation to reduce material consumption and environmental impact through efficiency measures. The well planned provision of water supply, waste disposal, power and other services will avoid slum conditions and increase the welfare of inhabitants.

Recommendation 6

In order to meet previously agreed goals for universal education, policy makers in countries with low school attendance need to work with international funders and organisations, such as UNESCO, UNFPA, UNICEF, IMF, World Bank and Education for All. **Financial and non-financial barriers must be overcome to achieve high-quality primary and secondary education for all the world's young, ensuring equal opportunities for girls and boys.**

Recommendation 7

Natural and social scientists need to increase their **research efforts on the interactions between consumption, demographic change and environmental impact.** They have a unique and vital role in developing a fuller picture of the problems, the uncertainties found in all such analyses, the efficacy of potential solutions, and providing an open, trusted source of information for policy makers and the public.

Recommendation 8

National Governments should accelerate the development of comprehensive wealth measures. This should include reforms to the system of national accounts, and improvement in natural asset accounting.

Recommendation 9

Collaboration between National Governments is needed to **develop socio-economic systems and institutions that are not dependent on continued material consumption growth.** This will inform the development and implementation of policies that allow both people and the planet to flourish.

Critical Thinking

1. Describe what is meant by *demographic change* and explain why it plays an important role in any discussions about planetary consumption.

2. The article states: "Population is not only about the growing numbers of people. . . ." Explain what this means.

3. What are the three pressing challenges confronting us today? Do you think the consuming the Earth/overconsumption issue has contributed to creating these challenges?

4. How do you think the recommendations of the summary report can be implemented (enforced)?

5. Can the recommendations be successful without first addressing the environmental overconsumption issue? Explain.

The Human Factor

E. O. Wilson talks to Elizabeth Kolbert about ants, habitat, and the threat of mass extinction.

ELIZABETH KOLBERT

If you could talk to any scientist in the world, who would it be? For me, the choice is pretty easy: E. O. Wilson. Wilson is one of the natural world's keenest observers and, at the same time, probably its most eloquent spokesman. As much as any one person can, he has tried to save what's left of the astonishing diversity of life.

Now 81, Wilson began his career as a naturalist when he was barely out of elementary school. As a 13-year-old Boy Scout, in 1942, he noticed some peculiar ant colonies in a vacant lot near his house, in Mobile, Alabama. In this way, he discovered that the imported red fire ant, a native of South America, had invaded the United States. (The Department of Agriculture's attempt to eradicate the ants in the 1950s would become a case study in misguided pesticide application, chronicled by Rachel Carson and dubbed by Wilson the "Vietnam of entomology.") Twenty-five years after his first boyhood discoveries, as a zoology professor at Harvard, Wilson cowrote one of the seminal books in ecology, *The Theory of Island Biogeography,* and in 1975, more or less singlehandedly, he created a whole new field of inquiry: sociobiology—the concept that behavior, including human behavior, has a basis in evolution. Wilson also became one of the first scientists to warn about what has since become known as the biodiversity crisis.

"The worst thing that can happen, will happen," he wrote in 1980.

> *Not energy depletion, economic collapse, limited nuclear war, or conquest by a totalitarian government. As terrible as these catastrophes would be for us, they can be repaired within a few generations. The one process ongoing in the 1980s that will take millions of years to correct is the loss of genetic and species diversity by the destruction of natural habitats. This is the folly our descendants are least likely to forgive us.*

In 1991, Wilson and his coauthor, Bert Hölldobler, won a Pulitzer Prize—Wilson's second—for their book *The Ants.* Though he is now retired from teaching, he is still writing; he published his first novel, *Anthill,* to generally positive reviews

in 2010. He is also still studying ants and still speaking out about the threat that one species—ours—poses to the millions of others on the planet.

I met up with Wilson in his office, which is above Harvard's Museum of Comparative Zoology and right across the hall from the university's ant collection. In the 1970s, Wilson's concept of sociobiology caused so much controversy at Harvard that he briefly considered decamping to another university; however, he has said, he could not bear to leave behind the school's ant collection, which contains nearly 1,000,000 individual ants, representing more than 6,000 species. After we had talked for a while, Wilson showed me around the collection. Many of the ants Wilson gathered himself on trips to, among other places, Brazil, Cuba, Fiji, and New Guinea. Each ant—some were so small I could barely make them out—was marked with a tiny label bearing its name in almost microscopic print. Wilson's office is decorated with pieces of ant-related art, which friends and colleagues have given him, and also with three hominid skulls, which, he likes to tell visitors, belonged to former grad students.

In his 1994 autobiography, *Naturalist,* Wilson described himself as "a happy man in a terrible century." By his own account, he places "great store in civility and good manners." Both in person and in his writing, he is courtly and good-humored. But he's blunt about the damage that humans are causing.

"It seems like almost on an annual basis now we have another really massive biodiversity problem to worry about," he told me. "We've reached the point where global catastrophes are getting to be the norm. We were just hacking down parts of the natural world in pieces here and there, but now things are getting global and they are coming right home." I spoke to Wilson about his early scientific career, his turn to environmental advocacy, and the future of biodiversity. One of the first topics we spoke about was Wilson's fieldwork in the 1950s and 1960s.

E. O. Wilson: When you went out into the field, say in Veracruz, Mexico, or Fiji, when I first got there, you didn't have somebody take you in a four-wheel drive to a field station that was all set up and give you a cot. I would get into a town somewhere and make myself comfortable and I almost always collected alone. I prefer that actually. Then someone would tell me there's a nice patch of rainforest if you go 30 kilometers up the road. And I would get a ride out there.

The next thing you'd do is get out of the car, pry apart a barbed wire fence, and step through. You'd walk among the cows, watching out not to get your feet gummy with cow manure. Then there it is: good-looking vegetation, and it's on an incline. So you know probably the reason it hasn't been cut was because it's on a slope. Before you get there the land dips down into a stream and a bog. You have to get through another barbed wire fence and cross the stream, getting yourself muddy, and then finally you are climbing your way up and starting to collect. That was so typical of those days. So many islands were like this. There were not many active field biologists in those days, but most of us became aware, especially those of us who were going into tropical areas, that entire ecosystems were being wiped out.

In 1967 Wilson and Robert MacArthur, a biology professor at Princeton, published The Theory of Island Biogeography, *their landmark work in ecology. The book attempted to explain why certain islands are species-rich, while others are species-poor. One of the key variables is size: all other things being equal, larger islands will be home to more species than smaller ones. The ratio, Wilson and MacArthur found, tends to follow a consistent mathematical formula: roughly speaking, the number of species doubles with every tenfold increase in area. The formula also works in reverse, so that if an island's area is reduced by 90 percent, the number of species that can survive will drop by half. I asked Wilson about the development of the theory and how it applied to "islands" of habitat in a fragmented landscape.*

Wilson: Islands are the laboratories of species formation and extinction. Islands are separate experiments in what species can get there and what species turn into new species, which then go extinct, and so on. That's why islands have always been so important. They were important to Darwin and much more important to Alfred Russel Wallace, who founded biogeography.

In 1959 I got together with Robert MacArthur. We were two young, ambitious biologists. I was much more the naturalist and data person and Robert was a brilliant mathematical modeler. The idea was to use islands as our laboratory to understand how species spread and how and why they become extinct. On islands the patterns became much clearer than on the mainland. It also turns out that islands have the highest rates of human-induced extinctions. They are the real disaster areas. Hawaii is the extinction capital of America and one of the hot spots of the world.

Then we realized that the same principles apply to anything that is broken into fragments—for example, bodies of fresh-water, from rivers to streams to springs to lakes to ponds.

In the 1970s Wilson's interest in extinction ceased to be academic. He began to speak out and to rethink his assumptions about what it means to be a scientist. I asked him about this transition.

Wilson: In the 1960s it was sort of just in my peripheral vision that what we were doing was relevant to understanding the extinction process. I thought that what we should be doing was finding out the information needed. At that time, oddly, in the scientific culture it was regarded as unseemly that scientists speak up or take an activist role. And being at Harvard, with all the devotion I felt to the purity of science, I certainly could see the problem, but my duty was to stay with science and just get the theory and facts right. And that I shouldn't say anything. After all, there was the World Wildlife Fund and the International Union for the Conservation of Nature, all those great people, and they were taking care of that.

By 1970 there was a significant burgeoning of conservation awareness and the early stages of the new conservation movement, which was also becoming increasingly scientific in its basis. I give credit to Paul Ehrlich [now president of Stanford University's Center for Conservation Biology] and to Peter Raven [now president emeritus of the Missouri Botanical Garden], who were first-rate scientists. They were yelling their heads off by that time, and nobody was calling them exhibitionists or loudmouths. So I thought probably they were an example to be followed. In 1974 I published a hard-hitting article in *Harvard Magazine* on species extinction and the desperate need to have a global conservation movement. That's how I got into it.

Wilson often compares humans with ants and termites. Social insects, like humans, cooperate in ways that are enormously empowering. I asked him why it is that ants don't, at least as far we know, drive other species to extinction.

Wilson: Ants and termites today make up very roughly three-quarters of the total insect biomass. A lot of people might think that's amazing because there are only some 15,000 ant species, whereas there are over 900,000 known species of insects. In terms of biomass, they own the world. But in terms of the damage they've done? Have the ants taken over the world? No. Why? Because—and this tells us the significance of the human story—ants came into existence in the Mesozoic, more than 100 million years ago. We have fossils now and we know what they looked like and how they appeared. In the beginning they were rather scarce. It was only after tens of millions of years of evolution, occasionally rushing out and stinging the feet of dinosaurs who were clumsy enough to step on their nests, that they diversified a good deal and probably had advanced societies. But they still were not the most abundant insects. That didn't occur until the beginning of the age of mammals.

We know that by 15 million years ago ants had taken over. Their habitats had tens of millions of years to evolve and accommodate the ants as they became more and more prevalent. There are vast numbers of insects that make their living off of ants, and the ants make their living off of other insects, particularly sap-sucking insects, with which they share nutrients. Many, many species of plants have special structures that they developed so that ants could live in them, and the ants in turn could offer them protection. After this long period of evolution, ants eased into their preeminent position, which they have held for 15 million years. Not bad.

Now considering humanity, good lord . . . When *Homo erectus* began to evolve a million years ago, they too were quite scarce. Then kaboom. *Homo sapiens* became prominent, maybe 200,000 years ago. Out of Africa they spread, maybe 60,000 years ago. And they began to have an impact. They wiped out the megafauna wherever they went, animals of maybe 100 pounds or more. Then kaboom, kaboom, kaboom. Neolithic agriculture. Villages and towns and technology. We hit the world's fauna and flora with a sucker punch, if I could use ordinary language like that. They just weren't ready for us. We really got on the scene big time 10,000 years ago. No way to coadapt. The ants gave them at least 30 million years, and we gave them 10,000 years. We're too darn good at it.

So what about us? I asked Wilson why it was that humans seem to be so good at driving other organisms extinct. Are we just very good competitors, or are we somehow qualitatively different from other organisms?

Wilson: This is a very rough figure, but before modern humans came along, especially before the Neolithic period 10,000 years ago, the extinction rate was, as close as we can tell, about one species going extinct per million per year. And the rate at which species were being created before humanity plunked down in the middle of things was the same. So you had very roughly an equilibrium. Even though species were coming and going, it was a very slow turnover. We have now upped the extinction rate by at least 1,000 times. It could easily go to 10,000. We have eliminated so much natural habitat that we now have places all around the world where we've cut everything down to 10 percent or 20 percent of the original cover. That's what you see in the Philippines, in Madagascar, in Hawaii. Substantial parts of Indonesia are going that way fast. Another very rough figure: if you reduce an area by 90 percent, so you have only 10 percent of the original forest left, for example, whatever the habitat is, you can expect to sustain only about half the species. Half of them, roughly, will go extinct. This is what we learned from our theory of island biogeography, and it happens pretty quick.

We are also reducing the birthrate—the cradles are disappearing. Now, why would I say that if we're not very careful we could easily take it from 1,000 up to 10,000 times the prehuman extinction rate? Because it's so easy to knock out the land that remains. That's why I think we can use those high figures. Some experts might be more conservative, but that's where we're at now. That's why we have to create preserves all around the world, make them as large as possible, and also connect them up with one another. Most scientists working in conservation and biology would agree. We need corridors, particularly in the face of climate change.

In other words, humans are unique as destroyers. There's never been anything like humans before on a global scale.

Wilson has said that "the pattern of human population growth in the twentieth century was more bacterial than primate." I asked him how this pattern has been sustained.

Wilson: We need to realize that up until this point we have saved our own species with technology, new developments in agriculture, opening new land—and therefore of course destroying large numbers of other species. We've always found a way around our exponential population growth through technology. When it comes to energy extraction, we've had to develop very high technology and complex systems, and they're getting more complex all the time. We've reached a point where what we have wrought is so complicated and ill planned that we can't handle a lot of it. That takes us to a point where we have to recognize that we're not going to have any kind of livable planet for ourselves unless we make our environment sustainable—and that includes the living environment. We have to slam on the brakes before we wreck the planet. Which we're about to do.

I'm exasperated by the professional optimists, who say, "These bad things are happening, but human genius and the resources this good earth has given us will allow us to keep on doing pretty much what we've been doing. So don't worry about it." There is a huge flaw in that reasoning. You may be familiar with the famous French riddle of the 29th day of lily pads, which everybody should know. Most of the things that should be worrying us are happening exponentially. With each interval that passes, the impact is more serious. There comes a time when the next interval is devastating, and here comes the riddle of the 29th day.

There is a pond with one lily pad. The number of lily pads doubles every day. The pond will fill up by the 30th day. On what day is the pond half full? The 29th day. That's the whole point. Okay, let's take the optimists' argument. Let's assume that the lily pads are getting really worried. They point out, Hey guys, we've got one short period left and then we're all going to be fighting over the same nutrients and choking each other out. Not to worry: some of the lily pads are technological geniuses. They'll figure out a way to get around this. They announce to every lily pad's joy that they have found an opening now to another empty pond of the same size, so you don't have to change your patterns and way of life, fellow lily pads, we can go on. Okay. When does that new pond fill up? On the 31st day. That's the problem with humanity today, and I don't care if we spend a huge amount of money to shoot people off into space toward Mars. I don't care if we figure out a way to do fusion or find some other energy source. We can't go on like this.

In his book The Diversity of Life, *Wilson writes about the five "mass extinction" events that have been identified in the geologic record. The most recent of these took place 65 million years ago, when an asteroid seven and a half miles wide hit the Earth, killing off not just the dinosaurs, but also the pterosaurs, plesiosaurs, and mosasaurs. I asked Wilson if he thought we were in the midst of another mass extinction event.*

Wilson: Sure we are. We're in the early stages but it's coming on fast. Compared with a giant meteorite strike, of course, it's taking longer, but that only means a few centuries, and we're coming up to where it will get up to those catastrophic levels within this century. Unless we can somehow stop it.

Population is going to peak at around 9 billion or 10 billion, 40 percent more than we have now. We can manage this if we make the necessary changes. We can't just go on using the usual nonrenewable energy and planting the same old crops. We have to develop dryland agriculture in areas that have run out of groundwater. That agriculture has to be based on—I will just say it right out—genetically modified crops that can produce high yields without sucking up the world's remaining groundwater. At the same time, we've got to save what's left—biological diversity. That's basically it. We can do it with reserves, with a lot more knowledge of what the species are and what they need. This is the reason, incidentally, why I put a lot of my time in the last 10 years into helping to get the Encyclopedia of Life started. [See Alan Burdick, "The (New) Web of Life," *OnEarth,* Fall 2009.] We need to be putting a lot more scientific and technical effort into mapping the world's biodiversity. We only know fewer than 10 percent of species. We're really flying blind.

Wilson has written that we are a product of our own evolutionary history, that we evolved to be shortsighted, kin-concerned creatures, and that this is no longer serving us very well. But how do you get beyond your own evolutionary history?

Wilson: We are ill equipped by instinct to control ourselves. Even with our tremendous intellect, we have a deep propensity for group conflict. Look at our defense expenditures, the way we glorify the constant expansion of human settlement and human growth, our archaic religions, which give us nothing but grief because they are essentially tribal. Our religions are ill equipped to handle our present problems, especially when they start trying to discredit what we can find out and prove with science.

However, we are adaptable. I think that when we get enough really serious knocks, we can start treating our problems as problems and not as the evil machinations of conspiracies—which is the way we tend to think about them. Our problems come from the fact that we are a *Star Wars* civilization. When you think about *Star Wars,* those movies reflect what we are: people blowing up whole planets. We have Paleolithic emotions, Stone Age emotions—we've inherited those nice and pure. We

have medieval institutions. And we have godlike technology. Put those three together and you have a very dangerous mix. So, somehow I think we ought to develop a new kind of self-understanding, self-reflection, and self-imaging. Then we might be able to actually get somewhere together.

NRDC: A Continent of Riches

Latin America is home to some of the most biodiverse ecosystems on the planet. Costa Rica immediately comes to mind. Its rainforests, mountains, and coasts simply pulse with life. Just a few minutes in a Costa Rican rainforest is enough to notice the complex networks among the variety of species there. At the southern end of the continent, in Chile's Patagonia, there are remarkable terrestrial and riparian ecosystems along the Baker River. This nutrient-rich river, which originates in Andean glaciers, ultimately feeds the biodiverse marine fjord system near the Pacific coast. The Baker is another tangible illustration of the connections in nature—in this case among the area's glaciers, temperate rainforests, swamps, and marine life.

Critical Thinking

1. According to E.O. Wilson, what may be the greatest environmental catastrophe looming ahead? Why is it so terrible?

2. Why is it, according to Wilson, that humans seem to be so good at driving other organisms extinct?

3. What linkage(s) might you see between what Wilson says about humanity's role in biodiversity loss and the "consuming the earth" theme in Unit I?

4. Wilson says, "We are ill-equipped by instinct to control ourselves." Compare this observation to the observations about human consumption in Articles 1, 2, 3, 4, and 5. Outline some of the similarities in the articles.

Global Aging and the Crisis of the 2020s

"The risk of social and political upheaval could grow throughout the developing world—even as the developed world's capacity to deal with such threats declines."

Neil Howe And Richard Jackson

From the fall of the Roman and the Mayan empires to the Black Death to the colonization of the New World and the youth-driven revolutions of the twentieth century, demographic trends have played a decisive role in many of the great invasions, political upheavals, migrations, and environmental catastrophes of history. By the 2020s, an ominous new conjuncture of demographic trends may once again threaten widespread disruption. We are talking about global aging, which is likely to have a profound effect on economic growth, living standards, and the shape of the world order.

For the world's wealthy nations, the 2020s are set to be a decade of rapid population aging and population decline. The developed world has been aging for decades, due to falling birthrates and rising life expectancy. But in the 2020s, this aging will get an extra kick as large postwar baby boom generations move fully into retirement. According to the United Nations Population Division (whose projections are cited throughout this article), the median ages of Western Europe and Japan, which were 34 and 33 respectively as recently as 1980, will soar to 47 and 52 by 2030, assuming no increase in fertility. In Italy, Spain, and Japan, more than half of all adults will be older than the official retirement age—and there will be more people in their 70s than in their 20s.

Falling birthrates are not only transforming traditional population pyramids, leaving them top-heavy with elders, but are also ushering in a new era of workforce and population decline. The working-age population has already begun to contract in several large developed countries, including Germany and Japan. By 2030, it will be stagnant or contracting in nearly all developed countries, the only major exception being the United States. In a growing number of nations, total population will begin a gathering decline as well. Unless immigration or birthrates surge, Japan and some European nations are on track to lose nearly one-half of their total current populations by the end of the century.

The working-age population has already begun to contract in several large developed countries, including Germany and Japan.

These trends threaten to undermine the ability of today's developed countries to maintain global security. To begin with, they directly affect population size and GDP size, and hence the manpower and economic resources that nations can deploy. This is what RAND scholar Brian Nichiporuk calls "the bucket of capabilities" perspective. But population aging and decline can also indirectly affect capabilities—or even alter national goals themselves.

Rising pension and health care costs will place intense pressure on government budgets, potentially crowding out spending on other priorities, including national defense and foreign assistance. Economic performance may suffer as workforces gray and rates of savings and investment decline. As societies and electorates age, growing risk aversion and shorter time horizons may weaken not just the ability of the developed countries to play a major geopolitical role, but also their will.

The weakening of the developed countries might not be a cause for concern if we knew that the world as a whole were likely to become more pacific. But unfortunately, just the opposite may be the case. During the 2020s, the developing world will be buffeted by its own potentially destabilizing demographic storms. China will face a massive age wave that could slow economic growth and precipitate political crisis just as that country is overtaking America as the world's leading economic power. Russia will be in the midst of the steepest and most protracted population implosion of any major power since the plague-ridden Middle Ages. Meanwhile, many other developing countries, especially in the Muslim world, will experience a sudden new resurgence of youth whose aspirations they are unlikely to be able to meet.

China will face a massive age wave that could slow economic growth and precipitate political crisis.

The risk of social and political upheaval could grow throughout the developing world—even as the developed world's capacity to deal with such threats declines. Yet, if the developed world seems destined to see its geopolitical stature diminish, there is one partial but important exception to the trend: the United States. While it is fashionable to argue that US power has peaked, demography suggests America will play as important a role in shaping the world order in this century as it did in the last.

Demography suggests America will play as important a role in shaping the world order in this century as it did in the last.

Graying Economies

Although population size alone does not confer geopolitical stature, no one disputes that population size and economic size together constitute a potent double engine of national power. A larger population allows greater numbers of young adults to serve in war and to occupy and pacify territory. A larger economy allows more spending on the hard power of national defense and the semi-hard power of foreign assistance. It can also enhance what political scientist Joseph Nye calls "soft power" by promoting business dominance, leverage with nongovernmental organizations and philanthropies, social envy and emulation, and cultural clout in the global media and popular culture.

The expectation that global aging will diminish the geopolitical stature of the developed world is thus based in part on simple arithmetic. By the 2020s and 2030s, the working-age population of Japan and many European countries will be contracting by between 0.5 and 1.5 percent per year. Even at full employment, growth in real GDP could stagnate or decline, since the number of workers may be falling faster than productivity is rising. Unless economic performance improves, some countries could face a future of secular economic stagnation—in other words, of zero real GDP growth from peak to peak of the business cycle.

Economic performance, in fact, is more likely to deteriorate than improve. Workforces in most developed countries will not only be stagnating or contracting, but also graying. A vast literature in the social and behavioral sciences establishes that worker productivity typically declines at older ages, especially in eras of rapid technological and market change.

Economies with graying workforces are also likely to be less entrepreneurial. According to the Global Entrepreneurship Monitor's 2007 survey of 53 countries, new business start-ups in high-income countries are heavily tilted toward the young. Of all "new entrepreneurs" in the survey (defined as owners of a business founded within the past three and one-half years), 40 percent were under age 35 and 69 percent under age 45. Only 9 percent were 55 or older.

At the same time, savings rates in the developed world will decline as a larger share of the population moves into the retirement years. If savings fall more than investment demand, as much macroeconomic modeling suggests is likely, either businesses will starve for investment funds or the developed economies' dependence on capital from higher-saving emerging markets will grow. In the first case, the penalty will be lower output. In the second, it will be higher debt service costs and the loss of political leverage, which history teaches is always ceded to creditor nations.

Even as economic growth slows, the developed countries will have to transfer a rising share of society's economic resources from working-age adults to nonworking elders. Graying means paying—more for pensions, more for health care, more for nursing homes for the frail elderly. According to projections by the Center for Strategic and International Studies, the cost of maintaining the current generosity of today's public old-age benefit systems would, on average across the developed countries, add an extra 7 percent of GDP to government budgets by 2030.

Yet the old-age benefit systems of most developed countries are already pushing the limits of fiscal and economic affordability. By the 2020s, political conflict over deep benefit cuts seems unavoidable. On one side will be young adults who face stagnant or declining after-tax earnings. On the other side will be retirees, who are often wholly dependent on pay-as-you-go public plans. In the 2020s, young people in developed countries will have the future on their side. Elders will have the votes on theirs.

Faced with the choice between economically ruinous tax hikes and politically impossible benefit cuts, many governments will choose a third option: cannibalizing other spending on everything from education and the environment to foreign assistance and national defense. As time goes by, the fiscal squeeze will make it progressively more difficult to pursue the obvious response to military manpower shortages—investing massively in military technology, and thereby substituting capital for labor.

Diminished Stature

The impact of global aging on the collective temperament of the developed countries is more difficult to quantify than its impact on their economies, but the consequences could be just as important—or even more so. With the size of domestic markets fixed or shrinking in many countries, businesses and unions may lobby for anticompetitive changes in the economy. We may see growing cartel behavior to protect market share and more restrictive rules on hiring and firing to protect jobs.

We may also see increasing pressure on governments to block foreign competition. Historically, eras of stagnant population and market growth—think of the 1930s—have been characterized by rising tariff barriers, autarky, corporatism, and other anticompetitive policies that tend to shut the door on free trade and free markets.

This shift in business psychology could be mirrored by a broader shift in social mood. Psychologically, older societies are likely to become more conservative in outlook and possibly more risk-averse in electoral and leadership behavior. Elder-dominated electorates may tend to lock in current public spending commitments at the expense of new priorities and shun decisive confrontations in favor of ad hoc settlements. Smaller families may be less willing to risk scarce youth in war.

We know that extremely youthful societies are in some ways dysfunctional—prone to violence, instability, and state failure. But extremely aged societies may also prove dysfunctional in some ways, favoring consumption over investment, the past over the future, and the old over the young.

Meanwhile, the rapid growth in ethnic and religious minority populations, due to ongoing immigration and higher-than-average minority fertility, could strain civic cohesion and foster a new diaspora politics. With the demand for low-wage labor rising, immigration (at its current rate) is on track by 2030 to double the percentage of Muslims in France and triple it in Germany. Some large European cities, including Amsterdam, Marseille, Birmingham, and Cologne, may be majority Muslim.

In Europe, the demographic ebb tide may deepen the crisis of confidence that is reflected in such best-selling books as *France Is Falling* by Nicolas Baverez, *Can Germany Be Saved?* by Hans-Werner Sinn, and *The Last Days of Europe* by Walter Laqueur. The media in Europe are already rife with dolorous stories about the closing of schools and maternity wards, the abandonment of rural towns, and the lawlessness of immigrant youths in large cities. In Japan, the government has half-seriously projected the date at which only one Japanese citizen will be left alive.

Over the next few decades, the outlook in the United States will increasingly diverge from that in the rest of the developed world. Yes, America is also graying, but to a lesser extent. Aside from Israel and Iceland, the United States is the only developed nation where fertility is at or above the replacement rate of 2.1 average lifetime births per woman. By 2030, its median age, now 37, will rise to only 39. Its working-age population, according to both US Census Bureau and UN projections, will also continue to grow through the 2020s and beyond, both because of its higher fertility rate and because of substantial net immigration, which America assimilates better than most other developed countries.

The United States faces serious structural challenges, including a bloated health care sector, a chronically low savings rate, and a political system that has difficulty making meaningful trade-offs among competing priorities. All of these problems threaten to become growing handicaps as the country's population ages. Yet, unlike Europe and Japan, the United States will still have the youth and the economic resources to play a major geopolitical role. The real challenge facing America by the 2020s may not be so much its inability to lead the developed world as the inability of the other developed nations to lend much assistance.

Perilous Transitions

Although the world's wealthy nations are leading the way into humanity's graying future, aging is a global phenomenon. Most of the developing world is also progressing through the so-called demographic transition—the shift from high mortality and high fertility to low mortality and low fertility that inevitably accompanies development and modernization. Since 1975, the average fertility rate in the developing world has dropped from 5.1 to 2.7 children per woman, the rate of population growth has decelerated from 2.2 to 1.3 percent per year, and the median age has risen from 21 to 28.

The demographic outlook in the developing world, however, is shaping up to be one of extraordinary diversity. In many of the poorest and least stable countries (especially in sub-Saharan Africa), the demographic transition has failed to gain traction, leaving countries burdened with large youth bulges. By contrast, in many of the most rapidly modernizing countries (especially in East Asia), the population shift from young and growing to old and stagnant or declining is occurring at a breathtaking pace—far more rapidly than it did in any of today's developed countries.

Notwithstanding this diversity, some demographers and political scientists believe that the unfolding of the transition is ushering in a new era in which demographic trends will promote global stability. This "demographic peace" thesis, as we dub it, begins with the observation that societies with rapidly growing populations and young age structures are often mired in poverty and prone to civil violence and state failure, while those with no or slow population growth and older age structures tend to be more affluent and stable. As the demographic transition progresses—and population growth slows, median ages rise, and child dependency burdens fall—the demographic peace thesis predicts that economic growth and social and political stability will follow.

We believe this thesis is deeply flawed. It fails to take into account the huge variation in the timing and pace of the demographic transition in the developing world. It tends to focus exclusively on the threat of state failure, which indeed is closely and negatively correlated with the degree of demographic transition, while ignoring the threat of "neo-authoritarian" state success, which is more likely to occur in societies in which the transition is well under way. We are, in other words, not talking just about a hostile version of the Somalia model, but also about a potentially hostile version of the China or Russia model, which appears to enjoy growing appeal among political leaders in many developing countries.

More fundamentally, the demographic peace thesis lacks any realistic sense of historical process. It is possible (though by no means assured) that the global security environment that emerges after the demographic transition has run its course will be safer than today's. It is very unlikely, however, that the transition will make the security environment progressively safer along the way. Journeys can be more dangerous than destinations.

Economists, sociologists, and historians who have studied the development process agree that societies, as they move from the traditional to the modern, are buffeted by powerful and disorienting social, cultural, and economic crosswinds. As countries are integrated into the global marketplace and global culture, traditional economic and social structures are overturned and traditional value systems are challenged.

Along with the economic benefits of rising living standards, development also brings the social costs of rapid urbanization, growing income inequality, and environmental degradation. When plotted against development, these stresses exhibit a hump-shaped or inverted-U pattern, meaning that they become most acute midway through the demographic transition.

The demographic transition can trigger a rise in extremism. Religious and cultural revitalization movements may seek to reaffirm traditional identities that are threatened by modernization and try to fill the void left when development uproots communities and fragments extended families. It is well documented that international terrorism, among the developing countries, is positively correlated with income, education, and urbanization. States that sponsor terrorism are rarely among the youngest and poorest countries; nor do the terrorists themselves usually originate in the youngest and poorest countries. Indeed, they are often disaffected members of the middle class in middle-income countries that are midway through the demographic transition.

Ethnic tensions can also grow. In many societies, some ethnic groups are more successful in the marketplace than others—which means that, as development accelerates and the market economy grows, rising inequality often falls along ethnic lines. The sociologist Amy Chua documents how the concentration of wealth among "market-dominant minorities" has triggered violent backlashes by majority populations in many developing countries, from Indonesia, Malaysia, and the Philippines (against the Chinese) to Sierra Leone (against the Lebanese) to the former Yugoslavia (against the Croats and Slovenes).

We have in fact only one historical example of a large group of countries that has completed the entire demographic transition—today's (mostly Western) developed nations. And their experience during that transition, from the late 1700s to the late 1900s, was filled with the most destructive revolutions, civil wars, and total wars in the history of civilization. The nations that engaged in World War II had a higher median age and a lower fertility rate—and thus were situated at a later stage of the transition—than most of today's developing world is projected to have over the next 20 years. Even if global aging breeds peace, in other words, we are not out of the woods yet.

Storms Ahead

A number of demographic storms are now brewing in different parts of the developing world. The moment of maximum risk still lies ahead—just a decade away, in the 2020s. Ominously, this is the same decade when the developed world will itself be experiencing its moment of greatest demographic stress.

Consider China, which may be the first country to grow old before it grows rich. For the past quarter-century, China has been "peacefully rising," thanks in part to a one-child-per-couple policy that has lowered dependency burdens and allowed both parents to work and contribute to China's boom. By the 2020s, however, the huge Red Guard generation, which was born before the country's fertility decline, will move into retirement, heavily taxing the resources of their children and the state.

China's coming age wave—by 2030 it will be an older country than the United States—may weaken the two pillars of the current regime's legitimacy: rapidly rising GDP and social stability. Imagine workforce growth slowing to zero while tens of millions of elders sink into indigence without pensions, without health care, and without large extended families to support them. China could careen toward social collapse—or, in reaction, toward an authoritarian clampdown. The arrival of China's age wave, and the turmoil it may bring, will coincide with its expected displacement of the United States as the world's largest economy in the 2020s. According to "power transition" theories of global conflict, this moment could be quite perilous.

By the 2020s, Russia, along with the rest of Eastern Europe, will be in the midst of an extended population decline as steep or steeper than any in the developed world. The Russian fertility rate has plunged far beneath the replacement level even as life expectancy has collapsed amid a widening health crisis. Russian men today can expect to live to 60—16 years less than American men and marginally less than their Red Army grandfathers at the end of World War II. By 2050, Russia is due to fall to 16th place in world population rankings, down from 4th place in 1950 (or third place, if we include all the territories of the former Soviet Union).

Prime Minister Vladimir Putin flatly calls Russia's demographic implosion "the most acute problem facing our country today." If the problem is not solved, Russia will weaken progressively, raising the nightmarish specter of a failing or failed state with nuclear weapons. Or this cornered bear may lash out in revanchist fury rather than meekly accept its demographic fate.

Of course, some regions of the developing world will remain extremely young in the 2020s. Sub-Saharan Africa, which is burdened by the world's highest fertility rates and is also ravaged by AIDS, will still be racked by large youth bulges. So will a scattering of impoverished and chronically unstable Muslim-majority countries, including Afghanistan, the Palestinian territories, Somalia, Sudan, and Yemen. If the correlation between extreme youth and violence endures, chronic unrest and state failure could persist in much of sub-Saharan Africa and parts of the Muslim world through the 2020s, or even longer if fertility rates fail to drop.

Meanwhile, many fast-modernizing countries where fertility has fallen very recently and very steeply will experience a sudden resurgence of youth in the 2020s. It is a law of demography that, when a population boom is followed by a bust, it causes a ripple effect, with a gradually fading cycle of echo booms and busts. In the 2010s, a bust generation will be coming of age in much of Latin America, South Asia, and the Muslim world. But by the 2020s, an echo boom will follow—dashing economic expectations and perhaps fueling political violence, religious extremism, and ethnic strife.

These echo booms will be especially large in Pakistan and Iran. In Pakistan, the decade-over-decade percentage growth in the number of people in the volatile 15- to 24-year-old age bracket is projected to drop from 32 percent in the 2000s to just 10 percent in the 2010s, but then leap upward again to 19 percent in the 2020s. In Iran, the swing in the size of the youth bulge population is projected to be even larger: minus 33 percent in the 2010s and plus 23 percent in the 2020s. These echo booms will be occurring in countries whose social fabric is already strained by rapid development. One country teeters on the brink of chaos, while the other aspires to regional hegemony. One already has nuclear weapons, while the other seems likely to obtain them.

Pax Americana Redux?

The demographer Nicholas Eberstadt has warned that demographic change may be "even more menacing to the security prospects of the Western alliance than was the cold war for the past generation." Although it would be fair to point out that such change usually presents opportunities as well as dangers, his basic point is incontestable: Planning national strategy for the next several decades with no regard for population projections is like setting sail without a map or a compass. It is likely to be an ill-fated voyage. In this sense, demography is the geopolitical cartography of the twenty-first century.

Although tomorrow's geopolitical map will surely be shaped in important ways by political choices yet to be made, the basic contours are already emerging. During the era of the Industrial Revolution, the population of what we now call the developed world grew faster than the rest of the world's population, peaking at 25 percent of the world total in 1930. Since then, its share has declined. By 2010, it stood at just 13 percent, and it is projected to decline still further, to 10 percent by 2050.

The collective GDP of the developed countries will also decline as a share of the world total—and much more steeply. According to new projections by the Carnegie Endowment for International Peace, the Group of 7 industrialized nations' share of the Group of 20 leading economies' total GDP will fall from 72 percent in 2009 to 40 percent in 2050. Driving this decline will be not just the slower growth of the developed world, as workforces age and stagnate or contract, but also the expansion of large, newly market-oriented economies, especially in East and South Asia.

Again, there is only one large country in the developed world that does not face a future of stunning relative demographic and economic decline: the United States. Thanks to its relatively high fertility rate and substantial net immigration, its current global population share will remain virtually unchanged in the coming decades. According to the Carnegie projections, the US share of total G-20 GDP will drop significantly, from 34 percent in 2009 to 24 percent in 2050. The combined share of Canada, France, Germany, Italy, Japan, and the United Kingdom, however, will plunge from 38 percent to 16 percent.

By the middle of the twenty-first century, the dominant strength of the US economy within the developed world will have only one historical parallel: the immediate aftermath of World War II, exactly 100 years earlier, at the birth of the "Pax Americana."

The UN regularly publishes a table ranking the world's most populous countries over time. In 1950, six of the top twelve were developed countries. In 2000, only three were. By 2050, only one developed country will remain—the United States, still in third place. By then, it will be the only country among the top twelve committed since its founding to democracy, free markets, and civil liberties.

All told, population trends point inexorably toward a more dominant US role in a world that will need America more, not less.

Critical Thinking

1. Outline briefly how global aging is likely to have a profound effect on economic growth, living standards, and the shape of the world order.

2. How might this global aging change natural resource consumption patterns? Will "new" resources come into demand and "old" resources see demand reductions?

3. What "demographic storms" are brewing in the world, and how might these "storms" impact demands for environmental resources?

4. Will global aging help or hinder efforts to achieve a sustainable relationship with the earth? Explain.

NEIL HOWE and **RICHARD JACKSON** are, respectively, a senior associate and a senior fellow at the Center for Strategic and International Studies. They are the authors of *The Graying of the Great Powers: Demography and Geopolitics in the 21st Century* (*CSIS,* 2008).

From *Current History*, January 2011, pp. 20–25. Copyright © 2011 by Current History, Inc. Reprinted by permission.

The New Population Bomb: The Four Megatrends That Will Change the World

Jack A. Goldstone

Forty-two years ago, the biologist Paul Ehrlich warned in The Population Bomb that mass starvation would strike in the 1970s and 1980s, with the world's population growth outpacing the production of food and other critical resources. Thanks to innovations and efforts such as the "green revolution" in farming and the widespread adoption of family planning, Ehrlich's worst fears did not come to pass. In fact, since the 1970s, global economic output has increased and fertility has fallen dramatically, especially in developing countries.

The United Nations Population Division now projects that global population growth will nearly halt by 2050. By that date, the world's population will have stabilized at 9.15 billion people, according to the "medium growth" variant of the UN's authoritative population database World Population Prospects: The 2008 Revision. (Today's global population is 6.83 billion.) Barring a cataclysmic climate crisis or a complete failure to recover from the current economic malaise, global economic output is expected to increase by two to three percent per year, meaning that global income will increase far more than population over the next four decades.

But twenty-first-century international security will depend less on how many people inhabit the world than on how the global population is composed and distributed: where populations are declining and where they are growing, which countries are relatively older and which are more youthful, and how demographics will influence population movements across regions.

These elements are not well recognized or widely understood. A recent article in The Economist, for example, cheered the decline in global fertility without noting other vital demographic developments. Indeed, the same UN data cited by The Economist reveal four historic shifts that will fundamentally alter the world's population over the next four decades: the relative demographic weight of the world's developed countries will drop by nearly 25 percent, shifting economic power to the developing nations; the developed countries' labor forces will substantially age and decline, constraining economic growth in the developed world and raising the demand for immigrant workers; most of the world's expected population growth will increasingly be concentrated in today's poorest, youngest, and most heavily Muslim countries, which have a dangerous lack of quality education, capital, and employment opportunities;

and, for the first time in history, most of the world's population will become urbanized, with the largest urban centers being in the world's poorest countries, where policing, sanitation, and health care are often scarce. Taken together, these trends will pose challenges every bit as alarming as those noted by Ehrlich. Coping with them will require nothing less than a major reconsideration of the world's basic global governance structures.

Europe's Reversal of Fortunes

At the beginning of the eighteenth century, approximately 20 percent of the world's inhabitants lived in Europe (including Russia). Then, with the Industrial Revolution, Europe's population boomed, and streams of European emigrants set off for the Americas. By the eve of World War I, Europe's population had more than quadrupled. In 1913, Europe had more people than China, and the proportion of the world's population living in Europe and the former European colonies of North America had risen to over 33 percent. But this trend reversed after World War I, as basic health care and sanitation began to spread to poorer countries. In Asia, Africa, and Latin America, people began to live longer, and birthrates remained high or fell only slowly. By 2003, the combined populations of Europe, the United States, and Canada accounted for just 17 percent of the global population. In 2050, this figure is expected to be just 12 percent—far less than it was in 1700. (These projections, moreover, might even understate the reality because they reflect the "medium growth" projection of the UN forecasts, which assumes that the fertility rates of developing countries will decline while those of developed countries will increase. In fact, many developed countries show no evidence of increasing fertility rates.) The West's relative decline is even more dramatic if one also considers changes in income. The Industrial Revolution made Europeans not only more numerous than they had been but also considerably richer per capita than others worldwide. According to the economic historian Angus Maddison, Europe, the United States, and Canada together produced about 32 percent of the world's GDP at the beginning of the nineteenth century. By 1950, that proportion had increased to a remarkable 68 percent of the world's total output (adjusted to reflect purchasing power parity).

This trend, too, is headed for a sharp reversal. The proportion of global GDP produced by Europe, the United States, and Canada fell from 68 percent in 1950 to 47 percent in 2003 and will decline even more steeply in the future. If the growth rate of per capita income (again, adjusted for purchasing power parity) between 2003 and 2050 remains as it was between 1973 and 2003—averaging 1.68 percent annually in Europe, the United States, and Canada and 2.47 percent annually in the rest of the world—then the combined GDP of Europe, the United States, and Canada will roughly double by 2050, whereas the GDP of the rest of the world will grow by a factor of five. The portion of global GDP produced by Europe, the United States, and Canada in 2050 will then be less than 30 percent—smaller than it was in 1820.

These figures also imply that an overwhelming proportion of the world's GDP growth between 2003 and 2050—nearly 80 percent—will occur outside of Europe, the United States, and Canada. By the middle of this century, the global middle class—those capable of purchasing durable consumer products, such as cars, appliances, and electronics—will increasingly be found in what is now considered the developing world. The World Bank has predicted that by 2030 the number of middle-class people in the developing world will be 1.2 billion—a rise of 200 percent since 2005. This means that the developing world's middle class alone will be larger than the total populations of Europe, Japan, and the United States combined. From now on, therefore, the main driver of global economic expansion will be the economic growth of newly industrialized countries, such as Brazil, China, India, Indonesia, Mexico, and Turkey.

Aging Pains

Part of the reason developed countries will be less economically dynamic in the coming decades is that their populations will become substantially older. The European countries, Canada, the United States, Japan, South Korea, and even China are aging at unprecedented rates. Today, the proportion of people aged 60 or older in China and South Korea is 12–15 percent. It is 15–22 percent in the European Union, Canada, and the United States and 30 percent in Japan. With baby boomers aging and life expectancy increasing, these numbers will increase dramatically. In 2050, approximately 30 percent of Americans, Canadians, Chinese, and Europeans will be over 60, as will more than 40 percent of Japanese and South Koreans.

Over the next decades, therefore, these countries will have increasingly large proportions of retirees and increasingly small proportions of workers. As workers born during the baby boom of 1945–65 are retiring, they are not being replaced by a new cohort of citizens of prime working age (15–59 years old).

Industrialized countries are experiencing a drop in their working-age populations that is even more severe than the overall slowdown in their population growth. South Korea represents the most extreme example. Even as its total population is projected to decline by almost 9 percent by 2050 (from 48.3 million to 44.1 million), the population of working-age South Koreans is expected to drop by 36 percent (from 32.9 million to 21.1 million), and the number of South Koreans aged 60 and older will increase by almost 150 percent (from 7.3 million to 18 million). By 2050, in other words, the entire working-age population will barely exceed the 60-and-older population. Although South Korea's case is extreme, it represents an increasingly common fate for developed countries. Europe is expected to lose 24 percent of its prime working-age population (about 120 million workers) by 2050, and its 60-and-older population is expected to increase by 47 percent. In the United States, where higher fertility and more immigration are expected than in Europe, the working-age population will grow by 15 percent over the next four decades—a steep decline from its growth of 62 percent between 1950 and 2010. And by 2050, the United States' 60-and-older population is expected to double.

All this will have a dramatic impact on economic growth, health care, and military strength in the developed world. The forces that fueled economic growth in industrialized countries during the second half of the twentieth century—increased productivity due to better education, the movement of women into the labor force, and innovations in technology—will all likely weaken in the coming decades. College enrollment boomed after World War II, a trend that is not likely to recur in the twenty-first century; the extensive movement of women into the labor force also was a one-time social change; and the technological change of the time resulted from innovators who created new products and leading-edge consumers who were willing to try them out—two groups that are thinning out as the industrialized world's population ages.

Overall economic growth will also be hampered by a decline in the number of new consumers and new households. When developed countries' labor forces were growing by 0.5–1.0 percent per year, as they did until 2005, even annual increases in real output per worker of just 1.7 percent meant that annual economic growth totaled 2.2–2.7 percent per year. But with the labor forces of many developed countries (such as Germany, Hungary, Japan, Russia, and the Baltic states) now shrinking by 0.2 percent per year and those of other countries (including Austria, the Czech Republic, Denmark, Greece, and Italy) growing by less than 0.2 percent per year, the same 1.7 percent increase in real output per worker yields only 1.5–1.9 percent annual overall growth. Moreover, developed countries will be lucky to keep productivity growth at even that level; in many developed countries, productivity is more likely to decline as the population ages.

A further strain on industrialized economies will be rising medical costs: as populations age, they will demand more health care for longer periods of time. Public pension schemes for aging populations are already being reformed in various industrialized countries—often prompting heated debate. In theory, at least, pensions might be kept solvent by increasing the retirement age, raising taxes modestly, and phasing out benefits for the wealthy. Regardless, the number of 80- and 90-year-olds—who are unlikely to work and highly likely to require nursing-home and other expensive care—will rise dramatically. And even if 60- and 70-year-olds remain active and employed, they will require procedures and medications—hip replacements, kidney transplants, blood-pressure treatments—to sustain their health in old age.

All this means that just as aging developed countries will have proportionally fewer workers, innovators, and consumerist young households, a large portion of those countries' remaining economic growth will have to be diverted to pay for the medical bills and pensions of their growing elderly populations. Basic services, meanwhile, will be increasingly costly because fewer young workers will be available for strenuous and labor-intensive jobs. Unfortunately, policymakers seldom reckon with these potentially disruptive effects of otherwise welcome developments, such as higher life expectancy.

Youth and Islam in the Developing World

Even as the industrialized countries of Europe, North America, and Northeast Asia will experience unprecedented aging this century, fast-growing countries in Africa, Latin America, the Middle East, and Southeast Asia will have exceptionally youthful populations. Today, roughly nine out of ten children under the age of 15 live in developing countries. And these are the countries that will continue to have the world's highest birthrates. Indeed, over 70 percent of the world's population growth between now and 2050 will occur in 24 countries, all of which are classified by the World Bank as low income or lower-middle income, with an average per capita income of under $3,855 in 2008.

Many developing countries have few ways of providing employment to their young, fast-growing populations. Would-be laborers, therefore, will be increasingly attracted to the labor markets of the aging developed countries of Europe, North America, and Northeast Asia. Youthful immigrants from nearby regions with high unemployment—Central America, North Africa, and Southeast Asia, for example—will be drawn to those vital entry-level and manual-labor jobs that sustain advanced economies: janitors, nursing-home aides, bus drivers, plumbers, security guards, farm workers, and the like. Current levels of immigration from developing to developed countries are paltry compared to those that the forces of supply and demand might soon create across the world.

These forces will act strongly on the Muslim world, where many economically weak countries will continue to experience dramatic population growth in the decades ahead. In 1950, Bangladesh, Egypt, Indonesia, Nigeria, Pakistan, and Turkey had a combined population of 242 million. By 2009, those six countries were the world's most populous Muslim-majority countries and had a combined population of 886 million. Their populations are continuing to grow and indeed are expected to increase by 475 million between now and 2050—during which time, by comparison, the six most populous developed countries are projected to gain only 44 million inhabitants. Worldwide, of the 48 fastest-growing countries today—those with annual population growth of two percent or more—28 are majority Muslim or have Muslim minorities of 33 percent or more.

It is therefore imperative to improve relations between Muslim and Western societies. This will be difficult given that many Muslims live in poor communities vulnerable to radical appeals and many see the West as antagonistic and militaristic.

In the 2009 Pew Global Attitudes Project survey, for example, whereas 69 percent of those Indonesians and Nigerians surveyed reported viewing the United States favorably, just 18 percent of those polled in Egypt, Jordan, Pakistan, and Turkey (all U.S. allies) did. And in 2006, when the Pew survey last asked detailed questions about Muslim-Western relations, more than half of the respondents in Muslim countries characterized those relations as bad and blamed the West for this state of affairs.

But improving relations is all the more important because of the growing demographic weight of poor Muslim countries and the attendant increase in Muslim immigration, especially to Europe from North Africa and the Middle East. (To be sure, forecasts that Muslims will soon dominate Europe are outlandish: Muslims compose just three to ten percent of the population in the major European countries today, and this proportion will at most double by midcentury.) Strategists worldwide must consider that the world's young are becoming concentrated in those countries least prepared to educate and employ them, including some Muslim states. Any resulting poverty, social tension, or ideological radicalization could have disruptive effects in many corners of the world. But this need not be the case; the healthy immigration of workers to the developed world and the movement of capital to the developing world, among other things, could lead to better results.

Urban Sprawl

Exacerbating twenty-first-century risks will be the fact that the world is urbanizing to an unprecedented degree. The year 2010 will likely be the first time in history that a majority of the world's people live in cities rather than in the countryside. Whereas less than 30 percent of the world's population was urban in 1950, according to UN projections, more than 70 percent will be by 2050.

Lower-income countries in Asia and Africa are urbanizing especially rapidly, as agriculture becomes less labor intensive and as employment opportunities shift to the industrial and service sectors. Already, most of the world's urban agglomerations—Mumbai (population 20.1 million), Mexico City (19.5 million), New Delhi (17 million), Shanghai (15.8 million), Calcutta (15.6 million), Karachi (13.1 million), Cairo (12.5 million), Manila (11.7 million), Lagos (10.6 million), Jakarta (9.7 million)—are found in low-income countries. Many of these countries have multiple cities with over one million residents each: Pakistan has eight, Mexico 12, and China more than 100. The UN projects that the urbanized proportion of sub-Saharan Africa will nearly double between 2005 and 2050, from 35 percent (300 million people) to over 67 percent (1 billion). China, which is roughly 40 percent urbanized today, is expected to be 73 percent urbanized by 2050; India, which is less than 30 percent urbanized today, is expected to be 55 percent urbanized by 2050. Overall, the world's urban population is expected to grow by 3 billion people by 2050.

This urbanization may prove destabilizing. Developing countries that urbanize in the twenty-first century will have far lower per capita incomes than did many industrial countries when they first urbanized. The United States, for example, did not reach 65 percent urbanization until 1950, when per capita income was nearly $13,000 (in 2005 dollars). By contrast, Nigeria, Pakistan, and the Philippines, which are approaching similar levels of urbanization, currently have per capita incomes of just $1,800–$4,000 (in 2005 dollars).

According to the research of Richard Cincotta and other political demographers, countries with younger populations are especially prone to civil unrest and are less able to create or sustain democratic institutions. And the more heavily urbanized, the more such countries are likely to experience Dickensian poverty and anarchic violence. In good times, a thriving economy might keep urban residents employed and governments flush with sufficient resources to meet their needs. More often, however, sprawling and impoverished cities are vulnerable to crime lords, gangs, and petty rebellions. Thus, the rapid urbanization of the developing world in the decades ahead might bring, in exaggerated form, problems similar to those that urbanization brought to nineteenth-century Europe. Back then, cyclical employment, inadequate policing, and limited sanitation and education often spawned widespread labor strife, periodic violence, and sometimes—as in the 1820s, the 1830s, and 1848—even revolutions.

International terrorism might also originate in fast-urbanizing developing countries (even more than it already does). With their neighborhood networks, access to the Internet and digital communications technology, and concentration of valuable targets, sprawling cities offer excellent opportunities for recruiting, maintaining, and hiding terrorist networks.

Defusing the Bomb

Averting this century's potential dangers will require sweeping measures. Three major global efforts defused the population bomb of Ehrlich's day: a commitment by governments and nongovernmental organizations to control reproduction rates; agricultural advances, such as the green revolution and the spread of new technology; and a vast increase in international trade, which globalized markets and thus allowed developing countries to export foodstuffs in exchange for seeds, fertilizers, and machinery, which in turn helped them boost production. But today's population bomb is the product less of absolute growth in the world's population than of changes in its age and distribution. Policymakers must therefore adapt today's global governance institutions to the new realities of the aging of the industrialized world, the concentration of the world's economic and population growth in developing countries, and the increase in international immigration.

During the Cold War, Western strategists divided the world into a "First World," of democratic industrialized countries; a "Second World," of communist industrialized countries; and a "Third World," of developing countries. These strategists focused chiefly on deterring or managing conflict between the First and the Second Worlds and on launching proxy wars and diplomatic initiatives to attract Third World countries into the First World's camp. Since the end of the Cold War, strategists have largely abandoned this three-group division and have tended to believe either that the United States, as the sole superpower, would maintain a Pax Americana or that the world

would become multipolar, with the United States, Europe, and China playing major roles.

Unfortunately, because they ignore current global demographic trends, these views will be obsolete within a few decades. A better approach would be to consider a different three-world order, with a new First World of the aging industrialized nations of North America, Europe, and Asia's Pacific Rim (including Japan, Singapore, South Korea, and Taiwan, as well as China after 2030, by which point the one-child policy will have produced significant aging); a Second World comprising fast-growing and economically dynamic countries with a healthy mix of young and old inhabitants (such as Brazil, Iran, Mexico, Thailand, Turkey, and Vietnam, as well as China until 2030); and a Third World of fast-growing, very young, and increasingly urbanized countries with poorer economies and often weak governments. To cope with the instability that will likely arise from the new Third World's urbanization, economic strife, lawlessness, and potential terrorist activity, the aging industrialized nations of the new First World must build effective alliances with the growing powers of the new Second World and together reach out to Third World nations. Second World powers will be pivotal in the twenty-first century not just because they will drive economic growth and consume technologies and other products engineered in the First World; they will also be central to international security and cooperation. The realities of religion, culture, and geographic proximity mean that any peaceful and productive engagement by the First World of Third World countries will have to include the open cooperation of Second World countries.

Strategists, therefore, must fundamentally reconsider the structure of various current global institutions. The G-8, for example, will likely become obsolete as a body for making global economic policy. The G-20 is already becoming increasingly important, and this is less a short-term consequence of the ongoing global financial crisis than the beginning of the necessary recognition that Brazil, China, India, Indonesia, Mexico, Turkey, and others are becoming global economic powers. International institutions will not retain their legitimacy if they exclude the world's fastest-growing and most economically dynamic countries. It is essential, therefore, despite European concerns about the potential effects on immigration, to take steps such as admitting Turkey into the European Union. This would add youth and economic dynamism to the EU—and would prove that Muslims are welcome to join Europeans as equals in shaping a free and prosperous future. On the other hand, excluding Turkey from the EU could lead to hostility not only on the part of Turkish citizens, who are expected to number 100 million by 2050, but also on the part of Muslim populations worldwide.

NATO must also adapt. The alliance today is composed almost entirely of countries with aging, shrinking populations and relatively slow-growing economies. It is oriented toward the Northern Hemisphere and holds on to a Cold War structure that cannot adequately respond to contemporary threats. The young and increasingly populous countries of Africa, the Middle East, Central Asia, and South Asia could mobilize insurgents much more easily than NATO could mobilize the troops it would need if it were called on to stabilize those countries. Long-standing

NATO members should, therefore—although it would require atypical creativity and flexibility—consider the logistical and demographic advantages of inviting into the alliance countries such as Brazil and Morocco, rather than countries such as Albania. That this seems far-fetched does not minimize the imperative that First World countries begin including large and strategic Second and Third World powers in formal international alliances.

The case of Afghanistan—a country whose population is growing fast and where NATO is currently engaged—illustrates the importance of building effective global institutions. Today, there are 28 million Afghans; by 2025, there will be 45 million; and by 2050, there will be close to 75 million. As nearly 20 million additional Afghans are born over the next 15 years, NATO will have an opportunity to help Afghanistan become reasonably stable, self-governing, and prosperous. If NATO's efforts fail and the Afghans judge that NATO intervention harmed their interests, tens of millions of young Afghans will become more hostile to the West. But if they come to think that NATO's involvement benefited their society, the West will have tens of millions of new friends. The example might then motivate the approximately one billion other young Muslims growing up in low-income countries over the next four decades to look more kindly on relations between their countries and the countries of the industrialized West.

Creative Reforms at Home

The aging industrialized countries can also take various steps at home to promote stability in light of the coming demographic trends. First, they should encourage families to have more children. France and Sweden have had success providing child care, generous leave time, and financial allowances to families with young children. Yet there is no consensus among policymakers—and certainly not among demographers—about what policies best encourage fertility.

More important than unproven tactics for increasing family size is immigration. Correctly managed, population movement can benefit developed and developing countries alike. Given the dangers of young, underemployed, and unstable populations in developing countries, immigration to developed countries can provide economic opportunities for the ambitious and serve as a safety valve for all. Countries that embrace immigrants, such as the United States, gain economically by having willing laborers and greater entrepreneurial spirit. And countries with high levels of emigration (but not so much that they experience so-called brain drains) also benefit because emigrants often send remittances home or return to their native countries with valuable education and work experience.

One somewhat daring approach to immigration would be to encourage a reverse flow of older immigrants from developed to developing countries. If older residents of developed countries took their retirements along the southern coast of the Mediterranean or in Latin America or Africa, it would greatly reduce the strain on their home countries' public entitlement systems. The developing countries involved, meanwhile, would benefit because caring for the elderly and providing retirement and leisure services is highly labor intensive. Relocating a portion of these activities to developing countries would provide employment and valuable training to the young, growing populations of the Second and Third Worlds.

This would require developing residential and medical facilities of First World quality in Second and Third World countries. Yet even this difficult task would be preferable to the status quo, by which low wages and poor facilities lead to a steady drain of medical and nursing talent from developing to developed countries. Many residents of developed countries who desire cheaper medical procedures already practice medical tourism today, with India, Singapore, and Thailand being the most common destinations. (For example, the international consulting firm Deloitte estimated that 750,000 Americans traveled abroad for care in 2008.)

Never since 1800 has a majority of the world's economic growth occurred outside of Europe, the United States, and Canada. Never have so many people in those regions been over 60 years old. And never have low-income countries' populations been so young and so urbanized. But such will be the world's demography in the twenty-first century. The strategic and economic policies of the twentieth century are obsolete, and it is time to find new ones.

Reference

Goldstone, Jack A. "The new population bomb: the four megatrends that will change the world." *Foreign Affairs* 89.1 (2010): 31. *General OneFile*. Web. 23 Jan. 2010. http://0-find.galegroup.com.www .consuls.org/gps/start.do?proId=IPS& userGroupName=a30wc.

Critical Thinking

1. What does the author contend will be the characteristics of future population growth?

2. The article argues that future population trends will have significant political and economic consequences. What do you see as the "environmental consequences"?

3. How might these environmental consequences vary for different populations around the world?

4. What impacts might these trends have on achieving the ideals of sustainability?

UNIT 3

The Geopolitical-Economy of Planetary Consumption

Unit Selections

Learning Outcomes

After reading this unit, you should be able to:

- Define *competitive exclusion principle* and explain how this biological principle may be associated with the idea of the geopolitical-economy of planetary consumption.

- Outline the reasons that extreme inequalities of income in the United States may contribute to future environmental problems.

- Explain why it is important to find ways to appraise and monitor the value of the Earth's ecosystem services.

- Identify the most critical uses of natural resources and their impacts.

- Describe the connection between the patterns of poverty and the pattern of environmental degradation.

- Provide examples of "environmental justice."

Student Website

www.mhhe.com/cls

Internet References

National Geographic Society
  www.nationalgeographic.com
Penn Library: Resources by Subject
  www.library.upenn.edu/cgi-bin/res/sr.cgi
Sustainable Development.Org
  www.sustainabledevelopment.org
United Nations
  www.unsystem.org

United Nations Environment Programme (UNEP)
  www.unep.ch
World Health Organization (WHO)
  www.who.ch
World Resources Institute (WRI)
  www.wri.org
WWW Virtual Library: Demography and Population Studies
  http://demography.anu.edu.au/VirtualLibrary

Fourth Law of Ecology: There's no such thing as a free lunch.

Barry Commoner, scientist, *The Closing Circle,* 1971

The five articles selected for Unit 3 are intended to explore the geopolitical-economy aspects of planetary consumption both directly and inferentially. But before introducing the articles, what do we mean by the "geopolitical-economy of planetary consumption"? It is the human and physical geographic, economic, and political elements that characterize a particular society, nation, or social group (demographic, ethnic, religious, socio-economic, etc.) and how those elements are linked to that particular society's, nation's, or social group's resource consumption behaviors.

When we hear and use such terms as *global population, resource consumption, sustainability, development, environmental impact,* or *consuming the earth* and the myriad other words associated with environmental discourse, they remain just that—words—vague, ambiguous, and oftentimes meaningless until the human actors imbedded in these concepts and implied in these words are identified (e.g., social, cultural, ethnic, religious, economic identities) and "placed." For example, wealthy, educated, politically powerful, well-fed, information-rich, stable people/groups/societies/ nations will play a significantly different role in the Earth resource consumption scenario than actors (people, societies, nations) characterized by poverty, illiteracy, political impotency, economic peripheralness, malnutrition, and civil conflict. Societies with tractors and dump trucks and titans of industry will certainly belly up to the Whole Earth Café and get their fill long before societies of ox carts, wicker baskets, and ineffectual leaders can rifle through the dumpster behind the Café.

Articles 13 and 14 examine this "placing" of actors and geopolitical economics. In Article 13, "The Competitive Exclusion Principle," the late Garret Harden, renowned ecologist who brought worldwide attention to his thesis of the "tragedy of the commons," discusses the ecology "competitive exclusion principle." In its simplest form, the principle states that *complete competitors cannot coexist.* If this is the case, what might that bode for nations and peoples who will be increasingly competing for Earth's resources? Will the wealthy, educated, strong, and aggressive of the Earth reign as top consumers?

In Article 14, "Of the 1%, by the 1%, for the 1%," Nobel Laureate economist Joseph Stiglitz observes that the increasing concentration of America's wealth (40 percent of it) into the hands of a small group of Americans (1 percent of them) is creating a serious inequality of income situation in the country. How might this income inequality translate into a component of Hardin's "exclusion principle" and how might it impact patterns of consumption in the United States? How might such wealth inequality at the global scale play into the exclusion principle and global resource access/consumption patterns?

Articles 15, 16, and 17 illustrate further that geopolitical-economy not only shapes the nature of resource access and consumption but also plays a significant role in ascribing accountability for the inevitable environmental externalities (waste, pollutants,

© Jack Star/PhotoLink/Getty Images

habitat loss, human health risks, and so on) that result from resource consumption. Resource consumption is fundamentally an economic production function that requires environmental decision making at nearly every step of the transformation process of converting raw materials (Earth's resources) into finished consumable products. For example, seeds, soil, and water are transformed into food, and environmental decisions have to be made; oil, coal, and natural gas are transformed into energy, and decisions have to be made; ores are transformed into machines, and decisions have to be made; precipitation is transformed into plastic bottles of drinking water, and more environmental decisions have to be made. However, the "best" environmental (transformation and consumption) decisions are made with the most information regarding the costs and benefits (referred to as *full cost accounting*) accruing to both people (shareholders) and the environment.

In Article 15, "The New Economy of Nature," Gretchen Daily and Katherine Ellison provide a clear example of how the geopolitical economy plays a significant role in ascribing accountability for the inevitable environmental externalities of consumption. The article argues that an ecosystem function (freshwater provision) failed because its "labor of nature" was badly appraised and undervalued, which can contribute to a lack of monitoring necessary to ensure against ecosystem service or loss. Ecosystem service valuation and protection are both activities that cannot be addressed independent of a "place's" geopolitical-economy structure.

Unit 3 concludes with Article 16, "Environmental Justice for All," which explores another "moral" aspect of the global environmental ethic challenge: how to address human poverty and its association with environmental degradation. The author believes today's environmental justice proponents are changing tactics. They are moving away from "reactive" responses to issues of environmental–social degradation to being "proactive." In a sense, it is an environmental ethic movement that is moving away from saving souls (victims of environmental degradation) to converting would-be sinners (agents of environmental degradation) into environmentally moral beings. In other words, some new advocates for environmental justice believe that by encouraging economic development with an environmental conscience (ethic), we can begin to change the cyclic association of environmental degradation–poverty–degradation. If such an approach proves to be effective in the urban American scene, could the next step be to go global?

The Competitive Exclusion Principle

An idea that took a century to be born has implications in ecology, economics, and genetics.

GARRETT HARDIN

On 21 March 1944 the British Ecological Society devoted a symposium to the ecology of closely allied species. There were about 60 members and guests present. In the words of an anonymous reporter[1], "a lively discussion . . . centred about Gause's contention (1934) that two species with similar ecology cannot live together in the same place. . . . A distinct cleavage of opinion revealed itself on the question of the validity of Gause's concept. Of the main speakers, Mr. Lack, Mr. Elton and Dr. Varley supported the postulate. . . . Capt. Diver made a vigorous attack on Gause's concept, on the grounds that the mathematical and experimental approaches had been dangerously over simplified. . . . Pointing out the difficulty of defining 'similar ecology' he gave examples of many congruent species of both plants and animals apparently living and feeding together."

Thus was born what has since been called "Gause's principle." I say "born" rather than "conceived" in order to draw an analogy with the process of mammalian reproduction, where the moment of birth, of exposure to the external world, of becoming a fully legal entity, takes place long after the moment of conception. With respect to the principle here discussed, the length of the gestation period is a matter of controversy: 10 years, 12 years, 18 years, 40 years, or about 100 years, depending on whom one takes to be the father of the child.

Statement of the Principle

For reasons given below, I here refer to the principle by a name already introduced[2]—namely, the "competitive exclusion principle," or more briefly, the "exclusion principle." It may be briefly stated thus: *Complete competitors cannot coexist.* Many published discussions of the principle revolve around the ambiguity of the words used in stating it. The statement given above has been very carefully constructed: every one of the four words is ambiguous. This formulation has been chosen not out of perversity but because of a belief that it is best to use that wording which is least likely to hide the fact that we still do not comprehend the exact limits of the principle. For the present, I think the "threat of clarity"[3] is a serious one that is best minimized by using a formulation that is *admittedly* unclear; thus can we keep in the forefront of our minds the unfinished work before us. The wording given has, I think, another point of superiority in that it seems brutal and dogmatic. By emphasizing the very aspects that might result in our denial of them were they less plain we can keep the principle explicitly present in our minds until we see if its implications are, or are not, as unpleasant as our subconscious might suppose. The meaning of these somewhat cryptic remarks should become clear further on in the discussion.

What does the exclusion principle mean? Roughly this: that (i) if two noninterbreeding populations "do the same thing"—that is, occupy precisely the same ecological niche in Elton's sense[4]—and (ii) if they are "sympatric"—that is, if they occupy the same geographic territory—and (iii) if population *A* multiplies even the least bit faster than population *B,* then ultimately *A* will completely displace *B,* which will become extinct. This is the "weak form" of the principle. Always in practice a stronger form is used, based on the removal of the hypothetical character of condition (iii). We do this because we adhere to what may be called the axiom of inequality, which states that no two things or processes, in a real world, are precisely equal. This basic idea is probably as old as philosophy itself but is usually ignored, for good reasons. With respect to the *things* of the world the axiom often leads to trivial conclusions. One postage stamp is as good as another. But with respect to competing *processes* (for example, the multiplication rates of competing species) the axiom is never trivial, as has been repeatedly shown[5-7]. No difference in rates of multiplication can be so slight as to negate the exclusion principle.

Demonstrations of the formal truth of the principle have been given in terms of the calculus (*5, 7*) and set theory[8]. Those to whom the mathematics does not appeal may prefer the following intuitive verbal argument (*2,* pp. 84–85), which is based on an economic analogy that is very strange economics but quite normal biology.

"Let us imagine a very odd savings bank which has only two depositors. For some obscure reason the bank pays one of the depositors 2 percent compound interest, while paying the other 2.01 percent. Let us suppose further (and here the analogy is really strained) that whenever the sum of the combined funds of the two depositors reaches two million dollars, the bank arbitrarily appropriates one million dollars of it, taking from each depositor in proportion to his holdings at that time. Then both accounts are allowed to grow until their sum again equals two million dollars, at which time the appropriation process is repeated. If this procedure is continued indefinitely, what will happen to the wealth of these two depositors? A little intuition shows us (and mathematics verifies) that the man who receives the greater rate of interest will, in time, have all the money, and the other man none (we assume a penny cannot be subdivided). No matter how small the difference between the two interest rates (so long as there is a difference) such will be the outcome.

"Translated into evolutionary terms, this is what competition in nature amounts to. The fluctuating limit of one million to two million represents the finite available wealth (food, shelter, etc.) of any natural environment, and the difference in interest rates represents the difference between the competing species in their efficiency in producing offspring. No matter how small this difference may be, one species will eventually replace the other. In the scale of geological time, even a small competitive difference will result in a rapid extermination of the less successful species. Competitive differences that are so small as to be unmeasurable by direct means will, by virtue of the compound-interest effect, ultimately result in the extinction of one competing species by another."

The Question of Evidence

So much for the theory. Is it true? This sounds like a straightforward question, but it hides subtleties that have, unfortunately, escaped a good many of the ecologists who have done their bit to make the exclusion principle a matter of dispute. There are many who have supposed that the principle is one that can be proved or disproved by empirical facts, among them[9-10] Gause himself. Nothing could be farther from the truth. The "truth" of the principle is and can be established only by theory, not being subject to proof or disproof by facts, as ordinarily understood. Perhaps this statement shocks you. Let me explain.

Suppose you believe the principle is true and set out to prove it empirically. First you find two noninterbreeding species that seem to have the same ecological characteristics. You bring them together in the same geographic location and await developments. What happens? Either one species extinguishes the other, or they coexist. If the former, you say, "The principle is proved." But if the species continue to coexist indefinitely, do you conclude the principle is false? Not at all. You decide there must have been some

subtle difference in the ecology of the species that escaped you at first, so you look at the species again to try to see how they differ ecologically, all the while retaining your belief in the exclusion principle. As Gilbert, Reynoldson, and Hobart[10] dryly remarked, "There is . . . a danger of a circular process here. . . ."

Indeed there is. Yet the procedure can be justified, both empirically and theoretically. First, empirically. On this point our argument is essentially an acknowledgement of ignorance. When we think of mixing two similar species that have previously lived apart, we realize that it is hardly possible to know enough about species to be able to say, in advance, which one will exclude the other in free competition. Or, as Darwin, at the close of chapter 4 of his *Origin of Species*[11] put it: "It is good thus to try in imagination to give any one species an advantage over another. Probably in no single instance should we know what to do. This ought to convince us of our ignorance on the mutual relations of all organic beings: a conviction as necessary, as it is difficult to acquire."

How profound our ignorance of competitive situations is has been made painfully clear by the extended experiments of Thomas Park and his collaborators[12]. For more than a decade Park has put two species of flour beetles (*Tribolium confusum* and *T. castaneum*) in closed universes under various conditions. In every experiment the competitive exclusion principle is obeyed—one of the species is completely eliminated, *but it is not always the same one*. With certain fixed values for the environmental parameters the experimenters have been unable to control conditions carefully enough to obtain an invariable result. Just how one is to interpret this is by no means clear, but in any case Park's extensive body of data makes patent our immense ignorance of the relations of organisms to each other and to the environment, even under the most carefully controlled conditions.

The theoretical defense for adhering come-hell-or-high-water to the competitive exclusion principle is best shown by apparently changing the subject. Consider Newton's first law: "Every body persists in a state of rest or of uniform motion in a straight line unless compelled by external force to change that state." How would one verify this law, by itself? An observer might (in principle) test Newton's first law by taking up a station out in space somewhere and then looking at all the bodies around him. Would any of the bodies be in a state of rest except (by definition) himself? Probably not. More important, would any of the bodies in motion be moving in a straight line? *Not one*. (We assume that the observer makes errorless measurements.) For the law says, ". . . in a straight line unless compelled by external force to change . . . ," and in a world in which another law says that "every body attracts every other body with a force that is inversely proportional to the square of the distance between them . . . ," the phrase in the first law that begins with the words *unless compelled* clearly indicates the hypothetical character of the law. So long as there are no sanctuaries from gravitation in space, every body is always "compelled." Our observer would claim that any body at rest or moving in a straight line verified the law; he would likewise claim that bodies moving in not-straight lines verified the law, too. In other words, any attempt to test Newton's first law *by itself* would lead to a circular argument of the sort encountered earlier in considering the exclusion principle.

The point is this: We do not test isolated laws, one by one. What we test is a whole conceptual model[13]. From the model we make predictions; these we test against empirical data. When we find that a prediction is not verifiable we then set about modifying the model. There is no procedural rule to tell us which element of the model is best abandoned or changed. (The scientific response to the results of the Michelson-Morley experiment was not in any sense *determined.*) Esthetics plays a part in such decisions.

The competitive exclusion principle is one element in a system of ecological thought. We cannot test it directly, by itself. What the whole ecological system is, we do not yet know. One immediate task is to discover the system, to find its elements, to work out their interactions, and to make the system as explicit as possible. (*Complete* explicitness can never be achieved.) The works of Lotka[14], Nicholson[15-16], and MacArthur[17] are encouraging starts toward the elaboration of such a theoretical system.

The Issue of Eponymy

That the competitive exclusion principle is often called "Gause's principle" is one of the more curious cases of eponymy in science (like calling human oviducts "Fallopian tubes," after a man who was not the first to see them and who misconstrued their significance). The practice was apparently originated by the English ecologists, among whom David Lack has been most influential. Lack made a careful study of *Geospiza* and other genera of finches in the Galápagos Islands, combining observational studies on location with museum work at the California Academy of Sciences. How his ideas of ecological principles matured during the process is evident from a passage in his little classic, *Darwin's Finches*[18].

"Snodgrass concluded that the beak differences between the species of *Geospiza* are not of adaptive significance in regard to food. The larger species tend to eat rather larger seeds, but this he considered to be an incidental result of the difference in the size of their beaks. This conclusion was accepted by Gilford (1919), Gulick (1932), Swarth (1934) and formerly by myself (Lack, 1945). Moreover, the discovery . . . that the beak differences serve as recognition marks, provided quite a different reason for their existence, and thus strengthened the view that any associated differences in diet are purely incidental and of no particular importance.

"My views have now completely changed, through appreciating the force of Gause's contention that two species with similar ecology cannot live in the same region (Gause, 1934).

This is a simple consequence of natural selection. If two species of birds occur together in the same habitat in the same region, eat the same types of food and have the same other ecological requirements, then they should compete with each other, and since the chance of their being equally well adapted is negligible, one of them should eliminate the other completely. Nevertheless, three species of ground-finch live together in the same habitat on the same Galapagos islands, and this also applies to two species of insectivorous tree-finch. There must be some factor which prevents these species from effectively competing."

Implicit in this passage is a bit of warm and interesting autobiography. It is touching to see how intellectual gratitude led Lack to name the exclusion principle after Gause, calling it, in successive publications, "Gause's contention," "Gause's hypothesis," and "Gause's principle." But the eponymy is scarcely justified. As Gilbert, Reynoldson, and Hobart point out (*10*, p. 312): "Gause . . . draws no general conclusions from his experiments, and moreover, makes no statement which resembles any wording of the hypothesis which has arisen bearing his name." Moreover, in the very publication in which he discussed the principle, Gause acknowledged the priority of Lotka in 1932 (*5*) and Volterra in 1926 (*6*). Gause gave full credit to these men, viewing his own work merely as an empirical testing of their theory—a quite erroneous view, as we have seen. How curious it is that the principle should be named after a man who did not state it clearly, who misapprehended its relation to theory, and who acknowledged the priority of others!

Recently Udvardy[19], in an admirably compact note, has pointed out that Joseph Grinnell, in a number of publications, expressed the exclusion principle with considerable clarity. In the earliest passage that Udvardy found, Grinnell, in 1904[20], said: "Every animal tends to increase at a geometric ratio, and is checked only by limit of food supply. It is only by adaptations to different sorts of food, or modes of food getting, that more than one species can occupy the same locality. Two species of approximately the same food habits are not likely to remain long enough evenly balanced in numbers in the same region. One will crowd out the other."

Udvardy quotes from several subsequent publications of Grinnell, from all of which it is quite clear that this well-known naturalist had a much better grasp of the exclusion principle than did Gause. Is this fact, however, a sufficiently good reason for now speaking (as Udvardy recommends) of "Grinnell's axiom?" On the basis of present evidence there seems to be justice in the proposal, but we must remember that the principle has already been referred to, in various publications, as "Gause's principle," the "Volterra-Gause principle," and the "Lotka-Volterra principle." What assurance have we that some diligent scholar will not tomorrow unearth a predecessor of Grinnell? And if this happens, should we then replace Grinnell's name with another's? Or should we, in a fine show of fairness, use all the names? (According to this system, the principle would, at present, be called the Grinnell-Volterra-Lotka-Gause-Lack principle—and, even so, injustice would be done to A. J. Nicholson, who, in his wonderful gold mine of unexploited aphorisms (15), wrote: "For the steady state [in the coexistence of two or more species] to exist, each species must possess some advantage over all other species with respect to some one, or group, of the control factors to which it is subject." This is surely a corollary of the exclusion principle.)

In sum, I think we may say that arguments for pinning an eponym on this idea are unsound. But it does need a name of some sort; its lack of one has been one of the reasons (though not the only one) why this basic principle has trickled out of the scientific consciousness after each mention during the last half century. Like Allee et al.[21] we should wish "to avoid further implementation of the facetious definition of ecology as being that phase of biology primarily abandoned to terminology." But, on the other side, it has been pointed out[22]: "Not many recorded facts are lost; the bibliographic apparatus of science is fairly equal to the problem of recording melting points, indices of refraction, etc., in such a way that they can be recalled when needed. Ideas, more subtle and more diffusely expressed present a bibliographic problem to which there is no present solution." To solve the bibliographic problem some sort of handle is needed for the idea here discussed; the name "the competitive exclusion principle" is correctly descriptive and will not be made obsolete by future library research.

The Exclusion Principle and Darwin

In our search for early statements of the principle we must not pass by the writings of Charles Darwin, who had so keen an appreciation of the ecological relationships of organisms. I have been unable to find any unambiguous references to the exclusion principle in the "Essays" of 1842 and 1844[23], but in the *Origin* itself there are several passages that deserve recording. All the following passages are quoted from the sixth edition (11).

"As the species of the same genus usually have, though by no means invariably, much similarity in habits and constitution, and always in structure, the struggle will generally be more severe between them, if they come into competition with each other, than between the species of distinct genera. We see this in the recent extension over parts of the United States of one species of swallow having caused the decrease of another species. The recent increase of the missel-thrush in parts of Scotland has caused the decrease of the song-thrush. How frequently we hear of one species of rat taking the place of another species under the most different climates! In Russia the small Asiatic cockroach has everywhere driven before it its great congener. In Australia the imported hive-bee is rapidly exterminating the small, stingless native bee. One species of charlock has been known to supplant another species; and so in other cases. We can dimly see why the competition should be most severe between allied forms, which fill nearly the same place in the economy of nature; but probably in no one case could we precisely say why one species has been victorious over another in the great battle of life" (p. 71).

"Owing to the high geometrical rate of increase of all organic beings, each area is already fully stocked with inhabitants; and it follows from this, that as the favored forms increase in number, so, generally, will the less favored decrease and become rare. Rarity, as geology tells us, is the precursor to extinction. We can see that any form which is represented by few individuals will run a good chance of utter extinction, during great fluctuations in the nature or the seasons, or from a temporary increase in the number of its enemies. But we may go further than this; for, as new forms are produced, unless we admit that specific forms can go on indefinitely increasing in number, many old forms must become extinct" (p. 102).

"From these several considerations I think it inevitably follows, that as new species in the course of time are formed through natural selection, others will become rarer and rarer, and finally extinct. The forms which stand in closest competition with those undergoing modification and improvement, will naturally suffer most. And we have seen in the chapter on the Struggle for Existence that it is the most closely-allied forms—varieties of the same species, and species of the same genus or related genera—which, from having nearly the same structure, constitution and habits, generally come into the severest competition with each other consequently, each new variety or species, during the progress of its formation, will generally press hardest on its nearest kindred, and tend to exterminate them. We see the same process of extermination among our domesticated productions, through the selection of improved forms by man. Many curious instances could be given showing how quickly new breeds of cattle, sheep and other animals, and varieties of flowers, take the place of older and inferior kinds. In Yorkshire, it is historically known that the ancient black cattle were displaced by the long-horns, and that these 'were swept away by the short-horns' (I quote the words of an agricultural writer) 'as if by some murderous pestilence'" (p. 103).

"For it should be remembered that the competition will generally be most severe between those forms which are most nearly related to each other in habits, constitution and structure. Hence all the intermediate forms between the earlier and later states, that is between the less and more improved states of the same species, as well as the original parent species itself, will generally tend to become extinct" (p. 114).

Those passages are, we must admit, typically Darwinian; by turn clear, obscure, explicit, cryptic, suggestive, they have in them all the characteristics that litterateurs seek in James Joyce. The complexity of Darwin's work, however, is unintended; it is the result partly of his limitations as an analytical thinker, but in part also it is the consequence of the magnitude, importance, and intrinsic difficulty of the ideas he grappled with. Darwin was not one to impose premature clarity on his writings.

Origins in Economic Theory?

In chapter 3 of *Nature and Man's Fate* I have argued for the correctness of John Maynard Keynes' view that the biological principle of natural selection is just a vast generalization of Ricardian economics. The argument is based on the isomorphism of theoretical systems in the two fields of human thought. Now that we have at last brought the competitive exclusion principle out of the periphery of our vision into focus on the *fovea centralis* it is natural to wonder if this principle, too, originated in economic thought. I think it is possible. At any rate, there is a passage by the French mathematician J. Bertrand[24], published in 1883, which shows an appreciation of the exclusion principle as it applies to economic matters. The passage occurs in a review of a book of Cournot, published much earlier, in which Cournot discussed the outcome of a struggle between two merchants engaged in selling identical products to the public. Bertrand says: "Their interest would be to unite or at least to agree on a common price so as to extract from the body of customers the greatest possible receipts. But this solution is avoided by Cournot who supposes that one of the competitors will lower his price in order to attract the buyers to himself, and that the other, trying to regain them, will set his price still lower. The two rivals will cease to pursue this path only when each has nothing more to gain by lowering his price.

"To this argument we make a peremptory objection. Given the hypothesis, no solution is possible: there is no limit to the lowering of the price. Whatever common price might be initially adopted, if one of the competitors were to lower the price unilaterally he would thereby attract the totality of the business to himself. . . ."

This passage clearly antedates Grinnell, Lack, *et al.,* but it comes long after the *Origin of Species.* Are there statements of the principle in the economic literature before Darwin? It would be nice to know. I have run across cryptic references to the work of Simonde de Sismondi (1773–1842) which imply that he had a glimpse of the exclusion principle, but I have not tracked them down. Perhaps some colleague in the history of economics will someday do so. If it is true that Sismondi understood the principle, this fact would add a nice touch to the interweaving of the history of ideas, for this famous Swiss economist was related to Emma Darwin by marriage; he plays a prominent role in the letters published under her name[25].

Utility of the Exclusion Principle

"The most important lesson to be learned from evolutionary theory," says Michael Scriven in a brilliant essay recently published[26], "is a negative one: the theory shows us what scientific explanations need not do. In particular it shows us that one cannot regard explanations as unsatisfactory when they are not such as to enable the event in question to have been predicted." The theory of evolution is not one with which we can predict exactly the future course of species formation and extinction; rather, the theory "explains" the past. Strangely enough, we take mental satisfaction in this ex post facto explanation. Scriven has done well in showing why we are satisfied.

Much of the theory of ecology fits Scriven's description of evolutionary theory. Told that two formerly separated species are to be introduced into the same environment and asked to predict exactly what will happen, we are generally unable to do so. We can only make certain predictions of this sort: either A will extinguish B, or B will extinguish A; or the two species are (or must become) ecologically different—that is, they must come to occupy different ecological niches. The general rule may be stated in either of two different ways: *Complete competitors cannot coexist*—as was said earlier; or, *Ecological differentiation is the necessary condition for coexistence.*

It takes little imagination to see that the exclusion principle, to date stated explicitly only in ecological literature, has applications in many academic fields of study. I shall now point out some of these, showing how the principle has been used (mostly unconsciously) in the past, and predicting some of its applications in the future.

Economics

The principle unquestionably plays an indispensable role in almost all economic thinking, though it is seldom explicitly stated. Any competitor knows that unrestrained competition will ultimately result in but one victor. If he is confident that he is that one, he may plump for "rugged individualism." If, on the other hand, he has doubts, then he will seek to restrain or restrict competition. He can restrain it by forming a cartel with his competitors, or by maneuvering the passage of "fair trade" laws. (Laboring men achieve a similar end—though the problem is somewhat different—by the formation of unions and the passage of minimum wage laws.) Or he may restrict competition by "ecological differentiation," by putting out a slightly different product (aided by restrictive patent and copyright laws). All this may be regarded as individualistic action.

Society as a whole may take action. The end of unrestricted competition is a monopoly. It is well known that monopoly breeds power which acts to insure and extend the monopoly; the system has "positive feedback" and hence is always a threat to those aspects of society still "outside" the monopoly. For this reason, men may, in the interest of "society" (rather than of themselves as individual competitors), band together to insure continued competition; this they do by passing anti-monopoly laws which prevent competition from proceeding to its "naturally" inevitable conclusion. Or "society" may permit monopolies but seek to remove the power element by the "socialization" of the monopoly (expropriation or regulation).

In their actions both as individuals and as groups, men show that they have an implicit understanding of the exclusion principle. But the failure to bring this understanding to the level of consciousness has undoubtedly contributed to the accusations of bad faith ("exploiter of the masses," "profiteer," "nihilist," "communist") that have characterized many of the interchanges between competing groups of society during the last century. F. A. Lange[27], thinking only of laboring men, spoke in most fervent terms of the necessity of waging a "struggle against the struggle for existence"—that is, a struggle against the unimpeded working out of the exclusion principle. Groups with interests opposed to those of "labor" are equally passionate about the same cause, though the examples they have in mind are different.

At the present time, one of the great fields of economics in which the application of the exclusion principle is resisted is international competition (nonbellicose). For emotional reasons, most discussion of problems in this field is restricted by the assumption (largely implicit) that Cournot's solution of the *intra*national competition problem is correct and applicable to the *inter*national problem. On the less frequent occasions when it is recognized that Bertrand's, not Cournot's, reasoning is correct, it is assumed that the consequences of the exclusion principle can be indefinitely postponed by a rapid and endless multiplication of "ecological niches" (largely unprotected though they are by copyright and patent). If some of these assumptions prove to be unrealistic, the presently fashionable stance toward tariffs and other restrictions of international competition will have to be modified.

Genetics

The application of the exclusion principle to genetics is direct and undeniable. The system of discrete alleles at the same gene locus competing for existence within a single population of organisms is perfectly isomorphic with the system of different species of organisms competing for existence in the same habitat and ecological niche. The consequences of this have frequently been acknowledged, usually implicitly, at least since J. B. S. Haldane's work of 1924[28]. But in this field, also, the consequences have often been denied, explicitly or otherwise, and again for emotional reasons. The denial has most often been coupled with a "denial" (in the psychological sense) of the priority of the inequality axiom. As a result of recent findings in the fields of physiological genetics and population genetics, particularly as concerns blood groups, the applicability of both the inequality axiom and the exclusion principle is rapidly becoming accepted. William C. Boyd has recorded, in a dramatic way[29], his escape from the bondage of psychological denial. The emotional restrictions of rational discussion in this field are immense. How "the struggle against the struggle for existence" will

be waged in the field of human genetics promises to make the next decade of study one of the most exciting of man's attempts to accept the implications of scientific knowledge.

Ecology

Once one has absorbed the competitive exclusion principle into one's thinking it is curious to note how one of the most popular problems of evolutionary speculation is turned upside down. Probably most people, when first taking in the picture of historical evolution, are astounded at the number of species of plants and animals that have become extinct. From Simpson's gallant "guesstimates"[30], it would appear that from 99 to 99.975 percent of all species evolved are now extinct, the larger percentage corresponding to 3999 million species. This seems like a lot. Yet it is even more remarkable that there should live at any one time (for example, the present) as many as a million species, more or less competing with each other. Competition is avoided between some of the species that coexist in time by separation in space. In addition, however, there are many ecologically more or less similar species that coexist. Their continued existence is a thing to wonder at and to study. As Darwin said (*11*, p. 363)—and this is one more bit of evidence that he appreciated the exclusion principle—"We need not marvel at extinction; if we must marvel, let it be at our own presumption in imagining for a moment that we understand the many complex contingencies on which the existence of each species depends."

I think it is not too much to say that in the history of ecology—which in the broadest sense includes the science of economics and the study of population genetics—we stand at the threshold of a renaissance of understanding, a renaissance made possible by the explicit acceptance of the competitive exclusion principle. This principle, like much of the essential theory of evolution, has (I think) long been psychologically denied, as the penetrating study of Morse Peckham[31] indicates. The reason for the denial is the usual one: admission of the principle to conciousness is painful. [Evidence for such an assertion is, in the nature of the case, difficult to find, but for a single clear-cut example see the letter by Krogman[32].] It is not sadism or masochism that makes us urge that the denial be brought to an end. Rather, it is a love of the reality principle, and recognition that only those truths that are admitted to the conscious mind are available for use in making sense of the world. To assert the truth of the competitive exclusion principle is not to say that nature is and always must be, everywhere, "red in tooth and claw." Rather, it is to point out that *every* instance of apparent coexistence must be accounted for. Out of the study of all such instances will come a fuller knowledge of the many prosthetic devices of coexistence, each with its own costs and its own benefits. On such a foundation we may set about the task of establishing a science of ecological engineering.

References

1. Anonymous, *J. Animal Ecol.* **13,** 176 (1944).
2. G. Hardin, *Nature and Man's Fate* (Rine-hart, New York, 1959).
3. ____, *Am. J. Psychiat.* **114,** 392 (1957).
4. C. Elton, *Animal Ecology* (Macmillan, New York, 1927).
5. A. J. Lotka, *J. Wash. Acad. Sci.* **22,** 469 (1932).
6. V. Volterra, *Mem. reale accad. nazl. Lincei, Classe sci. fis. mat. e nat. ser. 6, No. 2* (1926).
7. ____, *Leçons sur la Théorie Mathématique de la Lutte pour la Vie* (Gauthier-Villars, Paris, 1931).
8. G. E. Hutchinson, *Cold Spring Harbor Symposia Quant. Biol.* **22,** 415 (1957).
9. G. F. Gause, *The Struggle for Existence* (Williams and Wilkins, Baltimore, 1934); H. H. Ross, *Evolution* **11,** 113 (1957).
10. O. Gilbert, T. B. Reynoldson, J. Hobart, *J. Animal Ecol.* **21,** 310–312 (1952).
11. C. Darwin, *On the Origin of Species by Means of Natural Selection* (Macmillan, New York, new ed. 6, 1927).
12. T. Park and M. Lloyd, *Am. Naturalist* **89,** 235 (1955).
13. R. M. Thrall, C. H. Coombs, R. L. Davis, *Decision Processes* (Wiley, New York, 1954), pp. 22–23.
14. A. J. Lotka, *Elements of Physical Biology* (Williams and Wilkins, Baltimore, 1925).
15. A. J. Nicholson, *J. Animal Ecol.* **2,** suppl., 132–178 (1933).
16. ____, *Australian J. Zool.* **2,** 9 (1954).
17. R. H. MacArthur, *Ecology* **39,** 599 (1958).
18. D. Lack, *Darwin's Finches* (University Press, Cambridge, 1947).
19. M. F. D. Udvardy, *Ecology* **40,** 725 (1959).
20. J. Grinnell, *Auk* **21,** 364 (1904).
21. W. C. Allee, A. E. Emerson, O. Park, T. Park, K. P. Schmidt, *Principles of Ecology* (Saunders, Philadelphia, 1949).
22. G. Hardin, *Sci. Monthly* **70,** 178 (1950).
23. F. Darwin, *The Foundations of the Origin of Species* (University Press, Cambridge, 1909).
24. J. Bertrand, *J. savants* (Sept. 1883), pp. 499–508.
25. H. Litchfield, *Emma Darwin, A Century of Family Letters, 1792–1896* (Murray, London, 1915).
26. M. Scriven, *Science* **130,** 477 (1959).
27. F. A. Lange, *History of Materialism* (Harcourt Brace, New York, ed. 3, 1925).
28. J. B. S. Haldane, *Trans. Cambridge Phil. Soc.* **23,** 19 (1924).
29. W. C. Boyd, *Am. J. Human Genet.* **11,** 397 (1959).
30. G. G. Simpson, *Evolution* **6,** 342 (1952).
31. M. Peckham, *Victorian Studies* **3,** 19 (1959).
32. W. M. Krogman, *Science* **111,** 43 (1950).

Critical Thinking

1. Describe the competitive exclusion principle in its simplest form.

2. Identify several examples of how the principle evidences itself in human behavior at various levels—from the individual to the group to nation.

3. Describe how the principle may operate in the global geopolitical-economy.

4. What linkages can you see between the exclusion principle, environmental consumption patterns, and our success in achieving a sustainable world?

5. Would there be differences between the interpretation of the exclusion principle by wealthy people/nations and poor people/nations? Explain.

Of the 1%, by the 1%, for the 1%

Americans have been watching protests against oppressive regimes that concentrate massive wealth in the hands of an elite few. Yet in our own democracy, 1 percent of the people take nearly a quarter of the nation's income—an inequality even the wealthy will come to regret.

JOSEPH E. STIGLITZ

It's no use pretending that what has obviously happened has not in fact happened. The upper 1 percent of Americans are now taking in nearly a quarter of the nation's income every year. In terms of wealth rather than income, the top 1 percent control 40 percent. Their lot in life has improved considerably. Twenty-five years ago, the corresponding figures were 12 percent and 33 percent. One response might be to celebrate the ingenuity and drive that brought good fortune to these people, and to contend that a rising tide lifts all boats. That response would be misguided. While the top 1 percent have seen their incomes rise 18 percent over the past decade, those in the middle have actually seen their incomes fall. For men with only high-school degrees, the decline has been precipitous—12 percent in the last quarter-century alone. All the growth in recent decades—and more—has gone to those at the top. In terms of income equality, America lags behind any country in the old, ossified Europe that President George W. Bush used to deride. Among our closest counterparts are Russia with its oligarchs and Iran. While many of the old centers of inequality in Latin America, such as Brazil, have been striving in recent years, rather successfully, to improve the plight of the poor and reduce gaps in income, America has allowed inequality to grow.

Economists long ago tried to justify the vast inequalities that seemed so troubling in the mid-19th century—inequalities that are but a pale shadow of what we are seeing in America today. The justification they came up with was called "marginal-productivity theory." In a nutshell, this theory associated higher incomes with higher productivity and a greater contribution to society. It is a theory that has always been cherished by the rich. Evidence for its validity, however, remains thin. The corporate executives who helped bring on the recession of the past three years—whose contribution to our society, and to their own companies, has been massively negative—went on to receive large bonuses. In some cases, companies were so embarrassed about calling such rewards "performance bonuses" that they felt compelled to change the name to "retention bonuses" (even if the only thing being retained was bad performance). Those who have contributed great positive innovations to our society, from the pioneers of genetic understanding to the pioneers of the Information Age, have received a pittance compared with those responsible for the financial innovations that brought our global economy to the brink of ruin.

Some people look at income inequality and shrug their shoulders. So what if this person gains and that person loses? What matters, they argue, is not how the pie is divided but the size of the pie. That argument is fundamentally wrong. An economy in which *most* citizens are doing worse year after year—an economy like America's—is not likely to do well over the long haul. There are several reasons for this.

First, growing inequality is the flip side of something else: shrinking opportunity. Whenever we diminish equality of opportunity, it means that we are not using some of our most valuable assets—our people—in the most productive way possible. Second, many of the distortions that lead to inequality—such as those associated with monopoly power and preferential tax treatment for special interests—undermine the efficiency of the economy. This new inequality goes on to create new distortions, undermining efficiency even further. To give just one example, far too many of our most talented young people, seeing the astronomical rewards, have gone into finance rather than into fields that would lead to a more productive and healthy economy.

Third, and perhaps most important, a modern economy requires "collective action"—it needs government to invest in infrastructure, education, and technology. The United States and the world have benefited greatly from government-sponsored research that led to the Internet, to advances in public health, and so on. But America has long suffered from an under-investment in infrastructure (look at the condition of our highways and bridges, our railroads and airports), in basic research, and in education at all levels. Further cutbacks in these areas lie ahead.

None of this should come as a surprise—it is simply what happens when a society's wealth distribution becomes lopsided. The more divided a society becomes in terms of wealth, the more reluctant the wealthy become to spend money on common needs. The rich don't need to rely on government for parks or education or medical care or personal security—they can buy all these things for themselves. In the process, they become more distant from ordinary people, losing whatever empathy they may once have had. They also worry about strong government—one that could use its powers to adjust the balance, take some of their wealth, and invest it for the common good. The top 1 percent may complain about the kind of government we have in America, but in truth they like it just fine: too gridlocked to re-distribute, too divided to do anything but lower taxes.

Economists are not sure how to fully explain the growing inequality in America. The ordinary dynamics of supply and demand have certainly played a role: labor-saving technologies have reduced the demand for many "good" middle-class, blue-collar jobs. Globalization has created a worldwide marketplace, pitting expensive unskilled workers in America against cheap unskilled workers overseas. Social changes have also played a role—for instance, the decline of unions, which once represented a third of American workers and now represent about 12 percent.

But one big part of the reason we have so much inequality is that the top 1 percent want it that way. The most obvious example involves tax policy. Lowering tax rates on capital gains, which is how the rich receive a large portion of their income, has given the wealthiest Americans close to a free ride. Monopolies and near monopolies have always been a source of economic power—from John D. Rockefeller at the beginning of the last century to Bill Gates at the end. Lax enforcement of anti-trust laws, especially during Republican administrations, has been a godsend to the top 1 percent. Much of today's inequality is due to manipulation of the financial system, enabled by changes in the rules that have been bought and paid for by the financial industry itself—one of its best investments ever. The government lent money to financial institutions at close to 0 percent interest and provided generous bailouts on favorable terms when all else failed. Regulators turned a blind eye to a lack of transparency and to conflicts of interest.

When you look at the sheer volume of wealth controlled by the top 1 percent in this country, it's tempting to see our growing inequality as a quintessentially American achievement—we started way behind the pack, but now we're doing inequality on a world-class level. And it looks as if we'll be building on this achievement for years to come, because what made it possible is self-reinforcing. Wealth begets power, which begets more wealth. During the savings-and-loan scandal of the 1980s—a scandal whose dimensions, by today's standards, seem almost quaint—the banker Charles Keating was asked by a congressional committee whether the $1.5 million he had spread among a few key elected officials could actually buy influence. "I certainly hope so," he replied. The Supreme Court, in its recent *Citizens United* case, has enshrined the right of corporations to buy government, by removing limitations on campaign spending. The personal and the political are today in perfect alignment. Virtually all U.S. senators, and most of the representatives in the House, are members of the top 1 percent when they arrive, are kept in office by money from the top 1 percent, and know that if they serve the top 1 percent well they will be rewarded by the top 1 percent when they leave office. By and large, the key executive-branch policy-makers on trade and economic policy also come from the top 1 percent. When pharmaceutical companies receive a trillion-dollar gift—through legislation prohibiting the government, the largest buyer of drugs, from bargaining over price—it should not come as cause for wonder. It should not make jaws drop that a tax bill cannot emerge from Congress unless big tax cuts are put in place for the wealthy. Given the power of the top 1 percent, this is the way you would *expect* the system to work.

America's inequality distorts our society in every conceivable way. There is, for one thing, a well-documented lifestyle effect—people outside the top 1 percent increasingly live beyond their means. Trickle-down economics may be a chimera, but trickle-down behaviorism is very real. Inequality massively distorts our foreign policy. The top 1 percent rarely serve in the military—the reality is that the "all-volunteer" army does not pay enough to attract their sons and daughters, and patriotism goes only so far. Plus, the wealthiest class feels no pinch from higher taxes when the nation goes to war: borrowed money will pay for all that. Foreign policy, by definition, is about the balancing of national interests and national resources. With the top 1 percent in charge, and paying no price, the notion of balance and restraint goes out the window. There is no limit to the adventures we can undertake; corporations and contractors stand only to gain. The rules of economic globalization are likewise designed to benefit the rich: they encourage competition among countries for *business,* which drives down taxes on corporations, weakens health and environmental protections, and undermines what used to be viewed as the "core" labor rights, which include the right to collective bargaining. Imagine what the world might look like if the rules were designed instead to encourage competition among countries for *workers.* Governments would compete in providing economic security, low taxes on ordinary wage earners, good education, and a clean environment—things workers care about. But the top 1 percent don't need to care.

Or, more accurately, they think they don't. Of all the costs imposed on our society by the top 1 percent, perhaps the greatest is this: the erosion of our sense of identity, in which fair play, equality of opportunity, and a sense of community are so important. America has long prided itself on being a fair society, where everyone has an equal chance of getting ahead, but the statistics suggest otherwise: the chances of a poor citizen, or even a middle-class citizen, making it to the top in America are smaller than in many countries of Europe. The cards are stacked against them. It is this sense of an unjust system without opportunity that has given rise to the conflagrations in the Middle East: rising food prices and growing and persistent youth unemployment simply served as kindling. With youth unemployment in America at around 20 percent (and in some locations, and among some socio-demographic groups, at twice that); with one out of six Americans desiring a full-time job not able to get one; with one out of seven Americans on food stamps (and about the same number suffering from "food insecurity")—given all this, there is ample evidence that something has blocked the vaunted "trickling down" from the top 1 percent to everyone else. All of this is having the predictable effect of creating alienation—voter turnout among those in their 20s in the last election stood at 21 percent, comparable to the unemployment rate.

In recent weeks we have watched people taking to the streets by the millions to protest political, economic, and social conditions in the oppressive societies they inhabit. Governments have been toppled in Egypt and Tunisia. Protests have erupted in Libya, Yemen, and Bahrain. The ruling families elsewhere in the region look on nervously from their air-conditioned penthouses—will they be next? They are right to worry. These are societies where a minuscule fraction of the population—less than 1 percent—controls the lion's share of the wealth; where wealth is a main determinant of power; where entrenched corruption of one sort or another is a way of life; and where the wealthiest often stand actively in the way of policies that would improve life for people in general.

As we gaze out at the popular fervor in the streets, one question to ask ourselves is this: When will it come to America? In important ways, our own country has become like one of these distant, troubled places.

Alexis de Tocqueville once described what he saw as a chief part of the peculiar genius of American society—something he called "self-interest properly understood." The last two words were the key. Everyone possesses self-interest in a narrow sense: I want what's good for me right now! Self-interest "properly understood" is different. It means appreciating that paying attention to everyone else's self-interest—in other words, the common welfare—is in fact a precondition for one's own ultimate well-being. Tocqueville was not suggesting that there was anything noble or idealistic about this outlook—in fact, he was suggesting the opposite. It was a mark of American pragmatism. Those canny Americans understood a basic fact: looking out for the other guy isn't just good for the soul—it's good for business.

The top 1 percent have the best houses, the best educations, the best doctors, and the best lifestyles, but there is one thing that money doesn't seem to have bought: an understanding that their fate is bound up with how the other 99 percent live. Throughout history, this is something that the top 1 percent eventually do learn. Too late.

Critical Thinking

1. Identify some similarities between what Stiglitz is observing and what Hardin observes in Article 13.
2. How might the wealth gap depend on the exclusion principle?
3. Do you think the top 1 percent of Americans have different environmental sustainability priorities than say middle-income Americans? If so, give some examples.
4. Does maintaining extreme wealth require maintaining high levels of environmental consumption? Explain.
5. Can maintaining extreme levels of material wealth be compatible with the ideas of sustainability? Explain.

The New Economy of Nature

GRETCHEN C. DAILY AND KATHERINE ELLISON

When three-year-old Becky Furmann got the "poopies" and became dehydrated, her doctor urged her to drink water. He didn't know that water had caused the rare illness that would kill her. As the chubby blond child grew thin and pale, her sufferings were finally confirmed as the ravages of Cryptosporidium parvum, a parasite almost unheard of until April 1993, when it slipped through one of the two modern filtration plants in Milwaukee, Wisconsin, and entered the city's water supply. Becky had been born with human immunodeficiency virus (HIV), which weakened her immune system, yet she had seemed otherwise healthy until then. Cryptosporidiosis sealed her fate.

In all, Milwaukee's cryptosporidiosis epidemic led to more than one hundred deaths and four hundred thousand illnesses. The victims had been betrayed by their water—and by their faith in the technology keeping it safe. What's more, they had plenty of company throughout the world.

At the end of the twentieth century, more than three million people were dying every year of diseases spread by water, and another one billion were at risk, lacking access to water suitable to drink. As Milwaukee's disaster showed, the problem wasn't limited to developing countries. Some thirty-six million Americans were drinking water from systems violating Environmental Protection Agency standards. One million Americans were getting sick every year from the contamination, and as many as nine hundred were dying from it. And, as happened in Milwaukee, sometimes the highest-technology methods couldn't keep the contaminants out. Breakdowns were becoming a serious problem as the mechanical systems aged and many strapped local governments deferred maintenance, to the point that the American Water Works Association estimated it would cost $325 billion to rehabilitate the country's dilapidated mechanical systems to ensure safe drinking water for everyone.

This crisis, and particularly this specter of expense, led the city of New York in 1997 to embark on a bold experiment. With billions of dollars and the drinking water of nearly ten million people at stake, planners weighed the costs and benefits of two alternative solutions to their water problem—constructing a filtration plant or repairing the largely natural filtration system that had been purifying the city's water all along. Nature won. And in a turn of events that would have global implications, it won on economic grounds.

The battlefield on which this victory was achieved is the Catskill/Delaware Watershed, the heart of New York's purification and delivery system, named after the two major rivers flowing from it. This rural landscape is famed as a scene of great beauty, but it's also a highly efficient and valuable machine—its cogs two thousand square miles of crop-filled valleys and mountains blanketed in forest, all connected by meandering streams feeding into an extensive system of reservoirs. For nearly a century, the complex natural system had been delivering water of exceptional purity to the people of New York City and several upstate counties. In recent years, it produced as much as 1.8 billion gallons per day, serving New Yorkers with a healthy drink whose taste and clarity were the envy of mayors throughout the United States. And unlike most other large U.S. cities, New York's tap water has never passed through a filtration plant.

Instead, the water, born as rain and melted snow on mountaintops as far as 125 miles away from those who will ultimately drink it, is naturally cleansed as it makes its way downhill toward the reservoirs. Beneath the forest floor, soil and fine roots filter the water and hidden microorganisms break down contaminants. In the streams, plants absorb as much as half of the surplus nutrients running into the waterway, such as nitrogen from automobile emissions and fertilizer and manure used on nearby farms. In open stretches, wetlands continue the filtering as cattails and other plants voraciously take up nutrients while trapping sediment and heavy metals. After reaching the reservoirs, the water is further cleansed as it sits and waits. Dead algae, floating branches and leaves, and remaining particles of grit slowly sink to the bottom.

This natural process, supplemented by small doses of chlorine and fluoride at the end of the water's journey, worked beautifully for most of the twentieth century. But then signs appeared of some mechanical failures. The trouble was relentless new development: roads, subdivisions, and second homes were popping up all over the watershed, most of which is privately owned. Failing septic systems were leaking raw sewage into streams. Farming and forestry were also taking a toll, with lawn chemicals, fertilizers, pesticides, and manure all being washed into the reservoirs at an unprecedented rate.

By 1989, these problems could no longer be ignored. Congress that year amended the Safe Drinking Water Act, putting into motion a major review of the country's drinking water systems. New York City was faced with the potentially enormous cost of an artificial water filtration plant, estimated at as much as $6–$8 billion, plus yearly maintenance expenses amounting to $300–$500 million. That price tag meant potential catastrophe for New York's budget, and city officials were determined to avoid it. With vigorous lobbying, they won agreement from federal regulators to try an alternative: rather than pay for the costly new filtration plant, the city would spend the much smaller amount of about $1.5 billion to protect the upstate watershed, by buying land as buffers and upgrading polluting sewage treatment plants, among other tactics. The EPA, in turn, would grant a five-year reprieve of its order.

The scheme was seriously challenged from the start. Powerful developers filed suit, claiming that property values would plummet as the city imposed restrictions on new construction. Environmentalists criticized the city's efforts as too weak. Nonetheless, the unprecedented agreement was a milestone in a world in which nature's labor has too long been taken for granted. A major government body had acted as if an ecosystem—the watershed—were worth protecting in its natural state for the economic benefits it gives society. It had invested in its restoration as if it were in fact a precious piece of infrastructure.

Around the world, in city offices and university conference halls, among small groups of community activists and at the World Bank, scientists, legal scholars, bureaucrats, and professional environmentalists debated the implications of New York's experiment. Could it possibly work? Did scientists know enough about the mechanics of watersheds to give reliable advice on their management? And, assuming the approach turned out to be justified, how widely could it be replicated?

In fact, without clear answers to these questions, and in many cases without knowing much about New York, governments around the world—in Curitiba, Brazil; in Quito, Ecuador; and in more than 140 U.S. municipalities, from Seattle, Washington, to Dade County, Florida—were starting to calculate the costs of conserving watersheds and compare them with the costs of building mechanical plants. In a bold departure from business as usual, they were taking stock of their natural capital. In the process, they were learning how ecosystems—environments of interacting plants, animals, and microbes, from coastal tide pools to Loire Valley vineyards to expanses of Amazonian rainforest—can be seen as capital assets, supplying human beings with services that sustain and enhance our lives. These "ecosystem services" provide not only food and wine but also cleansing of the Earth's air and water, protection from the elements, and refreshment and serenity for human spirits.

Historically, the labor of nature has been thought of mostly as free. And with the exception of a few specific goods, such as farm crops and timber, the use of nature's services is startlingly unregulated. Despite our assiduous watch over other forms of capital—physical (homes, cars, factories), financial (cash, savings accounts, corporate stocks), and human (skills and knowledge)—we haven't even taken measure of the ecosystem capital stocks that produce these most vital of labors. We lack a formal system of appraising or monitoring the value of

natural assets, and we have few means of insuring them against damage or loss.

Although governments have negotiated a wide array of global and regional agreements to protect certain ecosystems from degradation and extinction—such as the Ramsar Convention on Wetlands, the Convention on Biological Diversity, and the Convention on the Law of the Sea—these agreements are mostly weak, lacking the participation, resources, and systems of incentives and enforcement they need to be effective.

Even more striking is how rarely investments in ecosystem capital are rewarded economically. Typically, the property owners—whether individuals, corporations, governments, or other institutions—are not compensated for the services the natural assets on their land provide to society. With rare exception, owners of coastal wetlands are not paid for the abundance of seafood the wetlands nurture, nor are owners of tropical forests compensated for that ecosystem's contribution to the pharmaceutical industry and climate stability. As a result, many crucial types of ecosystem capital are undergoing rapid degradation and depletion. Compounding the problem is that the importance of ecosystem services is often widely appreciated only upon their loss.

The source of this predicament is easy to comprehend. For most of humankind's experience on Earth, ecosystem capital was available in sufficient abundance, and human activities were sufficiently limited, so that it was reasonable to think of ecosystem services as free. Yet today, nature everywhere is under siege. Each year the world loses some thirty million acres of tropical forest, an area slightly larger than Pennsylvania. At this rate, the last rainforest tree will bow out—dead on arrival at a sawmill or in a puff of smoke—around the middle of the twenty-first century. Biodiversity is being reduced to the lowest levels in human history. Homo sapiens has already wiped out one-quarter of all bird species, and an estimated eleven percent more are on the path to extinction, along with twenty-four percent of mammal and eleven percent of plant species. One-quarter of the world's coral reefs have been destroyed, with many others undergoing serious decline. To top it off, we're taking fish out of the sea for consumption faster than they can reproduce.

The twenty-first century began with a growing sense among scientists that crucial thresholds had been reached and time to fix things was running out. This increasingly apparent deadline has begun to inspire a shift in thinking for many scholars, most notably economists. To be sure, economists have long been concerned with issues of resource scarcity and limits to human activities. That's why their field was dubbed "the dismal science." Yet throughout the 1960s, '70s, and '80s, most economists clashed with ecologists. Economists accused ecologists of being alarmist about adverse human effects on Earth and of proposing costly and unnecessary measures of protection. Meanwhile, ecologists charged economists with promoting "growth" at any price and misusing partial indicators of well-being, such as the gross national product, that are blind to wear and tear on the planet.

This conflict began to ease in the late 1980s, however, with efforts to forge a new discipline integrating ecology and economics. An early participant in this movement was Stanford professor and Nobel laureate Kenneth Arrow, who for decades has been disturbed by the way economics dismisses "externalities," activities of which there are two types. Positive externalities are activities that benefit people who don't pay for them; negative externalities harm people who don't receive compensation.

An example of a positive externality is modern Costa Rica's careful stewardship of its forests—a striking turnabout from the rampant deforestation that lasted into the 1980s. The new conservation policies contribute to sustainable development in the region while also helping to stabilize the global climate and maintain biodiversity. Yet for the most part, only Costa Ricans pay for these widely enjoyed benefits. In contrast, a negative externality occurs when Americans drive gas-guzzlers. This activity contributes to air pollution, potential climate change, and the risk of the U.S. being drawn into foreign conflicts over oil. Yet even though these negative consequences affect large numbers of people, the drivers—since U.S. gas is cheap and relatively untaxed—don't pay the costs.

"Internalization" of such externalities—enactment of a system of fair pricing and fair payment—is badly needed, but it will not be simple. Arrow has tried to meet the challenge in part by joining other economists and ecologists in a growing effort to "rethink economics," a process fortified by their yearly meetings in Sweden.

Another major player in these meetings has been Cambridge University professor Partha Dasgupta. Born in India, Dasgupta has devoted much of his career to studying the interplay of overpopulation, poverty, and environmental degradation. He remembers being stunned, at a United Nations meeting in 1981, when economists from developing countries stood up one by one and told him they couldn't afford to protect their environments. The encounter, he later said, showed him "how far we had yet to go. We must stop viewing the environment as an amenity, a luxury the poor can't afford." Quite the contrary, Dasgupta is convinced that the local environment is often the greatest asset for poor families because they have few alternatives for income if it fails. The rich, by contrast, have a global reach for all sorts of ecosystem goods and services, as revealed by their dinner tables laden with fresh fruit, fish, spring water, and flowers from all over the planet. Ultimately, though, the rich are also vulnerable to faltering ecosystem services and the social instability that can arise as a result.

Important as they clearly are to rich and poor alike, ecosystem services typically carry little or no formally recognized economic value. As Columbia University economist Geoffrey Heal points out, economics is concerned more with prices than with values or importance. "The price of a good"—say, a loaf of bread or a car or a piece of jewelry—"does not reflect its importance in any overall social or philosophical sense," says Heal. "Very unimportant goods can be valued more highly by the market—have higher prices—than very important goods."

This contradiction isn't new. Economists throughout the eighteenth and nineteenth centuries were perplexed by the paradox of diamonds and water. Why do diamonds command a much higher price than water, when water is obviously so much more key to human survival? The answer, proposed by Englishman Alfred Marshall, is now common knowledge: price is set by supply and demand. In the case of water, Heal explains, the supply (at least in Marshall's England) "was so large as to exceed the amount that could possibly be demanded at any price. Consequently the price was zero; water was free. Now, of course, the demand for water has increased greatly as a result of population growth and rising prosperity, while the supply has remained roughly constant, so that water is no longer free." Diamonds, by contrast, started out scarce: the desire for ownership always exceeded their supply. Their market price was thus high—set by rich people competing for the few diamonds available.

Ecosystem assets are gradually acquiring the scarcity of diamonds as the human population and its aspirations grow. As they become more like diamonds, they take on increasing potential value in economic terms. But major innovations to our economic and social institutions are needed to capture this value and incorporate it into day-to-day decision-making.

The main challenge in the pursuit of this goal is that most ecosystem services are currently treated as "public goods," which if provided for one are provided for all, no matter who pays. An example is air quality: if a government spends on reducing pollution, it helps taxpayers and non-taxpayers alike. That leads to a problem of "free riders," in which some people benefit without charge from services paid for by others. And this is particularly true with the services provided by nature. Although we've engineered a financial system so sophisticated as to include market values for feng shui masters and interest-rate derivatives, we've not yet managed to establish them for such vital and everyday services as water purification and flood protection.

The big challenge now is how to measure, capture, and protect these newly discovered values before they are lost. Since the late 1990s, there's been an urgent flurry of calls to do just that, yet not until New York made its historic decision to invest in its watershed did it seem possible that big governments would catch on, supporting the concrete results of nature's work with cash on the table. Replicating that endeavor to any great extent, by conserving not only watersheds for water purity but also wetlands for flood control and forests for climate stabilization and biodiversity conservation, would require a tremendous amount of new scientific understanding of ecosystems—of their functioning, of their susceptibility to adverse human effects and their amenability to repair, and of the pros and cons of replacing them with technological substitutes. More important, it would require a willingness to look at the world's economy in an entirely different way, starting with the assumption that ecosystems are assets whose output has concrete financial worth.

Next, we need to change the rules of the game so as to produce new incentives for environmental protection, geared to both society's long-term well-being and individuals' self-interest. One way to do this is with taxes and subsidies targeting major environmental externalities, a strategy widely employed in Europe. A tax on consumption of fossil fuels, for instance, makes users of a shared resource—in this case, the sky, being used as a dumping ground—reduce their consumption and the damage it causes. It also makes higher-priced alternative energy sources (with lower environmental costs) more financially attractive. Consumption taxes such as this can be offset by reductions in income tax rates. In the U.S., however, such taxes have been virtually impossible to pass through Congress.

Another tactic, sometimes more politically feasible, is to establish ownership of ecosystem assets and services. This can avert the famous "tragedy of the commons" that often occurs when there is open access to a natural resource. It happens because each individual has more to gain by, say, launching another fishing boat than to lose by depleting the fishery. But when ownership rights to nature's goods and services are assigned, the new owners—be they private citizens, communities, corporations, interest groups, or governments—face unshared risk of those rights diminishing in value. Thus, as explained by economist and Nobel laureate Ronald Coase, they are motivated to fight for the asset's protection.

Establishing ownership of natural capital and services enables the process of bargaining between those affected by an externality and those causing it. Creating a place where people can get together to bargain—a market, whether in the town square or on the internet—is an old approach being newly applied to capture the value of ecosystem assets. A premier example is the evolving legal concept of "carbon rights"—ownership of the capacity of forests to stabilize climate by absorbing carbon dioxide. Efforts are underway to establish such rights and develop international markets for the purchase and sale of this forest ecosystem service, which in turn would establish a "market value," or price.

"Without prices being set, nature becomes like an all-you-can-eat buffet—and I don't know anyone who doesn't overeat at a buffet," says Richard Sandor, an environmentally minded financial innovator based in Chicago. Sandor has been a leading pioneer in looking at the problem of our dwindling resources in a striking new way. He and others have begun to act, launching bold initiatives to find financial incentives for environmental conservation.

One thing is clear: private enterprise cannot substitute for governments, particularly in view of the increasing risk of climate change, a global problem requiring global cooperation if it's not to override all other environmental and economic worries in a matter of decades. Government regulation may be called for to kick-start and supervise the profound economic transformation needed to ward off this and other environmental threats. Yet this transformation can be speeded with the use of market mechanisms and other financial incentives, tactics that have been glaringly underemployed.

Whether they appeal to us or not, experiments in finding market values for such essential gifts of nature as clean water and fresh air are well underway. The great unanswered question in all of this is whether the drive for profits, which has done so much harm to the planet, can finally be harnessed to save it.

Critical Thinking

1. What ecosystem service failed and led to deaths and thousands of illnesses in Wisconsin? Why?

2. Why has the labor of nature historically been thought of as free? Is this true for all people in all regions of the world?

3. Would rural agrarian people who live "closer" to nature see the labor of nature differently than urban industrial people?

4. Why do diamonds command a much higher price than water when water is clearly so much more valuable in maintaining life? What does this paradox say about the consumption of the planet's resources and our valuation of its ecosystem services?

Environmental Justice for All

Leyla Kokmen

M anuel Pastor ran bus tours of Los Angeles a few years back. These weren't the typical sojourns to Disneyland or the MGM studios, though; they were expeditions to some of the city's most environmentally blighted neighborhoods—where railways, truck traffic, and refineries converge, and where people live 200 feet from the freeway.

The goal of the "toxic tours," explains Pastor, a professor of geography and of American studies and ethnicity at the University of Southern California (USC), was to let public officials, policy makers, and donors talk to residents in low-income neighborhoods about the environmental hazards they lived with every day and to literally see, smell, and feel the effects.

"It's a pretty effective forum," says Pastor, who directs USC's Program for Environmental and Regional Equity, noting that a lot of the "tourists" were eager to get back on the bus in a hurry. "When you're in these neighborhoods, your lungs hurt."

Like the tours, Pastor's research into the economic and social issues facing low-income urban communities highlights the environmental disparities that endure in California and across the United States. As stories about global warming, sustainable energy, and climate change make headlines, the fact that some neighborhoods, particularly low-income and minority communities, are disproportionately toxic and poorly regulated has, until recently, been all but ignored.

A new breed of activists and social scientists are starting to capitalize on the moment. In principle they have much in common with the environmental justice movement, which came of age in the late 1970s and early 1980s, when grassroots groups across the country began protesting the presence of landfills and other environmentally hazardous facilities in predominantly poor and minority neighborhoods.

In practice, though, the new leadership is taking a broader-based, more inclusive approach. Instead of fighting a proposed refinery here or an expanded freeway there, all along trying to establish that systematic racism is at work in corporate America, today's environmental justice movement is focusing on proactive responses to the social ills and economic roadblocks that if removed would clear the way to a greener planet.

The new movement assumes that society as a whole benefits by guaranteeing safe jobs, both blue-collar and white-collar, that pay a living wage. That universal health care would both decrease disease and increase awareness about the quality of everyone's air and water. That better public education and easier access to job training, especially in industries that are emerging to address the global energy crisis, could reduce crime, boost self-esteem, and lead to a homegrown economic boon.

That green rights, green justice, and green equality should be the environmental movement's new watchwords.

"This is the new civil rights of the 21st century," proclaims environmental justice activist Majora Carter.

A lifelong resident of Hunts Point in the South Bronx, Carter is executive director of Sustainable South Bronx, an eight-year-old nonprofit created to advance the environmental and economic future of the community. Under the stewardship of Carter, who received a prestigious MacArthur Fellowship in 2005, the organization has managed a number of projects, including a successful grassroots campaign to stop a planned solid waste facility in Hunts Point that would have processed 40 percent of New York City's garbage.

Her neighborhood endures exhaust from some 60,000 truck trips every week and has four power plants and more than a dozen waste facilities. "It's like a cloud," Carter says. "You deal with that, you're making a dent."

The first hurdle Carter and a dozen staff members had to face was making the environment relevant to poor people and people of color who have long felt disenfranchised from mainstream environmentalism, which tends to focus on important but distinctly nonurban issues, such as preserving Arctic wildlife or Brazilian rainforest. For those who are struggling to make ends meet, who have to cobble together adequate health care, education, and job prospects, who feel unsafe on their own streets, these grand ideas seem removed from reality.

That's why the green rights argument is so powerful: It spans public health, community development, and economic growth to make sure that the green revolution isn't just for those who can afford a Prius. It means cleaning up blighted communities like the South Bronx to prevent potential health problems and to provide amenities like parks to play in, clean trails to walk on, and fresh air to breathe. It also means building green industries into the local mix, to provide healthy jobs for residents in desperate need of a livable wage.

Historically, mainstream environmental organizations have been made up mostly of white staffers and have focused more on the ephemeral concept of the environment rather than on the people who are affected. Today, though, as climate change and gas prices dominate public discourse, the concepts driving the new environmental justice movement are starting to catch on. Just recently, for instance, *The New York Times* columnist Thomas Friedman dubbed the promise of public investment in the green economy the "Green New Deal."

Van Jones, whom Friedman celebrated in print last October, is president of the Ella Baker Center for Human Rights in Oakland, California. To help put things in context, Jones briefly sketches the history of environmentalism:

The first wave was conservation, led first by Native Americans who respected and protected the land, then later by Teddy Roosevelt, John Muir, and other Caucasians who sought to preserve green space.

The second wave was regulation, which came in the 1970s and 1980s with the establishment of the Environmental Protection Agency (EPA) and Earth Day. Increased regulation brought a backlash against poor people and people of color, Jones says. White, affluent communities sought to prevent environmental hazards from entering their neighborhoods. This "not-in-my-backyard" attitude spurred a new crop of largely grassroots environmental justice advocates who charged businesses with unfairly targeting low-income and minority communities. "The big challenge was NIMBY-ism," Jones says, noting that more toxins from power plants and landfills were dumped on people of color.

The third wave of environmentalism, Jones says, is happening today. It's a focus on investing in solutions that lead to "eco-equity." And, he notes, it invokes a central question: "How do we get the work, wealth, and health benefits of the green economy to the people who most need those benefits?"

There are a number of reasons why so many environmental hazards end up in the poorest communities.

Property values in neighborhoods with environmental hazards tend to be lower, and that's where poor people—and often poor people of color—can afford to buy or rent a home. Additionally, businesses and municipalities often choose to build power plants in or expand freeways through low-income neighborhoods because the land is cheaper and poor residents have less power and are unlikely to have the time or organizational infrastructure to evaluate or fight development.

"Wealthy neighborhoods are able to resist, and low-income communities of color will find their neighborhoods plowed down and [find themselves] living next to a freeway that spews pollutants next to their schools," USC's Manuel Pastor says.

Moreover, regulatory systems, including the EPA and various local and state zoning and environmental regulatory bodies, allow piecemeal development of toxic facilities. Each new chemical facility goes through an individual permit process, which doesn't always take into account the overall picture in the community. The regulatory system isn't equipped to address potentially dangerous cumulative effects.

In a single neighborhood, Pastor says, you might have toxins that come from five different plants that are regulated by five different authorities. Each plant might not be considered dangerous on its own, but if you throw together all the emissions from those static sources and then add in emissions from moving sources, like diesel-powered trucks, "you've created a toxic soup," he says.

In one study of air quality in the nine-county San Francisco Bay Area, Pastor found that race, even more than income, determined who lived in more toxic communities. That 2007 report, "Still Toxic After All These Years: Air Quality and Environmental Justice in the San Francisco Bay Area," published by the Center for Justice, Tolerance & Community at the University of California at Santa Cruz, explored data from the EPA's Toxic Release Inventory, which reports toxic air emissions from large industrial facilities. The researchers examined race, income, and the likelihood of living near such a facility.

More than 40 percent of African American households earning less than $10,000 a year lived within a mile of a toxic facility, compared to 30 percent of Latino households and fewer than 20 percent of white households.

As income rose, the percentages dropped across the board but were still higher among minorities. Just over 20 percent of African American and Latino households making more than $100,000 a year lived within a mile of a toxic facility, compared to just 10 percent of white households.

The same report finds a connection between race and the risk of cancer or respiratory hazards, which are both associated with environmental air toxics, including emissions both from large industrial facilities and from mobile sources. The researchers looked at data from the National Air Toxics Assessment, which includes estimates of such ambient air toxics as diesel particulate matter, benzene, and lead and mercury compounds. The areas with the highest risk for cancer had the highest proportion of African American and Asian residents, the lowest rate of home ownership, and the highest proportion of people in poverty. The same trends existed for areas with the highest risk for respiratory hazards.

According to the report, "There is a general pattern of environmental inequity in the Bay Area: Densely populated communities of color characterized by relatively low wealth and income and a larger share of immigrants disproportionately bear the hazard and risk burden for the region."

Twenty years ago, environmental and social justice activists probably would have presented the disparities outlined in the 2007 report as evidence of corporations deliberately targeting minority communities with hazardous waste. That's what happened in 1987, when the United Church of Christ released findings from a study that showed toxic waste facilities were more likely to be located near minority communities. At the 1991 People of Color Environmental Leadership Summit, leaders called the disproportionate burden both racist and genocidal.

In their 2007 book *Break Through: From the Death of Environmentalism to the Politics of Possibility,* authors Ted Nordhaus and Michael Shellenberger take issue with this strategy. They argue that some of the research conducted in the name of environmental justice was too narrowly focused and that activists have spent too much time looking for conspiracies of environmental racism and not enough time looking at the multifaceted problems facing poor people and people of color.

"Poor Americans of all races, and poor Americans of color in particular, disproportionately suffer from social ills of every kind," they write. "But toxic waste and air pollution are far from being the most serious threats to their health and well-being. Moreover, the old narratives of intentional discrimination fail to explain or address these disparities. Disproportionate environmental health outcomes can no more be reduced to intentional discrimination than can disproportionate economic and educational outcomes. They are due to a larger and more complex set of historic, economic, and social causes."

Today's environmental justice advocates would no doubt take issue with the finer points of Nordhaus and Shellenberger's criticism—in particular, that institutional racism is a red herring. Activists and researchers are acutely aware that they are facing a multifaceted spectrum of issues, from air pollution to a dire lack of access to regular health care. It's because of that complexity, however, that they are now more geared toward proactively addressing an array of social and political concerns.

"The environmental justice movement grew out of putting out fires in the community and stopping bad things from happening, like a landfill," says Martha Dina Argüello, executive director of Physicians for Social Responsibility—Los Angeles, an organization that connects environmental groups with doctors to promote public health. "The more this work gets done, the more you realize you have to go upstream. We need to stop bad things from happening."

"We can fight pollution and poverty at the same time and with the same solutions and methods," says the Ella Baker Center's Van Jones.

Poor people and people of color have borne all the burden of the polluting industries of today, he says, while getting almost none of the benefit from the shift to the green economy. Jones stresses that he is not an environmental justice activist, but a "social-uplift environmentalist." Instead of concentrating on the presence of pollution and toxins in low-income communities, Jones prefers to focus on building investment in clean, green, healthy industries that can help those communities. Instead of focusing on the burdens, he focuses on empowerment.

With that end in mind, the Ella Baker Center's Green-Collar Jobs Campaign plans to launch the Oakland Green Jobs Corps this spring. The initiative, according to program manager Aaron Lehmer, received $250,000 from the city of Oakland and will give people ages 18 to 35 with barriers to employment (contact with the criminal justice system, long-term unemployment) opportunities and paid internships for training in new energy skills like installing solar panels and making buildings more energy efficient.

The concept has gained national attention. It's the cornerstone of the Green Jobs Act of 2007, which authorizes $125 million annually for "green-collar" job training that could prepare 30,000 people a year for jobs in key trades, such as installing solar panels, weatherizing buildings, and maintaining

wind farms. The act was signed into law in December as part of the Energy Independence and Security Act.

While Jones takes the conversation to a national level, Majora Carter is focusing on empowerment in one community at a time. Her successes at Sustainable South Bronx include the creation of a 10-week program that offers South Bronx and other New York City residents hands-on training in brownfield remediation and ecological restoration. The organization has also raised $30 million for a bicycle and pedestrian greenway along the South Bronx waterfront that will provide both open space and economic development opportunities.

As a result of those achievements, Carter gets calls from organizations across the country. In December she traveled to Kansas City, Missouri, to speak to residents, environmentalists, businesses, and students. She mentions exciting work being done by Chicago's Blacks in Green collective, which aims to mobilize the African American community around environmental issues. Naomi Davis, the collective's founder, told Chicago Public Radio in November that the group plans to develop environmental and economic opportunities—a "green village" with greenways, light re-manufacturing, ecotourism, and energy-efficient affordable housing—in one of Chicago's most blighted areas.

Carter stresses that framing the environmental debate in terms of opportunities will engage the people who need the most help. It's about investing in the green economy, creating jobs, and building spaces that aren't environmentally challenged. It won't be easy, she says. But it's essential to dream big.

"It's about sacrifice," she says, "for something better and bigger than you could have possibly imagined."

Critical Thinking

1. The author argues that "given rights, green justice, and green equality should be the environmental movement's new watchwords." What does that mean?

2. How does the green rights movement engage poor people and people of color?

3. Can the green rights strategy work for all poor people and people of color around the world? Why or why not?

4. Referring to your text's world map, and your own knowledge, locate nations where ensuing environmental justice will be difficult. Why?

5. Identify in this article three key terms, concepts, or principles that are used in your textbook (environmental science, economics, sociology, history, geography, etc.) or employed in the discipline you are currently studying. (Note: The terms, concepts, or principles may be implicit, explicit, implied, or inferred.)

UNIT 4

The Whole Earth Café: Bellying Up to the Trough

Unit Selections

Learning Outcomes

After reading this unit, you will be able to:

- Explain the "cheeseburger footprint" and how that footprint contributes to environment pressures.

- Recognize the connections between land productivity, water resources, grain production consumption, and our future ability to feed 8 billion people.

- Compare/contrast arguments regarding the need for increased application of agricultural technologies/commercialization versus agricultural localizations and changing grain consumption patterns.

- Describe what the average American eats in a year.

- Identify where the world's water challenges may accelerate in the near future and explain your place selections.

- Explain why Americans have the illusion of water abundance and why we need to stop taking fresh water for granted.

- Define the concept of "water footprint" and explain what major factors determine a contry's water footprint.

- Articulate why the world should care about glacial shrinkage.

- Explain how it is that instead of eating food, in many cases we are really eating fossil fuels.

- Explain why mountaintop removal mining is so environmentally disruptive.

- Describe "Jevons" Paradox and discuss how increasing energy efficiency may actually carry environmental costs.

- Explain why we once thought the Earth's ocean biome was nearly infinite and inexhaustible, and explain how we are discovering it is not.

- Outline how wealthy countries and growing urban areas are consuming valuable agricultural and rural lands and asses the sustainability wisdom of such actions.

- Elaborate upon how most of the terrestrial biospheres and ecosystems have been invaded and altered by man, and describe what the environmental consequences have been.

- Explain how more quantitative knowledge about ecosystem responses to land-use change can help minimize potential environmental damage.

Student Website
www.mhhe.com/cls

Internet References

Alternative Energy Institute (AEI)
www.altenergy.org

Alternative Energy Institute, Inc.
www.altenergy.org

EnviroLink
www.envirolink.org

Endangered Species
www.endangeredspecie.com

Friends of the Earth
www.foe.co.uk/index.html

Freshwater Society
www.**freshwater**.org

Greenpeace
www.greenpeace.org

The Hunger Project
www.thp.org

National Geographic Society
www.nationalgeographic.com

Natural Resources Defense Council
http://nrdc.org

Smithsonian Institution Website
www.si.edu

The World's Water
www.world**water**.org

United Nations Environment Program (UNEP)
www.enep.ch

World Wildlife Federation (WWF)
www.wwf.org

The insufferable arrogance of human beings to think that Nature was made solely for their profit, as if it was conceivable that the sun had been set afire merely to ripen men's apples and head their cabbages.

Savinien deCryano de Bergerac, *États etempires de la lune, 1656*

Humans must consume the earth to survive and some of our most basic needs are food, water, energy, land, and functioning ecosystems. However, global population growth and changing consumption demands for Earth's resources over the last century are putting increasing strains on the planet's ability to provide for the needs, and desires, of the human species. Our consumption patterns of food, water, energy and land are changing from the patterns of our early ancestors. The use of Earth's resources is moving from need to desire, functionality

David Frazier/Corbis

to fashion. As societies transition from rural agrarian to urban industrial, lifestyles transition as well. These new lifestyles bring with them new needs, and more importantly, they bring the luxury of new desires. These new lifestyles also increase the physical, social and psychological distance between people and the Earth's resources they consume. This combination of desire and distance are changing more and more of humanity's relationship to the earth. It is a relationship plagued with growing unfamiliarity, unawareness and under-appreciation of the planetary ecosystems that humbly provide for our needs and desires.

The articles in Unit IV explore these issues of changing consumption patterns and changing lifestyles, and the impact these changes are having on our food, water, energy, land and ecosystem resources. This unit has been dived into five sections, referred to as "menu items" at the Whole Earth Café and they are: Item #1—Farmer's Omelet (food); Item #2—Sparkling Water (water); Item #3—Electrical Banana (energy); Item #4—Mud Pie (land); and Item #5—The Gaia Kabob (ecosystems).

Six articles are presented under **Menu Item #1—Farmer's Omelet** which discuss our changing food consumption patterns and its relationship to human lifestyles, energy, geopolitics, future trends in demand, availability and consumption, our penchant for meat and a illustrated look at what we eat in a year.

Article 17 focuses on encouraging and implementing more "hard science" to meet our agricultural needs. The authors argue that success in feeding the world's billions will depend on acceptance and use of our agricultural science knowledge: genetic engineering, pesticides, herbicides, irrigation techniques. They believe we need to scale up and build on existing technologies and innovations, and do so immediately. And, at the same time, we need to reevaluate existing agricultural regulatory frameworks that are based on scientific data and not fear. And on the other end of the spectrum is the aspect of lifestyle, the "carbon facts" of one of America's most beloved pastimes (and growing numbers of our global citizenry) are provided by Jamais Cascio in Article 19, "The Cheeseburger Footprint." Article 21, "Rethinking the Meat-Guzzler," expands upon Cascio's "cheeseburger footprint" pointing out that Americans consume close to 200 pounds of meat a year. The problem is that meat production demands so many resources and results in so many environmental negative externalities.

The possibility of the changing consumption geopolitics of food supply and scarcity and how it may create a new food economy is fast approaching according to Lester Brown in Article 20, "The New Geopolitics of Food." This new food economy may lead to global political instability and increasing environmental pressures. Article 22, "This Is What You Eat in a Year," offers a summary chart of the "ingredients" of this unit's Farmer's Omelet.

In Menu Item #2—Sparkling Water, four articles examine some of the fresh water consumption issues. Water may become a crisis of such magnitude that unless we begin to address it now, the food debate may become a moot point, and attending to global conflict resulting from fresh water scarcity and changing geographic patterns of fresh water access will be the order of the day. Unlike food, fresh water provides for a myriad of Human-Earth ecosystems needs. If water were simply called upon to slake our thirst (thus meeting our requisite biological functions) as food is essentially called upon to fill our bellies and provide for cellular respiration (give us the energy needed to consume the Earth), our fresh water issue may be simpler.

Article 24, "The World's Water Challenge," provides a summary view of the world's water challenge for now and into the future. Again the themes of the issue remain consistent: consumption, availability, access, use and resource competition. However, the authors argue that despite recognition of this issue and its themes, focusing on managing for the sustainability of the resource may be advanced by establishing a value for the resource. Data is available that suggests water can be ascribed a market value against which more sustainable consumption decisions and policies can be made. Without better understanding and appreciation of the linkages between access to and availability of fresh water, such a valuation can lead to inappropriate pricing and denied access. Our illusions of abundant water, likewise, contribute toward the tendency to undervalue it. Article 25, "Wet Dreams: Water Consumption in America," argues for the need to stop ignoring our water.

The "water footprint" of a country is described as the volume of water needed for the production of goods and services consumed by a nation's people. Article 26, "Water Footprints of Nations: Water Use By People as a Function of their Consumption Patterns," observes that there are four main factors determining the water footprint of a country: consumption volume, consumption pattern, climate, and agricultural practices.

The last article in this section, (#27) "The Big Melt," provides an illustration of the complex "environmental science" linkages between the earth's hydrological cycle, climate change, fresh water availability, agriculture, politics and conflict. Glacial shrinkage may become a very serious geopolitical issue. While places like the United States, Europe, and Latin America may not currently depend on glaciers for their water, other place likes India, China and Pakistan do. And to make matters worse, these regions also have population growth stresses, which mean higher demands for food resulting in an increasing demand for water. Locate this situation in an already contentious sociopolitical arena, and the majesty of glaciers takes on a whole new meaning.

Menu Item #3—Electrical Banana, takes a look at some of our energy consumption behaviors. The four essays in this section include two aspects of how we "consume" energy and two examples of how we might even consume more as our efficiency to produce it increases. Article 28, "Eating Fossil Fuels," discusses how we now expend 400 gallons of oil equivalent per person to annually feed America to make up for our 40% loss of land-based photosynthetic capability.

On the other hand, Article 29, "The Myth of Mountaintop Mining," provides an example of one of our more environmentally disruptive ways of producing and consuming one form of energy: coal. Coal is touted as a viable way to reduce our energy dependence on foreign supplies and we have lots of it. But to consume it, the coal industry says we will have to accept a certain degree of environmental impact. The author asks if it really has to be this kind of a trade-off: jobs and energy independence or a pristine environment.

The consequences of trading oil for food or coal for energy independence will influence our energy consumption patterns. Will increased energy production efficiency influence it as well? Articles 30 and 31 look at that very question with its investigation of the "Jevons' Paradox." In Article 30, "The Efficiency Dilemma," David Owen describes the paradox as such: *increased energy efficiency* leads to *increased productivity* which *reduces price* and *demand/consumption increases.* Now the question becomes can squeezing more consumption from less fuel carry an environmental cost? And in Article 31, "Jevons' Paradox and

Perils of Efficient Energy Use," writer Grey Lindsey offers a few more insights into the paradox. "The Piggy Principle" is one of those insights.

Humans also consume land, but first it must be transformed. And human transformation of the land has been taking place for millennia to the point where studies indicate that now nearly 83% of the planet's ice-free land surface is influenced by humans in some way or another. Our two most popular kinds of transformation are urbanization and agriculture, and demand for these two "products" of our transformation is growing each year. We now have half the world's population living in cities and over the next twenty-five years we will see another one million square kilometers of land transformed into urban space. The demands for agricultural land are rising as well. The paradox is that the land we need for agriculture is typically the land most prized for urban sprawl. This paradox, however, has not gone unrecognized. Rich countries are now realizing the need for agricultural land and are buying up the agricultural rights to millions of hectares, mostly in developing countries.

Menu Item #4—Mud Pie provides three articles that explore this situation of land consumption via urbanization and agriculture. Article 32, "Rich Countries Launch Great Land Grab to Safeguard Food Supply," argues that agricultural land-buying activities by wealthy nations could create a form of neo-colonialism in which poor nation's sacrifice the food needs of thier own people to produce food for the rich. On the other hand, sustainable urban farming in developed and emerging nations may help ease the demand for agricultural land abroad. According to Article 33, "Global Urbanization: Can Ecologists Identify a Sustainable Way Forward," urban farms can help reduce stress on ecosystems, control energy use, and help feed people. Finally, Article 34, "Development at the Urban Fringe and Beyond: Impacts on Agricultural and Rural Land," presents insights into the forces driving the kinds of development that are threatening agricultural and rural communities, and offers ways to channel and control such growth.

Menu Item #5—The Gaia Kabob is intended to provide an overview of this unit by looking at the broader picture of planetary consumption and its impact on the dish that provides us with our most complex and essential meal of Whole Earth Café menu—ecosystems. When did humans begin consuming ecosystems—ecophagy? Many scholars believe it was the period following modern agriculturalization and industrialization when humans finally became a global geoenvironmental force. The Earth was being tamed and transformed for even more explosive growth of the human population. To meet the growing demands from these revolutions and population growth, we needed more products of the earth's ecosystems: food, fresh water, timber, fiber, minerals, ores and fuel. Earth was consumed at faster and faster

rates, and humanity spread into more and more ecosystems and biomes. In fact, one region of our earth we once thought to be infinite, inexhaustible and out or our reach is no longer. Article 35, "The End of a Myth," looks at one our last remaining, relatively intact biomes—the ocean. It is a biome composed of myriad ecosystems which have remained essentially un-degraded by humans. But this is changing as we intrude further and further into its virgin domains, nibbling away at its edges. If our past consumption behaviors are any indication, it won't be long until we learn how to transform the ocean's bounty of ecosystems into palatable dishes for human consumption.

Ecosystems have contributed to our making substantial gains in human well-being and socioeconomic development, but gains have not come without their costs. For example, Article 36, "Land-Use Choices: Balancing Human Needs and Ecosystem Function," explains how our consumption needs and desires for ecosystem products often require land-use change which frequently results in alterations to the ecosystem itself.

Loss of biodiversity is another cost as Article 37, "Economic Report into Biodiversity Crisis Reveals Price of Consuming the Planet," points out. However, biodiversity is linked directly to habitat, and habitat to ecosystem. When we lose biodiversity we lose the other components as well. Unfortunately, the harmful effects of ecosystem degradation are too often initiated in one *place* by people who are not intimately connected to the ecosystems but who benefit from its services; while at the same time the harmful effects are borne disproportionately in *another place* by people who, more often than not are not just "benefiting" from an ecosystem, but have their lives depend on it. The irony is that the people who are most impacted by ecosystem degradation are the ones who can least afford it. For example, wealthy, industrialized societies have the capital and technology to build extensive and complex linkages to capture the benefits of ecosystem services located great distances from their own degraded habitats and/or faltering ecosystems. Poor, agrarian, non-industrialized economies (ecosystem peoples) have no such capabilities to reach that far for ecosystem services should their systems destabilize; and they have even far less agency to protect those systems.

The last article included in this section attempts to help the reader visualize to what extent humans have impacted global systems of biodiversity and ecosystem. If ever there was an ultimate invasive species, humans would have to take the title. The authors of Article 38, "Putting People in the Map," create a map of the earth's "anthropogenic biomes" which they derived from empirical analyses of global population, land use and land cover in an attempt to help us better understand the patterns of our planetary consumption and the human/environment relationship.

Radically Rethinking Agriculture for the 21st Century

N. V. FEDOROFF ET AL.

opulation experts anticipate the addition of another roughly 3 billion people to the planet's population by the mid-21st century. However, the amount of arable land has not changed appreciably in more than half a century. It is unlikely to increase much in the future because we are losing it to urbanization, salinization, and desertification as fast as or faster than we are adding it.[1] Water scarcity is already a critical concern in parts of the world.[2]

Climate change also has important implications for agriculture. The European heat wave of 2003 killed some 30,000 to 50,000 people.[3] The average temperature that summer was only about 3.5°C above the average for the last century. The 20 to 36% decrease in the yields of grains and fruits that summer drew little attention. But if the climate scientists are right, summers will be that hot on average by midcentury, and by 2090 much of the world will be experiencing summers hotter than the hottest summer now on record.

The yields of our most important food, feed, and fiber crops decline precipitously at temperatures much above 30°C.[4] Among other reasons, this is because photosynthesis has a temperature optimum in the range of 20° to 25°C for our major temperate crops, and plants develop faster as temperature increases, leaving less time to accumulate the carbohydrates, fats, and proteins that constitute the bulk of fruits and grains.[5] Widespread adoption of more effective and sustainable agronomic practices can help buffer crops against warmer and drier environments,[6] but it will be increasingly difficult to maintain, much less increase, yields of our current major crops as temperatures rise and drylands expand.[7]

Climate change will further affect agriculture as the sea level rises, submerging low-lying cropland, and as glaciers melt, causing river systems to experience shorter and more intense seasonal flows, as well as more flooding.[7]

Recent reports on food security emphasize the gains that can be made by bringing existing agronomic and food science technology and know-how to people who do not yet have it,[8,9] as well as by exploring the genetic variability in our existing food crops and developing more ecologically sound farming practices.[10] This requires building local educational, technical, and research capacity, food processing capability, storage capacity, and other aspects of agribusiness, as well as rural transportation and water and communications infrastructure. It also necessitates addressing the many trade, subsidy, intellectual property, and regulatory issues that interfere with trade and inhibit the use of technology.

What people are talking about today, both in the private and public research sectors, is the use and improvement of conventional and molecular breeding, as well as molecular genetic modification (GM), to adapt our existing food crops to increasing temperatures, decreased water availability in some places and flooding in others, rising salinity,[8,9] and changing pathogen and insect threats.[11] Another important goal of such research is increasing crops' nitrogen uptake and use efficiency, because nitrogenous compounds in fertilizers are major contributors to waterway eutrophication and greenhouse gas emissions.

There is a critical need to get beyond popular biases against the use of agricultural biotechnology and develop forward-looking regulatory frameworks based on scientific evidence. In 2008, the most recent year for which statistics are available, GM crops were grown on almost 300 million acres in 25 countries, of which 15 were developing countries.[12] The world has consumed GM crops for 13 years without incident. The first few GM crops that have been grown very widely, including insect-resistant and herbicide-tolerant corn, cotton, canola, and soybeans, have increased agricultural productivity and farmers' incomes. They have also had environmental and health benefits, such as decreased use of pesticides and herbicides and increased use of no-till farming.[13]

Despite the excellent safety and efficacy record of GM crops, regulatory policies remain almost as restrictive as they were when GM crops were first introduced. In the United States, case-by-case review by at least two and sometimes three regulatory agencies (USDA, EPA, and FDA) is still commonly the rule rather than the exception. Perhaps the most detrimental effect of this complex, costly, and time-intensive regulatory apparatus is the virtual exclusion of public-sector researchers from the use of molecular methods to improve crops for farmers. As a result, there are still only a few GM crops, primarily those for which there is a large seed market,[12] and the benefits of biotechnology have not been realized for the vast majority of food crops.

What is needed is a serious reevaluation of the existing regulatory framework in the light of accumulated evidence and experience. An authoritative assessment of existing data on GM crop safety is timely and should encompass protein safety, gene stability, acute toxicity, composition, nutritional value, allergenicity, gene flow, and effects on nontarget organisms. This would establish a foundation for reducing the complexity of the regulatory process without affecting the integrity of the safety assessment. Such an evolution of the regulatory process in the United States would be a welcome precedent globally.

It is also critically important to develop a public facility within the USDA with the mission of conducting the requisite safety testing of GM crops developed in the public sector. This would make it possible for university and other public-sector researchers to use contemporary molecular knowledge and techniques to improve local crops for farmers.

However, it is not at all a foregone conclusion that our current crops can be pushed to perform as well as they do now at much higher temperatures and with much less water and other agricultural inputs. It will take new approaches, new methods, new technology—indeed, perhaps even new crops and new agricultural systems.

Aquaculture is part of the answer. A kilogram of fish can be produced in as little as 50 liters of water,[14] although the total water requirements depend on the feed source. Feed is now commonly derived from wild-caught fish, increasing pressure on marine fisheries. As well, much of the growing aquaculture industry is a source of nutrient pollution of coastal waters, but self-contained and isolated systems are increasingly used to buffer aquaculture from pathogens and minimize its impact on the environment.[15]

Another part of the answer is in the scale-up of dryland and saline agriculture.[16] Among the research leaders are several centers of the Consultative Group on International Agricultural Research, the International Center for Biosaline Agriculture, and the Jacob Blaustein Institutes for Desert Research of the Ben-Gurion University of the Negev.

Systems that integrate agriculture and aquaculture are rapidly developing in scope and sophistication. A 2001 United Nations Food and Agriculture Organization report[17] describes the development of such systems in many Asian countries. Today, such systems increasingly integrate organisms from multiple trophic levels[18]. An approach particularly well suited for coastal deserts includes inland seawater ponds that support aquaculture, the nutrient efflux from which fertilizes the growth of halophytes, seaweed, salt-tolerant grasses, and mangroves useful for animal feed, human food, and biofuels, and as carbon sinks.[19] Such integrated systems can eliminate today's flow of agricultural nutrients from land to sea. If done on a sufficient scale, inland seawater systems could also compensate for rising sea levels.

The heart of new agricultural paradigms for a hotter and more populous world must be systems that close the loop of nutrient flows from microorganisms and plants to animals and back, powered and irrigated as much as possible by sunlight and seawater. This has the potential to decrease the land, energy,

and freshwater demands of agriculture, while at the same time ameliorating the pollution currently associated with agricultural chemicals and animal waste. The design and large-scale implementation of farms based on nontraditional species in arid places will undoubtedly pose new research, engineering, monitoring, and regulatory challenges, with respect to food safety and ecological impacts as well as control of pests and pathogens. But if we are to resume progress toward eliminating hunger, we must scale up and further build on the innovative approaches already under development, and we must do so immediately.

References and Notes

1. *The Land Commodities Global Agriculture & Farmland Investment Report 2009* (Land Commodities Asset Management AG, Baar, Switzerland, 2009; www.landcommodities.com).

2. *Water for Food, Water for Life: A Comprehensive Assessment of Water Management* (International Water Management Institute, Colombo, Sri Lanka, 2007).

3. D. S. Battisti, R. L. Naylor, Historical warnings of future food insecurity with unprecedented seasonal heat. *Science* **323,** 240 (2009). [Abstract/Free Full Text].

4. W. Schlenker, M. J. Roberts, Nonlinear temperature effects indicate severe damages to U.S. crop yields under climate change. *Proc. Natl. Acad. Sci. U.S.A.* **106,** 15594 (2009). [Abstract/Free Full Text].

5. M. M. Qaderi, D. M. Reid, in *Climate Change and Crops,* S. N. Singh, Ed. (Springer-Verlag, Berlin, 2009), pp. 1–9.

6. J. I. L. Morison, N. R. Baker, P. M. Mullineaux, W. J. Davies, Improving water use in crop production. *Philos. Trans. R. Soc. London Ser. B* **363,** 639 (2008). [Abstract/Free Full Text].

7. Intergovernmental Panel on Climate Change, *Climate Change 2007: Impacts, Adaptation and Vulnerability* (Cambridge Univ. Press, Cambridge, 2007; www.ipcc.ch/publications_and_data/publications_ipcc_fourth_assessment_report_wg2_report_impacts_adaptation_and_vulnerability.htm).

8. *Agriculture for Development* (World Bank, Washington, DC, 2008; http://siteresources.worldbank.org/INTWDR2008/Resources/WDR_00_book.pdf).

9. *Reaping the Benefits: Science and the Sustainable Intensification of Global Agriculture* (Royal Society, London, 2009; http://royalsociety.org/Reapingthebenefits).

10. *The Conservation of Global Crop Genetic Resources in the Face of Climate Change* (Summary Statement from a Bellagio Meeting, 2007; http://iis-db.stanford.edu/pubs/22065/Bellagio_final1.pdf).

11. P. J. Gregory, S. N. Johnson, A. C. Newton, J. S. I. Ingram, Integrating pests and pathogens into the climate change/food security debate. *J. Exp. Bot.* **60,** 2827 (2009). [Abstract/Free Full Text].

12. C. James, *Global Status of Commercialized Biotech/GM Crops: 2008* (International Service for the Acquisition of Agri-biotech Applications, Ithaca, NY, 2008).

13. G. Brookes, P. Barfoot, *AgBioForum* **11,** 21 (2008).

14. S. Rothbard, Y. Peretz, in *Tilapia Farming in the 21st Century,* R. D. Guerrero III, R. Guerrero-del Castillo, Eds. (Philippines Fisheries Associations, Los Baños, Philippines, 2002), pp. 60–65.

15. *The State of World Fisheries and Aquaculture 2008* (United Nations Food and Agriculture Organization, Rome, 2009; www.fao.org/docrep/011/i0250e/i0250e00.HTM).

16. M. A. Lantican, P. L. Pingali, S. Rajaram, Is research on marginal lands catching up? The case of unfavourable wheat growing environments. *Agric. Econ.* **29,** 353 (2003). [CrossRef].

17. *Integrated Agriculture-Aquaculture* (United Nations Food and Agriculture Organization, Rome, 2001; www.fao.org/DOCREP/005/Y1187E/y1187e00.htm).

18. T. Chopin *et al.,* in *Encyclopedia of Ecology,* S. E. Jorgensen, B. Fath, Eds. (Elsevier, Amsterdam, 2008), pp. 2463–2475.

19. The Seawater Foundation, www.seawaterfoundation.org.

20. The authors were speakers in a workshop titled "Adapting Agriculture to Climate Change: What Will It Take?" held 14 September 2009 under the auspices of the Office of the Science and Technology Adviser to the Secretary of State. The views expressed here should not be construed as representing those of the U.S. government. N.V.F. is on leave from Pennsylvania State University. C.N.H. is co-chair of Global Seawater, which promotes creation of Integrated Seawater Farms.

Critical Thinking

1. What does it mean to "radically re-examine" agriculture?

2. Who is doing the "re-examining"; who will "implement" the "conclusions" of the re-examination; who will "pay" for the "application" of the conclusions; who will benefit?

3. Compare the agriculture ideas presented in Article 9 to the ideas presented in Article 12. Where do they differ?

4. Locate on the map provided in this text the following: regions of food production; regions of consumption; regions where risks of global food security are highest.

5. Referring to question 4 above, list some general characteristics like socioeconomic, cultural, and environmental of the places you have identified. Describe what you see.

6. Identify in this article three key terms, concepts, or principles that are used in your textbook (environmental science, economics, sociology, history, geography, etc.) or employed in the discipline you are currently studying. (Note: The terms, concepts, or principles may be implicit, explicit, implied, or inferred.)

The Cheeseburger Footprint

JAMAIS CASCIO

We're growing accustomed to thinking about the greenhouse gas impact of transportation and energy production, but nearly everything we do leaves a carbon footprint. If it requires energy to make or do, chances are, some carbon was emitted along the way. But these are the early days of the climate awareness era, and it's not yet habit to consider the greenhouse implications of otherwise prosaic actions.

So as an exercise, let's examine the carbon footprint of something commonplace—a cheeseburger. There's a good chance you've eaten one this week, perhaps even today. What was its greenhouse gas impact? Do you have any idea? This is the kind of question we'll be forced to ask more often as we pay greater attention to our individual greenhouse gas emissions.

Burgers are common food items for most people in the US—surprisingly common. Estimates for the average American diet range from an average of about one per week, or about 50/year (*Fast Food Nation*) to as many as three burgers per week, or roughly 150/year (the *Economist,* **among other sources**). So what's the global warming impact of all those cheeseburgers? I don't just mean cooking the burger; I mean the gamut of energy costs associated with a hamburger—including growing the feed for the cattle for beef and cheese, growing the produce, storing and transporting the components, as well as cooking.

The first step in answering this question requires figuring out the life cycle energy of a cheeseburger, and it turns out we're in luck. *Energy Use in the Food Sector* (PDF), a 2000 report from Stockholm University and the Swiss Federal Institute of Technology, does just that. This highly-detailed report covers the myriad elements going into the production of the components of a burger, from growing and milling the wheat to make bread, to feeding, slaughtering and freezing the cattle for meat—even the energy costs of pickling cucumbers. The report is fascinating in its own right, but it also gives us exactly what we need to make a relatively decent estimation of the carbon footprint of a burger.

Overall, the researchers conclude that the total energy use going into a single cheeseburger amounts to somewhere between about 7 and 20 megajoules (the range comes from the variety of methods available to the food industry).

The researchers break this down by process, but not by energy type. Here, then, is a first approximation: we can split

Carbon Facts	
Product Size 1 Cheeseburger (130g)	
Amount per Serving	
Kilograms co_2 Equivalent 5.18	
kilograms co_2 243 kilograms CH_4 215	
Total C: Energy Sources	243g
Transportation	
Fossil Fuel (Diesel)	120g
Fossil Fuel (Gasoline)	48g
Electricity Production	
Fossil Fuel (Natural Gas)	75g
Fossil Fuel (Coal)	0g
Other	
Total C: Non-Energy Sources	4939gCO_2E
Enteric Fermentation	181.0g(4163gCO_2E)
Manure	25.8g(656gCO_2E)
Other	5.2g(120gCO_2E)
Carbon/Product Ratio	39.9
Localism Rating	0+
Sustainable production Rating	D+
overall carbon code:	orange

the food production and transportation uses into a diesel category, and the food processing (milling, cooking, storage) uses into an electricity category. Split this way, the totals add up thusly:

Diesel—4.7 to 10.8 MJ per burger
Electricity—2.6 to 8.4 MJ per burger

With these ranges in hand, we can then convert the energy use into carbon dioxide emissions, based on fuel. **Diesel is straightforward.** For **electricity,** we should calculate the footprint using both natural gas and coal, as their carbon emissions vary considerably. (If you're lucky enough to have your local cattle ranches, farms and burger joints powered by wind farm, you can drop that part of the footprint entirely.) The results:

Diesel—350 to 800 grams of carbon dioxide per burger
Gas—416 to 1340 grams of carbon dioxide per burger
Coal—676 to 2200 grams of carbon dioxide per burger

. . . for a combined carbon dioxide footprint of a cheeseburger of 766 grams of CO_2 (at the low end, with gas) to 3000 grams of CO_2 (at the high end, with coal). Adding in the carbon from operating the restaurant (and driving to the burger shop in the first place), we can reasonably call it somewhere between 1 kilogram and 3.5 kilograms of energy-based carbon dioxide emissions per cheeseburger.

But that's not the whole story. There's a little thing called methane. It's a greenhouse gas that is, pound for pound, about 23 times more effective a greenhouse gas than carbon dioxide. It's also something that cattle make, in abundance.

By regulation, a beef cow must be at least 21 months old before going to the slaughterhouse; let's call it two years. A single cow produces about 110 kilos of methane per year in manure and what the EPA delicately calls "enteric fermentation," so over its likely lifetime, a beef cow produces 220 kilos of methane. Since a single kilo of methane is the equivalent of 23 kilos of carbon dioxide, a single beef cow produces a bit more than 5,000 CO_2-equivalent kilograms of methane over its life.

A typical beef cow produces **approximately 500 lbs of meat** for boneless steaks and ground beef. If we assume that the typical burger is a quarter-pound of pre-cooked meat, that's 2,000 burgers per cow. Dividing the methane total by the number of burgers, then, we get about 2.6 CO_2-equivalent kilograms of additional greenhouse gas emissions from methane, per burger, or roughly as much greenhouse gas produced from cow burps (etc.) as from all of the energy used to raise, feed or produce all of the components of a completed cheeseburger!

That's a total of 3.6-6.1 kg of CO_2-equivalent per burger. If we accept the ~3/week number, that's 540-915 kg of greenhouse gas per year for an average American's burger consumption. And for the nation as a whole?

300,000,000	citizens
* 150	burgers/year
* 4.35	kilograms of CO_2-equivalent per burger
/ 1000	kilograms per metric ton
= 195,750,000	annual metric tons of CO_2-equivalent for all US burgers

That's at a lower-than-average level of kg/burger.

Even with the lower claim of one cheeseburger per week, for an average American, the numbers remain sobering.

300,000,000	citizens
* 50	burgers/year (~ Fast Food Nation)
* 4.35	kilograms of CO_2-equivalent per burger
/ 1000	kilograms per metric ton
= 65,250,000	annual metric tons of CO_2-equivalent for all US burgers

Those numbers are big, impressive, and probably meaningless. So let's convert that to something more visceral. Let's compare to the output from a more familiar item: an SUV.

A Hummer H3 SUV emits 11.1 tons (imp.) of CO_2 over a year; this converts to about 10.1 metric tons, so we'll call it 10 to make the math easy.

195,750,000 annual metric tons of CO_2-equivalent for all US burgers
/10 metric tons of CO_2-equivalent per SUV

= 19.6 million SUVs

—or—

65,250,000 annual metric tons of CO_2-equivalent for all US burgers
/10 metric tons of CO_2-equivalent per SUV

= 6.5 million SUVs

To make it clear, then: **the greenhouse gas emissions arising every year from the production and consumption of cheeseburgers is roughly the amount emitted by 6.5 million to 19.6 million SUVs.** There are now approximately 16 million SUVs currently on the road in the US.

Will this information alone make a difference? Probably not; after all, nutrition info panels on packaged foods didn't turn us all into health food consumers. But they will allow us more informed choices, with no appeals to not knowing the consequences of our actions.

This was, ultimately, an attempt to take a remarkably prosaic activity and parse out its carbon aspects. After all, we're all increasingly accustomed to recognizing obvious, direct carbon emissions, but we're still wrapping our heads around the secondary and tertiary sources. Exercises like this one help to reveal the less-obvious ways that our behaviors and choices impact the planet and our civilization.

I doubt that we'll have to go through this process with everything we eat, from now until the end of the world. As our societies become more conscious of the impact of greenhouse gases, and the need for very tight and careful controls on just how much carbon we dump into the air, we'll need to create mechanisms for carbon transparency. Be they labels, icons, colorcodes, or [RFIDs], we'll need to be able to see, at a glance, just how much of a hit our personal carbon budgets take with each purchase.

The Cheeseburger Footprint is about much more than raw numbers. It's about how we live our lives, and the recognition that every action we take, even the most prosaic, can have unexpectedly profound consequences. The article was meant to poke us in our collective ribs, waking us up to the effects of our choices.

Critical Thinking

1. How might you reduce a cheeseburger's carbon footprint?

2. Provide a brief example of how you could apply full cost accounting to the production of a cheeseburger. How might that encourage reduced consumption and ultimately a smaller carbon footprint?

3. Put together a project for which you and other students construct a carbon footprint of a dinner, a tailgate party, or the food sold in a restaurant in one day.

The New Geopolitics of Food

From the Middle East to Madagascar, high prices are spawning land grabs and ousting dictators. Welcome to the 21st-century food wars.

Lester R. Brown

In the United States, when world wheat prices rise by 75 percent, as they have over the last year, it means the difference between a $2 loaf of bread and a loaf costing maybe $2.10. If, however, you live in New Delhi, those skyrocketing costs really matter: A doubling in the world price of wheat actually means that the wheat you carry home from the market to hand-grind into flour for chapatis costs twice as much. And the same is true with rice. If the world price of rice doubles, so does the price of rice in your neighborhood market in Jakarta. And so does the cost of the bowl of boiled rice on an Indonesian family's dinner table.

Welcome to the new food economics of 2011: Prices are climbing, but the impact is not at all being felt equally. For Americans, who spend less than one-tenth of their income in the supermarket, the soaring food prices we've seen so far this year are an annoyance, not a calamity. But for the planet's poorest 2 billion people, who spend 50 to 70 percent of their income on food, these soaring prices may mean going from two meals a day to one. Those who are barely hanging on to the lower rungs of the global economic ladder risk losing their grip entirely. This can contribute—and it has—to revolutions and upheaval.

Already in 2011, the U.N. Food Price Index has eclipsed its previous all-time global high; as of March it had climbed for eight consecutive months. With this year's harvest predicted to fall short, with governments in the Middle East and Africa teetering as a result of the price spikes, and with anxious markets sustaining one shock after another, food has quickly become the hidden driver of world politics. And crises like these are going to become increasingly common. The new geopolitics of food looks a whole lot more volatile—and a whole lot more contentious—than it used to. Scarcity is the new norm.

Until recently, sudden price surges just didn't matter as much, as they were quickly followed by a return to the relatively low food prices that helped shape the political stability of the late 20th century across much of the globe. But now both the causes and consequences are ominously different.

In many ways, this is a resumption of the 2007–2008 food crisis, which subsided not because the world somehow came together to solve its grain crunch once and for all, but because

the Great Recession tempered growth in demand even as favorable weather helped farmers produce the largest grain harvest on record. Historically, price spikes tended to be almost exclusively driven by unusual weather—a monsoon failure in India, a drought in the former Soviet Union, a heat wave in the U.S. Midwest. Such events were always disruptive, but thankfully infrequent. Unfortunately, today's price hikes are driven by trends that are both elevating demand and making it more difficult to increase production: among them, a rapidly expanding population, crop-withering temperature increases, and irrigation wells running dry. Each night, there are 219,000 additional people to feed at the global dinner table.

More alarming still, the world is losing its ability to soften the effect of shortages. In response to previous price surges, the United States, the world's largest grain producer, was effectively able to steer the world away from potential catastrophe. From the mid-20th century until 1995, the United States had either grain surpluses or idle cropland that could be planted to rescue countries in trouble. When the Indian monsoon failed in 1965, for example, President Lyndon Johnson's administration shipped one-fifth of the U.S. wheat crop to India, successfully staving off famine. We can't do that anymore; the safety cushion is gone.

That's why the food crisis of 2011 is for real, and why it may bring with it yet more bread riots cum political revolutions. What if the upheavals that greeted dictators Zine el-Abidine Ben Ali in Tunisia, Hosni Mubarak in Egypt, and Muammar al-Qaddafi in Libya (a country that imports 90 percent of its grain) are not the end of the story, but the beginning of it? Get ready, farmers and foreign ministers alike, for a new era in which world food scarcity increasingly shapes global politics.

THE DOUBLING OF WORLD grain prices since early 2007 has been driven primarily by two factors: accelerating growth in demand and the increasing difficulty of rapidly expanding production. The result is a world that looks strikingly different from the bountiful global grain economy of the last century. What will the geopolitics of food look like in a new era dominated by scarcity? Even at this early stage, we can see at least the broad outlines of the emerging food economy.

On the demand side, farmers now face clear sources of increasing pressure. The first is population growth. Each year the world's farmers must feed 80 million additional people, nearly all of them in developing countries. The world's population has nearly doubled since 1970 and is headed toward 9 billion by midcentury. Some 3 billion people, meanwhile, are also trying to move up the food chain, consuming more meat, milk, and eggs. As more families in China and elsewhere enter the middle class, they expect to eat better. But as global consumption of grain-intensive livestock products climbs, so does the demand for the extra corn and soybeans needed to feed all that livestock. (Grain consumption per person in the United States, for example, is four times that in India, where little grain is converted into animal protein. For now.)

At the same time, the United States, which once was able to act as a global buffer of sorts against poor harvests elsewhere, is now converting massive quantities of grain into fuel for cars, even as world grain consumption, which is already up to roughly 2.2 billion metric tons per year, is growing at an accelerating rate. A decade ago, the growth in consumption was 20 million tons per year. More recently it has risen by 40 million tons every year. But the rate at which the United States is converting grain into ethanol has grown even faster. In 2010, the United States harvested nearly 400 million tons of grain, of which 126 million tons went to ethanol fuel distilleries (up from 16 million tons in 2000). This massive capacity to convert grain into fuel means that the price of grain is now tied to the price of oil. So if oil goes to $150 per barrel or more, the price of grain will follow it upward as it becomes ever more profitable to convert grain into oil substitutes. And it's not just a U.S. phenomenon: Brazil, which distills ethanol from sugar cane, ranks second in production after the United States, while the European Union's goal of getting 10 percent of its transport energy from renewables, mostly biofuels, by 2020 is also diverting land from food crops.

This is not merely a story about the booming demand for food. Everything from falling water tables to eroding soils and the consequences of global warming means that the world's food supply is unlikely to keep up with our collectively growing appetites. Take climate change: The rule of thumb among crop ecologists is that for every 1 degree Celsius rise in temperature above the growing season optimum, farmers can expect a 10 percent decline in grain yields. This relationship was borne out all too dramatically during the 2010 heat wave in Russia, which reduced the country's grain harvest by nearly 40 percent.

While temperatures are rising, water tables are falling as farmers overpump for irrigation. This artificially inflates food production in the short run, creating a food bubble that bursts when aquifers are depleted and pumping is necessarily reduced to the rate of recharge. In arid Saudi Arabia, irrigation had surprisingly enabled the country to be self-sufficient in wheat for more than 20 years; now, wheat production is collapsing because the non-replenishable aquifer the country uses for irrigation is largely depleted. The Saudis soon will be importing all their grain.

Saudi Arabia is only one of some 18 countries with water-based food bubbles. All together, more than half the world's people live in countries where water tables are falling. The politically troubled Arab Middle East is the first geographic region where grain production has peaked and begun to decline because of water shortages, even as populations continue to grow. Grain production is already going down in Syria and Iraq and may soon decline in Yemen. But the largest food bubbles are in India and China. In India, where farmers have drilled some 20 million irrigation wells, water tables are falling and the wells are starting to go dry. The World Bank reports that 175 million Indians are being fed with grain produced by overpumping. In China, overpumping is concentrated in the North China Plain, which produces half of China's wheat and a third of its corn. An estimated 130 million Chinese are currently fed by overpumping. How will these countries make up for the inevitable shortfalls when the aquifers are depleted?

Even as we are running our wells dry, we are also mismanaging our soils, creating new deserts. Soil erosion as a result of overplowing and land mismanagement is undermining the productivity of one-third of the world's cropland. How severe is it? Look at satellite images showing two huge new dust bowls: one stretching across northern and western China and western Mongolia; the other across central Africa. Wang Tao, a leading Chinese desert scholar, reports that each year some 1,400 square miles of land in northern China turn to desert. In Mongolia and Lesotho, grain harvests have shrunk by half or more over the last few decades. North Korea and Haiti are also suffering from heavy soil losses; both countries face famine if they lose international food aid. Civilization can survive the loss of its oil reserves, but it cannot survive the loss of its soil reserves.

Beyond the changes in the environment that make it ever harder to meet human demand, there's an important intangible factor to consider: Over the last half-century or so, we have come to take agricultural progress for granted. Decade after decade, advancing technology underpinned steady gains in raising land productivity. Indeed, world grain yield per acre has tripled since 1950. But now that era is coming to an end in some of the more agriculturally advanced countries, where farmers are already using all available technologies to raise yields. In effect, the farmers have caught up with the scientists. After climbing for a century, rice yield per acre in Japan has not risen at all for 16 years. In China, yields may level off soon. Just those two countries alone account for one-third of the world's rice harvest. Meanwhile, wheat yields have plateaued in Britain, France, and Germany—Western Europe's three largest wheat producers.

IN THIS ERA OF TIGHTENING world food supplies, the ability to grow food is fast becoming a new form of geopolitical leverage, and countries are scrambling to secure their own parochial interests at the expense of the common good.

The first signs of trouble came in 2007, when farmers began having difficulty keeping up with the growth in global demand for grain. Grain and soybean prices started to climb, tripling by mid-2008. In response, many exporting countries tried to control the rise of domestic food prices by restricting exports. Among them were Russia and Argentina, two leading wheat exporters. Vietnam, the No. 2 rice exporter, banned exports entirely for several months in early 2008. So did several other smaller exporters of grain.

With exporting countries restricting exports in 2007 and 2008, importing countries panicked. No longer able to rely on the market to supply the grain they needed, several countries took the novel step of trying to negotiate long-term grain-supply agreements with exporting countries. The Philippines, for instance, negotiated a three-year agreement with Vietnam for 1.5 million tons of rice per year. A delegation of Yemenis traveled to Australia with a similar goal in mind, but had no luck. In a seller's market, exporters were reluctant to make long-term commitments.

Fearing they might not be able to buy needed grain from the market, some of the more affluent countries, led by Saudi Arabia, South Korea, and China, took the unusual step in 2008 of buying or leasing land in other countries on which to grow grain for themselves. Most of these land acquisitions are in Africa, where some governments lease cropland for less than $1 per acre per year. Among the principal destinations were Ethiopia and Sudan, countries where millions of people are being sustained with food from the U.N. World Food Program. That the governments of these two countries are willing to sell land to foreign interests when their own people are hungry is a sad commentary on their leadership.

By the end of 2009, hundreds of land acquisition deals had been negotiated, some of them exceeding a million acres. A 2010 World Bank analysis of these "land grabs" reported that a total of nearly 140 million acres were involved—an area that exceeds the cropland devoted to corn and wheat combined in the United States. Such acquisitions also typically involve water rights, meaning that land grabs potentially affect all downstream countries as well. Any water extracted from the upper Nile River basin to irrigate crops in Ethiopia or Sudan, for instance, will now not reach Egypt, upending the delicate water politics of the Nile by adding new countries with which Egypt must negotiate.

The potential for conflict—and not just over water—is high. Many of the land deals have been made in secret, and in most cases, the land involved was already in use by villagers when it was sold or leased. Often those already farming the land were neither consulted about nor even informed of the new arrangements. And because there typically are no formal land titles in many developing-country villages, the farmers who lost their land have had little backing to bring their cases to court. Reporter John Vidal, writing in Britain's *Observer*, quotes Nyikaw Ochalla from Ethiopia's Gambella region: "The foreign companies are arriving in large numbers, depriving people of land they have used for centuries. There is no consultation with the indigenous population. The deals are done secretly. The only thing the local people see is people coming with lots of tractors to invade their lands."

Local hostility toward such land grabs is the rule, not the exception. In 2007, as food prices were starting to rise, China signed an agreement with the Philippines to lease 2.5 million acres of land slated for food crops that would be shipped home. Once word leaked, the public outcry—much of it from Filipino farmers—forced Manila to suspend the agreement. A similar uproar rocked Madagascar, where a South Korean firm, Daewoo Logistics, had pursued rights to more than 3 million acres of land. Word of the deal helped stoke a political furor that toppled the government and forced cancellation of the agreement. Indeed, few things are more likely to fuel insurgencies than taking land from people. Agricultural equipment is easily sabotaged. If ripe fields of grain are torched, they burn quickly.

Not only are these deals risky, but foreign investors producing food in a country full of hungry people face another political question of how to get the grain out. Will villagers permit trucks laden with grain headed for port cities to proceed when they themselves may be on the verge of starvation? The potential for political instability in countries where villagers have lost their land and their livelihoods is high. Conflicts could easily develop between investor and host countries.

These acquisitions represent a potential investment in agriculture in developing countries of an estimated $50 billion. But it could take many years to realize any substantial production gains. The public infrastructure for modern market-oriented agriculture does not yet exist in most of Africa. In some countries it will take years just to build the roads and ports needed to bring in agricultural inputs such as fertilizer and to export farm products. Beyond that, modern agriculture requires its own infrastructure: machine sheds, grain-drying equipment, silos, fertilizer storage sheds, fuel storage facilities, equipment repair and maintenance services, well-drilling equipment, irrigation pumps, and energy to power the pumps. Overall, development of the land acquired to date appears to be moving very slowly.

So how much will all this expand world food output? We don't know, but the World Bank analysis indicates that only 37 percent of the projects will be devoted to food crops. Most of the land bought up so far will be used to produce biofuels and other industrial crops.

Even if some of these projects do eventually boost land productivity, who will benefit? If virtually all the inputs—the farm equipment, the fertilizer, the pesticides, the seeds—are brought in from abroad and if all the output is shipped out of the country, it will contribute little to the host country's economy. At best, locals may find work as farm laborers, but in highly mechanized operations, the jobs will be few. At worst, impoverished countries like Mozambique and Sudan will be left with less land and water with which to feed their already hungry populations. Thus far the land grabs have contributed more to stirring unrest than to expanding food production.

And this rich country-poor country divide could grow even more pronounced—and soon. This January, a new stage in the scramble among importing countries to secure food began to unfold when South Korea, which imports 70 percent of its grain, announced that it was creating a new public-private entity that will be responsible for acquiring part of this grain. With an initial office in Chicago, the plan is to bypass the large international trading firms by buying grain directly from U.S. farmers. As the Koreans acquire their own grain elevators, they may well sign multiyear delivery contracts with farmers, agreeing to buy specified quantities of wheat, corn, or soybeans at a fixed price.

Other importers will not stand idly by as South Korea tries to tie up a portion of the U.S. grain harvest even before it gets to market. The enterprising Koreans may soon be joined by China, Japan, Saudi Arabia, and other leading importers. Although South

Korea's initial focus is the United States, far and away the world's largest grain exporter, it may later consider brokering deals with Canada, Australia, Argentina, and other major exporters. This is happening just as China may be on the verge of entering the U.S. market as a potentially massive importer of grain. With China's 1.4 billion increasingly affluent consumers starting to compete with U.S. consumers for the U.S. grain harvest, cheap food, seen by many as an American birthright, may be coming to an end.

No one knows where this intensifying competition for food supplies will go, but the world seems to be moving away from the international cooperation that evolved over several decades following World War II to an every-country-for-itself philosophy. Food nationalism may help secure food supplies for individual affluent countries, but it does little to enhance world food security. Indeed, the low-income countries that host land grabs or import grain will likely see their food situation deteriorate.

AFTER THE CARNAGE of two world wars and the economic missteps that led to the Great Depression, countries joined together in 1945 to create the United Nations, finally realizing that in the modern world we cannot live in isolation, tempting though that might be. The International Monetary Fund was created to help manage the monetary system and promote economic stability and progress. Within the U.N. system, specialized agencies from the World Health Organization to the Food and Agriculture Organization (FAO) play major roles in the world today. All this has fostered international cooperation.

But while the FAO collects and analyzes global agricultural data and provides technical assistance, there is no organized effort to ensure the adequacy of world food supplies. Indeed, most international negotiations on agricultural trade until recently focused on access to markets, with the United States, Canada, Australia, and Argentina persistently pressing Europe and Japan to open their highly protected agricultural markets. But in the first decade of this century, access to supplies has emerged as the overriding issue as the world transitions from an era of food surpluses to a new politics of food scarcity. At the same time, the U.S. food aid program that once worked to fend off famine wherever it threatened has largely been replaced by the U.N. World Food Program (WFP), where the United States is the leading donor. The WFP now has food-assistance operations in some 70 countries and an annual budget of $4 billion. There is little international coordination otherwise. French President Nicolas Sarkozy—the reigning president of the G-20—is proposing to deal with rising food prices by curbing speculation in commodity markets. Useful though this may be, it treats the symptoms of growing food insecurity, not the causes, such as population growth and climate change. The world now needs to focus not only on agricultural policy, but on a structure that integrates it with energy, population, and water policies, each of which directly affects food security.

But that is not happening. Instead, as land and water become scarcer, as the Earth's temperature rises, and as world food security deteriorates, a dangerous geopolitics of food scarcity is emerging. Land grabbing, water grabbing, and buying grain directly from farmers in exporting countries are now integral parts of a global power struggle for food security.

With grain stocks low and climate volatility increasing, the risks are also increasing. We are now so close to the edge that a breakdown in the food system could come at any time. Consider, for example, what would have happened if the 2010 heat wave that was centered in Moscow had instead been centered in Chicago. In round numbers, the 40 percent drop in Russia's hoped-for harvest of roughly 100 million tons cost the world 40 million tons of grain, but a 40 percent drop in the far larger U.S. grain harvest of 400 million tons would have cost 160 million tons. The world's carryover stocks of grain (the amount in the bin when the new harvest begins) would have dropped to just 52 days of consumption. This level would have been not only the lowest on record, but also well below the 62-day carryover that set the stage for the 2007–2008 tripling of world grain prices.

Then what? There would have been chaos in world grain markets. Grain prices would have climbed off the charts. Some grain-exporting countries, trying to hold down domestic food prices, would have restricted or even banned exports, as they did in 2007 and 2008. The TV news would have been dominated not by the hundreds of fires in the Russian countryside, but by footage of food riots in low-income grain-importing countries and reports of governments falling as hunger spread out of control. Oil-exporting countries that import grain would have been trying to barter oil for grain, and low-income grain importers would have lost out. With governments toppling and confidence in the world grain market shattered, the global economy could have started to unravel.

We may not always be so lucky. At issue now is whether the world can go beyond focusing on the symptoms of the deteriorating food situation and instead attack the underlying causes. If we cannot produce higher crop yields with less water and conserve fertile soils, many agricultural areas will cease to be viable. And this goes far beyond farmers. If we cannot move at wartime speed to stabilize the climate, we may not be able to avoid runaway food prices. If we cannot accelerate the shift to smaller families and stabilize the world population sooner rather than later, the ranks of the hungry will almost certainly continue to expand. The time to act is now—before the food crisis of 2011 becomes the new normal.

Critical Thinking

1. Describe the geopolitics of the "new food economy" the author warns is coming.

2. What will be the variable global impacts of this new food economy? Who will be negatively impacted most? Who the least? Why?

3. What does the author mean by "food is the new oil"?

4. What does the article say about global climate change and food production? How might this impact current food consumption patterns?

5. Can this trend to a new food economy be averted? If so, how?

Reprinted in entirety by McGraw-Hill with permission from *Foreign Policy*, May/June 2011, www.foreignpolicy.com. © 2011 Washingtonpost.Newsweek Interactive, LLC.

How to Feed 8 Billion People

Record grain shortages are threatening global food security in the immediate future. A noted environmental analyst shows how nations can better manage their limited resources.

LESTER R. BROWN

The world is entering a new food era. It will be marked by higher food prices, rapidly growing numbers of hungry people, and an intensifying competition for land and water resources that crosses national boundaries when food-importing countries buy or lease vast tracts of land in other countries. Because some of the countries where land is being acquired do not have enough land to adequately feed their own people, the stage is being set for future conflicts.

The sharp rise of grain prices in recent years underlines the gravity of the situation. From mid-2006 to mid-2008, world prices of wheat, rice, corn, and soybeans roughly tripled, reaching historic highs. It was not until the global economic crisis beginning in 2008 that grain prices began to level off and recede slightly.

The world has experienced several grain price surges over the last half century, but none like this. Earlier surges were event-driven, weather-related, and temporary—caused by monsoons, droughts, heat waves, etc.

The recent record surge in grain prices has been trend-driven. Working our way out of this tightening food situation means reversing the trends that are causing it, such as soil erosion, falling water tables, and rising carbon emissions.

As a result of persistently high food prices, hunger is spreading. In the mid-1990s, the number of hungry people had fallen to 825 million. But instead of continuing to decline, the number of people facing chronic food insecurity and undernourishment started to edge upward, jumping to more than 1 billion in 2009.

Rising food prices and the swelling ranks of the hungry are among the early signs of a tightening world food situation. More and more, food is looking like the weak link in our civilization, much as it was for the earlier ones whose archaeological sites we now study.

Food: The Weak Link

As the world struggles to feed all its people, farmers are facing some worrying trends. On the demand side of the food equation are three consumption-boosting trends: population growth, the growing consumption of grain-based animal protein, and, most recently, the massive use of grain to fuel cars.

Each year there are 79 million more people at the dinner table, and the overwhelming majority of these individuals are being added in countries where soils are eroding, water tables are falling, and irrigation wells are going dry.

Even as our numbers are multiplying, some 3 billion people are trying to add to their diets, consuming more meat and dairy products. At the top of the food-consumption ranking are the United States and Canada, where people consume on average 800 kilograms of grain per year, most of it indirectly as beef, pork, poultry, milk, and eggs. Near the bottom of this ranking is India, where people have less than 200 kilograms of grain each, and thus must consume nearly all of it directly, leaving little for conversion into animal protein.

The orgy of investment in ethanol fuel distilleries that followed the 2005 surge in U.S. gas prices doubled grain consumption to 40 million tons by 2008.

On the supply side, ongoing environmental trends are making it very difficult to expand food production fast enough. These include soil erosion, aquifer depletion, crop-shrinking heat waves, melting ice sheets and rising sea levels, and the melting of the mountain glaciers that feed major rivers and irrigation systems. In addition, three resource trends are affecting our food supply: the loss of cropland to non-farm uses, the diversion of irrigation water to cities, and the coming reduction in oil supplies.

Soil erosion is currently lowering the inherent productivity of some 30% of the world's cropland. In some countries, it has reduced grain production by half or more over the last three decades. Vast dust storms coming out of sub-Saharan Africa, northern China, western Mongolia, and Central Asia remind us that the loss of topsoil is not only continuing but expanding. Advancing deserts in China—the result of overgrazing, overplowing, and deforestation—have forced the complete or partial abandonment of some 24,000 villages and the cropland surrounding them.

The loss of topsoil began with the first wheat and barley plantings, but falling water tables are historically quite recent, simply because the pumping capacity to deplete aquifers has evolved only in recent decades. Water tables are now falling in countries that together contain half the world's people. An estimated 400 million people (including 175 million in India and 130 million in China) are being fed by overpumping, a process that is by definition short term. Saudi Arabia has announced that, because its major aquifer, a nonreplenishable fossil aquifer, is largely depleted, it will be phasing out wheat production entirely by 2016.

An estimated 400 million people are being fed by overpumping water, a process that is by definition short term.

Climate change also threatens food security. For each 1°C rise in temperature above the norm during the growing season, farmers can expect a 10% decline in wheat, rice, and corn yields. Since 1970, the earth's average surface temperature has increased by 0.6°C. And the Intergovernmental Panel on Climate Change projects that the temperature will rise by up to 6°C during this century.

As the earth's temperature continues to rise, mountain glaciers are melting throughout the world. The projected melting of the glaciers on which China and India depend presents the most massive threat to food security that humanity has ever faced. China and India are the world's leading wheat producers and also dominate the world rice harvest. Whatever happens to the wheat and rice harvests in these two population giants will affect food prices everywhere.

The accelerating melting of the Greenland and West Antarctic ice sheets combined with thermal expansion of the oceans could raise sea level by up to six feet during this century. Every rice-growing river delta in Asia is threatened by the melting of these ice sheets. Even a three-foot rise would devastate the rice harvest in the Mekong Delta, which produces more than half the rice in Vietnam, the world's number-two rice exporter.

Three-fourths of oceanic fisheries are now being fished at or beyond capacity or are recovering from over-exploitation. If we continue with business as usual, many of these fisheries will collapse. We are taking fish from the oceans faster than they can reproduce.

With additional water no longer available in many countries, growing urban thirst can be satisfied only by taking irrigation water from farmers. Thousands of farmers in California find it more profitable to sell their irrigation water to Los Angeles and San Diego and leave their land idle. China's farmers are also losing irrigation water to the country's fast-growing cities.

If we paid the full cost of producing it—including the true cost of the oil used in producing it, the future costs of overpumping aquifers, the destruction of land through erosion, and the carbon-dioxide emissions from land clearing—food would cost far more than we now pay for it in the supermarket.

The question—at least for now—is: Will the world grain harvest expand fast enough to keep pace with steadily growing demand? Food security will deteriorate further unless leading countries collectively mobilize to stabilize population, stabilize climate, stabilize aquifers, conserve soils, protect cropland, and restrict the use of grain to produce fuel for cars.

The Emerging Geopolitics of Food Scarcity

As world food security deteriorates, individual countries, acting in their narrowly defined self-interest, are banning or limiting grain exports.

In response, other countries have been trying to nail down long-term bilateral trade agreements that would lock up future grain supplies. Several have succeeded. Egypt, for example, has reached a long-term agreement with Russia for more than 3 million tons of wheat each year.

The more affluent food-importing countries have sought to buy or lease for the long term large blocks of land to farm in other countries. Libya, which imports 90% of its grain and has been worried about

access to supplies, was one of the first to look abroad for land. After more than a year of negotiations, it reached an agreement to farm 100,000 hectares (250,000 acres) of land in Ukraine to grow wheat for its own people.

Countries selling or leasing their land are often low-income countries and, more often than not, those where chronic hunger and malnutrition are commonplace. A major acquisition site for Saudi Arabia and several other countries is Sudan—the site of the World Food Programme's largest famine relief effort.

The growing competition for land across national boundaries is also an indirect competition for water. In effect, land acquisitions are also water acquisitions. Land acquisitions in Sudan that tap water from the Nile, which is already fully utilized, may mean that Egypt will get less water from the river—making it even more dependent on imported grain.

Such bilateral land acquisitions raise many questions. To begin with, negotiations and the agreements they lead to tend to lack transparency. Typically, only a few high-ranking officials are involved, and the terms are confidential. Not only are many stakeholders such as farmers not at the table when the agreements are negotiated, they do not even learn about the deals until after they have been signed. And since there is rarely idle productive land in the countries where the land is being purchased or leased, the agreements suggest that many local farmers will be displaced. Their land may be confiscated or bought from them at a price over which they have little say.

This helps explain the public hostility that often arises within host countries. China, for example, signed an agreement with the Philippine government to lease more than a million hectares of land on which to produce crops that would be shipped home. Once word leaked out, the public outcry—much of it from Filipino farmers—forced the government to suspend the agreement. A similar situation developed in Madagascar, where South Korea's Daewoo Logistics had pursued rights to an area half the size of Belgium. The political furor led to a change in government and cancellation of the agreement.

Raising Land Productivity

There are many things that can be done in agriculture to raise land and water productivity. The challenge is for each country to fashion agricultural and economic policies that enable it to realize its unique potential.

Prior to 1950, expansion of the food supply came almost entirely from expanding cropland area. Then as frontiers disappeared and population growth accelerated after World War II, the world quickly shifted to raising land productivity. After several decades of rapid rise, however, it is now becoming more difficult to continue increasing productivity.

Gains in land productivity have come primarily from three sources: the growing use of fertilizer, the spread of irrigation, and the development of higher-yielding varieties of wheat, rice, and corn.

Among the three grains, corn is the only one where the yield is continuing to rise in high-yield countries. Even though fertilizer use has not increased since 1980, corn yields continue to edge upward as seed companies invest huge sums in corn breeding.

Despite dramatic past leaps in grain yields, it is becoming more difficult to expand world food output. There is little productive new land to bring under the plow. Expanding the irrigated area is difficult. Returns on the use of additional fertilizer are mostly diminishing. In the more arid countries of Africa, there is not enough rainfall to raise yields dramatically.

One way is to breed crops that are more tolerant of drought and cold. Another way to raise land productivity, where soil moisture permits, is to expand the area of land that produces more than one crop per year. Indeed, the tripling in the world grain harvest from 1950 to 2000 was due in part to widespread increases in multiple cropping in Asia. The spread of double cropping of winter wheat and corn on the North China Plain helped boost China's grain production to where it now rivals that of the United States.

A concerted U.S. effort to both breed earlier-maturing varieties and develop cultural practices that would facilitate multiple cropping could boost crop output. If China's farmers can extensively double crop wheat and corn, then U.S. farmers—at a similar latitude and with similar climate patterns—could do more if agricultural research and farm policy were reoriented to support it. Western Europe, with its mild winters and high-yielding winter wheat, might also be able to double crop more with a summer grain, such as corn, or an oilseed crop. Brazil and Argentina, which have extensive frost-free growing seasons, commonly multicrop wheat or corn with soybeans.

One encouraging effort to raise cropland productivity in Africa is the simultaneous planting of grain and leguminous trees. At first the trees grow slowly, permitting the grain crop to mature and be harvested; then the saplings grow quickly to several feet in height, dropping leaves that provide nitrogen and organic matter, both sorely needed in African soils. The wood is then cut and used for fuel. This simple, locally adapted technology, developed by scientists at the International Centre for Research in Agroforestry in Nairobi, has enabled farmers to double their grain yields within a matter of years as soil fertility builds.

Raising Water Productivity

Since it takes 1,000 tons of water to produce one ton of grain, it is not surprising that 70% of world water use is devoted to irrigation. Thus, raising irrigation efficiency is central to raising water productivity overall.

Data on the efficiency of surface water irrigation projects—that is, dams that deliver water to farmers through a network of canals—show that crop usage of irrigation water never reaches 100% because some irrigation water evaporates, some percolates downward, and some runs off. Water policy analysts have found that surface water irrigation efficiency is well below 50% in a number of countries, including India and Thailand.

Irrigation water efficiency is affected not only by the type and condition of irrigation systems but also by soil type, temperature, and humidity. In hot, arid regions, the evaporation of irrigation water is far higher than in cooler, humid regions. In a May 2004 meeting, China's Minister of Water Resources Wang Shucheng outlined for me in some detail the plans to raise China's irrigation efficiency from 43% in 2000 to 51% in 2010 and 55% in 2030. The steps he described included raising the price of water, providing incentives for adopting more irrigation-efficient technologies, and developing the local institutions to manage this process. Reaching these goals, he believes, would assure China's future food security.

Raising irrigation efficiency typically means shifting to overhead sprinklers or drip irrigation, the gold standard of irrigation efficiency. Switching to low-pressure sprinkler systems reduces water use by an estimated 30%, while switching to drip irrigation typically cuts water use in half.

A drip system also raises yields because it provides a steady supply of water with minimal losses to evaporation. Since drip systems are both labor-intensive and water-efficient, they are well suited to countries with a surplus of labor and a shortage of water. Israel (where the method was pioneered) and neighboring Jordan both rely heavily on drip irrigation. In contrast, among the big three agricultural producers, this more-efficient technology is used on roughly 3% of irrigated land in India and China and on roughly 4% in the United States.

In recent years, small-scale drip-irrigation systems—literally a bucket or drum with flexible plastic tubing to distribute the water—have been developed to irrigate small vegetable gardens. The containers are elevated slightly so that gravity distributes the water. Large-scale drip systems using plastic lines that can be moved easily are also becoming popular. These simple systems can pay for themselves in one year. By simultaneously reducing water costs and raising yields, they can dramatically raise incomes.

Shifting to more water-efficient crops wherever possible also boosts water productivity. Rice production is being phased out around Beijing because rice is such a thirsty crop. Similarly, Egypt restricts rice production in favor of wheat.

Strategic Reductions in the Demand for Grain

Although we seldom consider the climate effect of various dietary options, they are substantial, to say the least. A plant-based diet requires roughly one-fourth as much energy as a diet rich in red meat. Shifting to a vegetarian diet cuts greenhouse gas emissions almost as much as shifting from an SUV to a hybrid vehicle does. Shifting to less grain-intensive forms of animal protein such as poultry or certain types of fish can also reduce pressure on the earth's land and water resources.

Shifting to a vegetarian diet cuts greenhouse gas emissions almost as much as shifting from an SUV to a hybrid vehicle does.

When considering how much animal protein to consume, it is useful to distinguish between grass-fed and grain-fed products. For example, most of the world's beef is produced with grass. Even in the United States, with an abundance of feedlots, over half of all beef cattle weight gain comes from grass rather than grain. Grasslands are usually too steeply sloping or too arid to plow, and can contribute to the food supply only if used for grazing.

Beyond the role of grass in providing high-quality protein in our diets, it is sometimes assumed that we can increase the efficiency of land and water use by shifting from animal protein to high-quality plant protein, such as that from soybeans. It turns out, however, that since corn yields in the U.S. Midwest are three to four times those of soybeans, it may be more resource-efficient to produce corn and convert it into poultry or catfish at a ratio of two to one than to have everyone heavily reliant on soy.

The massive conversion of grain into biofuel began just a few years ago. If we are to reverse the spread of hunger, we will almost certainly have to cut back on ethanol production. Removing the incentives for converting food into fuel will help ensure that everyone has enough to eat. It will also lessen the pressures that lead to overpumping of groundwater and the clearing of tropical rain forests. If the U.S. government were to abolish the subsidies and mandates that are driving the conversion of grain into fuel, it would help stabilize grain prices and set the stage for relaxing the political tensions that have emerged within importing countries.

The Localization of Agriculture

In the United States, there has been a surge of interest in eating fresh local foods, corresponding with mounting concerns about the climate effects of consuming food from distant places. This is reflected in the rise in urban gardening, school gardening, and farmers' markets.

Food from more distant locations boosts carbon emissions while losing flavor and nutrition. A localized food economy reduces fossil fuel usage. Supermarkets are increasingly contracting with local farmers, and upscale restaurants are emphasizing locally grown food on their menus.

In school gardens, children learn how food is produced, a skill often lacking in urban settings, and they may get their first taste of freshly picked peas or vine-ripened tomatoes. School gardens also provide fresh produce for school lunches. California, a leader in this area, has 6,000 school gardens.

Many universities are now making a point of buying local food as well. Some universities compost kitchen and cafeteria food waste and make the compost available to the farmers who supply them with fresh produce.

Community gardens can be used by those who would otherwise not have access to land for gardening. Providing space for community gardens is seen by many local governments as an essential service.

Many market outlets are opening up for local produce. Perhaps the best known of these are the farmers' markets where local farmers bring their produce for sale. Many farmers' markets also now take food stamps, giving low-income consumers access to fresh produce that they might not otherwise be able to afford.

A survey of food consumed in Iowa showed conventional produce traveled on average 1,500 miles, not including food imported from other countries. In contrast, locally grown produce traveled on average 56 miles—a huge difference in fuel investment.

Concerns about the climate effects of transporting food long distances has led Tesco, the leading U.K. supermarket chain, to begin labeling products with their carbon footprint, indicating the greenhouse gas contribution of food items from the farm to supermarket shelf.

The shift from factory farm production of milk, meat, and eggs to mixed crop—livestock operations also facilitates nutrient recycling as local farmers return livestock manure to the land. The combination of high prices of natural gas, which is used to make nitrogen fertilizer, and of phosphate, as reserves are depleted, suggests a much greater future emphasis on nutrient recycling—an area where small farmers producing for local markets have a distinct advantage over massive feeding operations.

Costs and Solutions

If we cannot quickly cut carbon emissions, the world will face crop-shrinking heat waves that can massively and unpredictably reduce harvests. A hotter world will mean melting ice sheets, rising sea levels, and the inundation of the highly productive rice-growing river deltas of Asia. The loss of glaciers in the Himalayas and on the Tibetan Plateau will shrink wheat and rice harvests in both India and China, the world's most populous countries. Both are already facing water shortages driven by aquifer depletion and melting glaciers.

Since hunger is almost always the result of poverty, eradicating hunger depends on eradicating poverty. And where people are outrunning their land and water resources, this means stabilizing population.

Given that a handful of the more affluent grain-importing countries are reportedly investing some $20–30 billion in land acquisition, there is no shortage of capital to invest in agricultural development. Why not invest it across the board in helping low-income countries develop their unrealized potential for expanding food production, enabling them to export more grain?

We have a role to play as individuals. Whether we bike, bus, or drive to work will affect carbon emissions, climate change, and food security. The size of the car we drive to the supermarket and its effect on climate may indirectly affect the size of the bill at the supermarket checkout counter. If we are living high on the food-consumption chain, we can move down, improving our health while helping to stabilize climate. Food security is something in which we all have a stake—and a responsibility.

Critical Thinking

1. Sketch out a concept map illustrating the linkages between the three consumption-boosting trends the author suggests. Now describe the geographic location of the (1) population growth, (2) grain-based animal protein consumption, and (3) automobile driving. What are your observations?

2. The author states that food security will deteriorate further unless leading countries collectively mobilize to do several things. What are those actions, and how do these recommendations differ from those argued in Article 9?

3. Describe what is meant by the "localization of agriculture."

4. How might the promotion of "agricultural localization" have more global benefit and better environmental resiliency than large, commercial agricultural production operations?

5. Identify in this article three key terms, concepts, or principles that are used in your textbook (environmental science, economics, sociology, history, geography, etc.) or employed in the discipline you are currently studying. (Note: The terms, concepts, or principles may be implicit, explicit, implied, or inferred.)

LESTER R. BROWN is the founder and president of the Washington, D.C.-based nonprofit Earth Policy Institute. This article draws from his most recent book, *Plan B 4.0: Mobilizing to Save Civilization* (W.W. Norton and Co., 2009). For additional information, visit www.earth-policy.org.

Originally published in the January/February 2010, issue of *The Futurist*. Copyright © 2010 by World Future Society, 7910 Woodmont Avenue, Suite 450, Bethesda, MD 20814; phone: 301/656-8274. Used with permission. www.wfs.org

Rethinking the Meat-Guzzler

MARK BITTMAN

A SEA change in the consumption of a resource that Americans take for granted may be in store—something cheap, plentiful, widely enjoyed and a part of daily life. And it isn't oil.

It's meat.

The two commodities share a great deal: Like oil, meat is subsidized by the federal government. Like oil, meat is subject to accelerating demand as nations become wealthier, and this, in turn, sends prices higher. Finally—like oil—meat is something people are encouraged to consume less of, as the toll exacted by industrial production increases, and becomes increasingly visible.

Global demand for meat has multiplied in recent years, encouraged by growing affluence and nourished by the proliferation of huge, confined animal feeding operations. These assembly-line meat factories consume enormous amounts of energy, pollute water supplies, generate significant greenhouse gases and require ever-increasing amounts of corn, soy and other grains, a dependency that has led to the destruction of vast swaths of the world's tropical rain forests.

Just this week, the president of Brazil announced emergency measures to halt the burning and cutting of the country's rain forests for crop and grazing land. In the last five months alone, the government says, 1,250 square miles were lost.

The world's total meat supply was 71 million tons in 1961. In 2007, it was estimated to be 284 million tons. Per capita consumption has more than doubled over that period. (In the developing world, it rose twice as fast, doubling in the last 20 years.) World meat consumption is expected to double again by 2050, which one expert, Henning Steinfeld of the United Nations, says is resulting in a "relentless growth in livestock production."

Americans eat about the same amount of meat as we have for some time, about eight ounces a day, roughly twice the global average. At about 5 percent of the world's population, we "process" (that is, grow and kill) nearly 10 billion animals a year, more than 15 percent of the world's total.

Growing meat (it's hard to use the word "raising" when applied to animals in factory farms) uses so many resources that it's a challenge to enumerate them all. But consider: an estimated 30 percent of the earth's ice-free land is directly or indirectly involved in livestock production, according to the United Nations' Food and Agriculture Organization, which also estimates that livestock production generates nearly a fifth of the world's greenhouse gases—more than transportation.

To put the energy-using demand of meat production into easy-to-understand terms, Gidon Eshel, a geophysicist at the Bard Center, and Pamela A. Martin, an assistant professor of geophysics at the University of Chicago, calculated that if Americans were to reduce meat consumption by just 20 percent it would be as if we all switched from a standard sedan—a Camry, say—to the ultra-efficient Prius. Similarly, a study last year by the National Institute of Livestock and Grassland Science in Japan estimated that 2.2 pounds of beef is responsible for the equivalent amount of carbon dioxide emitted by the average European car every 155 miles, and burns enough energy to light a 100-watt bulb for nearly 20 days.

Grain, meat and even energy are roped together in a way that could have dire results. More meat means a corresponding increase in demand for feed, especially corn and soy, which some experts say will contribute to higher prices.

This will be inconvenient for citizens of wealthier nations, but it could have tragic consequences for those of poorer ones, especially if higher prices for feed divert production away from food crops. The demand for ethanol is already pushing up prices, and explains, in part, the 40 percent rise last year in the food price index calculated by the United Nations' Food and Agricultural Organization.

Though some 800 million people on the planet now suffer from hunger or malnutrition, the majority of corn and soy grown in the world feeds cattle, pigs and chickens. This despite the inherent inefficiencies: about two to five times more grain is required to produce the same amount of calories through livestock as through direct grain consumption, according to Rosamond Naylor, an associate professor of economics at Stanford University. It is as much as 10 times more in the case of grain-fed beef in the United States.

The environmental impact of growing so much grain for animal feed is profound. Agriculture in the United States—much of which now serves the demand for meat—contributes to nearly three-quarters of all water-quality problems in the nation's rivers and streams, according to the Environmental Protection Agency.

Because the stomachs of cattle are meant to digest grass, not grain, cattle raised industrially thrive only in the sense that they gain weight quickly. This diet made it possible to remove cattle from their natural environment and encourage the efficiency of mass confinement and slaughter. But it causes enough health problems that administration of antibiotics is routine, so much so that it can result in antibiotic-resistant bacteria that threaten the usefulness of medicines that treat people.

Those grain-fed animals, in turn, are contributing to health problems among the world's wealthier citizens—heart disease, some types of cancer, diabetes. The argument that meat provides useful protein makes sense, if the quantities are small. But the "you gotta eat meat" claim collapses at American levels. Even if the amount of meat we eat weren't harmful, it's way more than enough.

Americans are downing close to 200 pounds of meat, poultry and fish per capita per year (dairy and eggs are separate, and hardly insignificant), an increase of 50 pounds per person from 50 years ago. We each consume something like 110 grams of protein a day, about twice the federal government's recommended allowance; of that, about 75 grams come from animal protein. (The recommended level is itself considered by many dietary experts to be higher than it needs to be.) It's likely that most of us would do just fine on around 30 grams of protein a day, virtually all of it from plant sources.

What can be done? There's no simple answer. Better waste management, for one. Eliminating subsidies would also help; the United Nations estimates that they account for 31 percent of global farm income. Improved farming practices would help, too. Mark W. Rosegrant, director of environment and production technology at the nonprofit International Food Policy Research Institute, says, "There should be investment in livestock breeding and management, to reduce the footprint needed to produce any given level of meat."

Then there's technology. Israel and Korea are among the countries experimenting with using animal waste to generate electricity. Some of the biggest hog operations in the United States are working, with some success, to turn manure into fuel.

Longer term, it no longer seems lunacy to believe in the possibility of "meat without feet"—meat produced in vitro, by growing animal cells in a super-rich nutrient environment before being further manipulated into burgers and steaks.

Another suggestion is a return to grazing beef, a very real alternative as long as you accept the psychologically difficult and politically unpopular notion of eating less of it. That's because grazing could never produce as many cattle as feedlots do. Still, said Michael Pollan, author of the recent book "In Defense of Food," "In places where you can't grow grain, fattening cows on grass is always going to make more sense."

But pigs and chickens, which convert grain to meat far more efficiently than beef, are increasingly the meats of choice for producers, accounting for 70 percent of total meat production, with industrialized systems producing half that pork and three-quarters of the chicken.

Once, these animals were raised locally (even many New Yorkers remember the pigs of Secaucus), reducing transportation costs and allowing their manure to be spread on nearby fields. Now hog production facilities that resemble prisons more than farms are hundreds of miles from major population centers, and their manure "lagoons" pollute streams and groundwater. (In Iowa alone, hog factories and farms produce more than 50 million tons of excrement annually.)

These problems originated here, but are no longer limited to the United States. While the domestic demand for meat has leveled off, the industrial production of livestock is growing more than twice as fast as land-based methods, according to the United Nations.

Perhaps the best hope for change lies in consumers' becoming aware of the true costs of industrial meat production. "When you look at environmental problems in the U.S.," says Professor Eshel, "nearly all of them have their source in food production and in particular meat production. And factory farming is 'optimal' only as long as degrading waterways is free. If dumping this stuff becomes costly—even if it simply carries a non-zero price tag—the entire structure of food production will change dramatically."

Animal welfare may not yet be a major concern, but as the horrors of raising meat in confinement become known, more animal lovers may start to react. And would the world not be a better place were some of the grain we use to grow meat directed instead to feed our fellow human beings?

Real prices of beef, pork and poultry have held steady, perhaps even decreased, for 40 years or more (in part because of grain subsidies), though we're beginning to see them increase now. But many experts, including Tyler Cowen, a professor of economics at George Mason University, say they don't believe meat prices will rise high enough to affect demand in the United States.

"I just don't think we can count on market prices to reduce our meat consumption," he said. "There may be a temporary spike in food prices, but it will almost certainly be reversed and then some. But if all the burden is put on eaters, that's not a tragic state of affairs."

If price spikes don't change eating habits, perhaps the combination of deforestation, pollution, climate change, starvation, heart disease and animal cruelty will gradually encourage the simple daily act of eating more plants and fewer animals.

Mr. Rosegrant of the food policy research institute says he foresees "a stronger public relations campaign in the reduction of meat consumption—one like that around cigarettes—emphasizing personal health, compassion for animals, and doing good for the poor and the planet."

It wouldn't surprise Professor Eshel if all of this had a real impact. "The good of people's bodies and the good of the planet are more or less perfectly aligned," he said.

The United Nations' Food and Agriculture Organization, in its detailed 2006 study of the impact of meat consumption on the planet, "Livestock's Long Shadow," made a similar point: "There are reasons for optimism that the conflicting demands for animal products and environmental services can be reconciled. Both demands are exerted by the same group of people . . . the relatively affluent, middle- to high-income class, which is no longer confined to industrialized countries. . . . This group of consumers is probably ready to use its growing voice to exert pressure for change and may be willing to absorb the inevitable price increases."

In fact, Americans are already buying more environmentally friendly products, choosing more sustainably produced meat, eggs and dairy. The number of farmers' markets has more than doubled in the last 10 years or so, and it has escaped no one's notice that the organic food market is growing fast. These all represent products that are more expensive but of higher quality.

If those trends continue, meat may become a treat rather than a routine. It won't be uncommon, but just as surely as the S.U.V. will yield to the hybrid, the half-pound-a-day meat era will end.

Maybe that's not such a big deal. "Who said people had to eat meat three times a day?" asked Mr. Pollan.

Critical Thinking

1. How is meat like oil?

2. While Although higher prices for meat are expected and can be an inconvenience for citizens of wealthy nations, why could this have tragic consequence for the people of poorer nations?

3. Why is the environmental impact of growing so much grain for animal feed so significant?

4. List the negative externalities of meat production; then list the countries within which these externalities are located.

Chart: This Is What You Eat in a Year (Including 42 Pounds of Corn Syrup)

What do you call 200 pounds of meat, 31 pounds of cheese, 16 pounds of fish, and 415 pounds of veggies? Just a year in the life of the American stomach.

DEREK THOMPSON

What Are We Eating?

What the Average American Consumes in a Year

Fats & Oils
85.5 lbs

Red Meat
110 lbs

Poultry
73.6 lbs

Fish & Shellfish
16.1 lbs

Fruits
273.2 lbs

Chicken
60.4 lbs

Eggs
32.7 lbs

Cheese
31.4 lbs

Coffee, Cocoa & Nuts
24 lbs

The Average American

Age: 36.6

Height:
5'9" (m)
5'4" (f)

Dairy Products
(non-cheese)
600.5 lbs

Vegetables
415.4 lbs

Weight:
190 lbs (m)
164 lbs (f)

Corn
58 lbs

Corn
Syrup
42 lbs

Beverage Milks
181 lbs

Wheat flour
134.1 lbs

Caloric Sweeteners
141.6 lbs

Flour & Cereal Products
194.1 lbs

That Includes:
(every year)

French Fries	Pizza	Ice Cream	Soda	Artificial Sweeteners	Sodium	Caffeine	And 2,700 calories
29 lbs	23 lbs	24 lbs	53 gallons (about a gallon/week)	24 lbs	2.736 lbs (47% more than recommended)	0.2 lbs (90.700 mg)	a day

Critical Thinking

1. Are there other countries that would have a similar food chart?

2. With a group of students, construct similar food charts for three of the world's poorest countries.

3. Take each food group on the chart and identify countries other than the United States that supply the food or the ingredients for its production.

The World's Water Challenge

If oil is the key geopolitical resource of today, water will be as important—if not more so—in the not-so-distant future.

Erik R. Peterson and Rachel A. Posner

Historically, water has meant the difference between life and death, health and sickness, prosperity and poverty, environmental sustainability and degradation, progress and decay, stability and insecurity. Societies with the wherewithal and knowledge to control or "smooth" hydrological cycles have experienced more rapid economic progress, while populations without the capacity to manage water flows—especially in regions subject to pronounced flood-drought cycles—have found themselves confronting tremendous social and economic challenges in development.

Tragically, a substantial part of humanity continues to face acute water challenges. We now stand at a point at which an obscenely large portion of the world's population lacks regular access to fresh drinking water or adequate sanitation. Water-related diseases are a major burden in countries across the world. Water consumption patterns in many regions are no longer sustainable. The damaging environmental consequences of water practices are growing rapidly. And the complex and dynamic linkages between water and other key resources—especially food and energy—are inadequately understood. These factors suggest that even at current levels of global population, resource consumption, and economic activity, we may have already passed the threshold of water sustainability.

An obscenely large portion of the world's population lacks regular access to fresh drinking water or adequate sanitation.

A major report recently issued by the 2030 Water Resources Group (whose members include McKinsey & Company, the World Bank, and a consortium of business partners) estimated that, assuming average economic growth and no efficiency gains, the gap between global water demand and reliable supply could reach 40 percent over the next 20 years. As serious as this world supply-demand gap is, the study notes, the dislocations will be even more concentrated in developing regions that account for one-third of the global population, where the water deficit could rise to 50 percent.

It is thus inconceivable that, at this moment in history, no generally recognized "worth" has been established for water to help in its more efficient allocation. To the contrary, many current uses of water are skewed by historical and other legacy practices that perpetuate massive inefficiencies and unsustainable patterns.

The Missing Links

In addition, in the face of persistent population pressures and the higher consumption implicit in rapid economic development among large populations in the developing world, it is noteworthy that our understanding of resource linkages is so limited. Our failure to predict in the spring of 2008 a spike in food prices, a rise in energy prices, and serious droughts afflicting key regions of the world—all of which occurred simultaneously—reveals how little we know about these complex interrelationships.

Without significant, worldwide changes—including more innovation in and diffusion of water-related technologies; fundamental adjustments in consumption patterns; improvements in efficiencies; higher levels of public investment in water infrastructures; and an integrated approach to governance based on the complex relationships between water and food, water and economic development, and water and the environment—the global challenge of water resources could become even more severe.

Also, although global warming's potential effects on watersheds across the planet are still not precisely understood, there can be little doubt that climate change will in a number of regions generate serious dislocations in water supply. In a June 2008 technical paper, the Inter governmental Panel on Climate Change (IPCC) concluded that "globally, the negative impacts of climate change on freshwater systems are expected to outweigh the benefits." It noted that "higher water temperatures and changes in extremes, including droughts and floods, are projected to affect water quality and exacerbate many forms of water pollution."

Climate change will in a number of regions generate serious dislocations in water supply.

As a result, we may soon be entering unknown territory when it comes to addressing the challenges of water in all their dimensions, including public health, economic development, gender equity, humanitarian crises, environmental degradation, and global security. The geopolitical consequences alone could be profound.

Daunting Trends

Although water covers almost three-quarters of the earth's surface, only a fraction of it is suitable for human consumption. According to the United Nations, of the water that humans consume, approximately 70 percent is used in agricultural production, 22 percent in industry, and 8 percent in domestic use. This consumption—critical as it is for human health, economic development, political and social stability, and security—is unequal, inefficient, and unsustainable.

Indeed, an estimated 884 million people worldwide do not have access to clean drinking water, and 2.5 billion lack adequate sanitation. A staggering 1.8 million people, 90 percent of them children, lose their lives each year as a result of diarrheal diseases resulting from unsafe drinking water and poor hygiene. More generally, the World Health Organization (WHO) estimates that inadequate water, sanitation, and hygiene are responsible for roughly half the malnutrition in the world.

In addition, we are witnessing irreparable damage to ecosystems across the globe. Aquifers are being drawn down faster than they can naturally be recharged. Some great lakes are mere fractions of what they once were.

And water pollution is affecting millions of people's lives. China typifies this problem. More than 75 percent of its urban river water is unsuitable for drinking or fishing, and 90 percent of its urban groundwater is contaminated. On the global scale, according to a recent UN report on world water development, every day we dump some 2 million tons of industrial waste and chemicals, human waste, and agricultural waste (fertilizers, pesticides, and pesticide residues) into our water supply.

Over the past century, as the world's population rose from 1.7 billion people in 1900 to 6.1 billion in 2000, global fresh water consumption increased six-fold—more than double the rate of population growth over the same period. The latest "medium" projections from the UN's population experts suggest that we are on the way to 8 billion people by the year 2025 and 9.15 billion by the middle of the century.

The contours of our predicament are clear-cut: A finite amount of water is available to a rapidly increasing number of people whose activities require more water than ever before. The UN Commission on Sustainable Development has indicated that we may need to double the amount of freshwater available today to meet demand at the middle of the century—after which time demand for water will increase by 50 percent with each additional generation.

Why is demand for water rising so rapidly? It goes beyond population pressures. According to a recent report from the UN Food and Agriculture Organization, the world will require 70 percent more food production over the next 40 years to meet growing per capita demand. This rising agricultural consumption necessarily translates into higher demand for water. By 2025, according to the water expert Sandra Postel, meeting projected global agricultural demand will require additional irrigation totaling some 2,000 cubic kilometers—roughly the equivalent of the annual flow of 24 Nile Rivers or 110 Colorado Rivers.

Consumption patterns aside, climate change will accelerate and intensify stress on water systems. According to the IPCC, in coming decades the frequency of extreme droughts will double while the average length of droughts will increase six times. This low water flow, combined with higher temperatures, not only will create devastating shortages. It will also increase pollution of fresh water by sediments, nutrients, pesticides, pathogens, and salts. On the other hand, in some regions, wet seasons will be more intense (but shorter).

In underdeveloped communities that lack capture and storage capacity, water will run off and will be unavailable when it is needed in dry seasons, thus perpetuating the cycle of poverty.

Climatic and demographic trends indicate that the regions of the world with the highest population growth rates are precisely those that are already the "driest" and that are expected to experience water stress in the future. The Organization for Economic Cooperation and Development has suggested that the number of people in water-stressed countries—where governments encounter serious constraints on their ability to meet household, industrial, and agricultural water demands—could rise to nearly 4 billion by the year 2030.

The Geopolitical Dimension

If oil is the key geopolitical resource of today, water will be as important—if not more so—in the not-so-distant future. A profound mismatch exists between the distribution of the human population and the availability of fresh water. At the water-rich extreme of the spectrum is the Amazon region, which has an estimated 15 percent of global runoff and less than 1 percent of the world's people. South America as a whole has only 6 percent of the world's population but more than a quarter of the world's runoff.

At the other end of the spectrum is Asia. Home to 60 percent of the global population, it has a freshwater endowment estimated at less than 36 percent of the world total. It is hardly surprising that some water-stressed countries in the region have pursued agricultural trade mechanisms to gain access to more water—in the form of food. Recently, this has taken the form of so-called "land grabs," in which governments and state companies have invested in farmland overseas to meet their countries' food security needs. *The Economist* has estimated that, to date, some 50 million acres have been remotely purchased or leased under these arrangements in Africa and Asia.

Although freshwater management has historically represented a means of preventing and mitigating conflict between countries with shared water resources, the growing scarcity of water will likely generate new levels of tension at the local, national, and even international levels. Many countries with limited water availability also depend on shared water, which increases the risk of friction, social tensions, and conflict.

The Euphrates, Jordan, and Nile Rivers are obvious examples of places where frictions already have occurred. But approximately 40 percent of the world's population lives in more than 260 international river basins of major social and economic importance, and 13 of these basins are shared by five or more countries. Interstate tensions have already escalated and could easily intensify as increasing water scarcity raises the stakes.

Within countries as well, governments in water-stressed regions must effectively and transparently mediate the concerns and demands of various constituencies. The interests of urban and rural populations, agriculture and industry, and commercial and domestic sectors often conflict. If allocation issues are handled inappropriately, subnational disputes and unrest linked to water scarcity and poor water quality could arise, as they already have in numerous cases.

Addressing the Challenge

Considering the scope and gravity of these water challenges, responses by governments and nongovernmental organizations have fallen short of what is needed. Despite obvious signs that we overuse water, we continue to perpetuate gross inefficiencies. We continue to skew consumption on the basis of politically charged subsidies or other supports. And we continue to pursue patently unsustainable practices whose costs will grow more onerous over time.

The Colorado River system, for example, is being overdrawn. It supplies water to Las Vegas, Los Angeles, San Diego, and other growing communities in the American Southwest. If demand on this river system is not curtailed, there is a 50 percent chance that Lake Mead will be dry by 2021, according to experts from the Scripps Institution of Oceanography.

Despite constant reminders of future challenges, we continue to be paralyzed by short-term thinking and practices. What is especially striking about water is the extent to which the world's nations are unprepared to manage such a vital resource sustainably. Six key opportunities for solutions stand out.

First, the global community needs to do substantially more to address the lack of safe drinking water and sanitation. Donor countries, by targeting water resources, can simultaneously address issues associated with health, poverty reduction, and environmental stewardship, as well as stability and security concerns. It should be stressed in this regard that rates of return on investment in water development—financial, political, and geopolitical—are all positive. The WHO estimates that the global return on every dollar invested in water and sanitation programs is $4 and $9, respectively.

Consider, for example, how water problems affect the earning power of women. Typically in poor countries, women and girls are kept at home to care for sick family members inflicted with water-related diseases. They also spend hours each day walking to collect water for daily drinking, cooking, and washing. According to the United Nations Children's Fund, water and sanitation issues explain why more than half the girls in sub-Saharan Africa drop out of primary school.

Second, more rigorous analyses of sustainability could help relevant governments and authorities begin to address the conspicuous mismanagement of water resources in regions across the world. This would include reviewing public subsidies—for water-intensive farming, for example—and other supports that tend to increase rather than remove existing inefficiencies.

Priced to Sell

Third, specialists, scholars, practitioners, and policy makers need to make substantial progress in assigning to water a market value against which more sustainable consumption decisions and policies can be made. According to the American Water Works Association, for example, the average price of water in the United States is $1.50 per 1,000 gallons—or less than a single penny per gallon. Yet, when it comes to the personal consumption market, many Americans do not hesitate to pay prices for bottled water that are higher than what they pay at the pump for a gallon of gasoline. What is clear, both inside and outside the United States, is that mechanisms for pricing water on the basis of sustainability have yet to be identified.

Fourth, rapid advances in technology can and should have a discernible effect on both the supply and demand sides of the global water equation. The technology landscape is breathtaking—from desalination, membrane, and water-reuse technologies to a range of cheaper and more efficient point-of-use applications (such as drip irrigation and rainwater harvesting). It remains to be seen, however, whether the acquisition and use of such technologies can be accelerated and dispersed so that they can have an appreciable effect in offsetting aggregate downside trends.

From a public policy perspective, taxation and regulatory policies can create incentives for the development and dissemination of such technologies, and foreign assistance projects can promote their use in developing countries. Also, stronger links with the private sector would help policy makers improve their understanding of technical possibilities, and public-private partnerships can be effective mechanisms for distributing technologies in the field.

Fifth, although our understanding of the relationship between climate change and water will continue to be shaped by new evidence, it is important that we incorporate into our approach to climate change our existing understanding of water management and climate adaptation issues.

Sixth, the complex links among water, agriculture, and energy must be identified with greater precision. An enormous amount of work remains to be done if we are to appreciate these linkages in the global, basin, and local contexts.

In the final analysis, our capacity to address the constellation of challenges that relate to water access, sanitation, ecosystems, infrastructure, adoption of technologies, and the mobilization of resources will mean the difference between rapid economic development and continued poverty, between healthier populations and continued high exposure to water-related diseases, between a more stable world and intensifying geopolitical tensions.

Critical Thinking

1. In what way will the geopolitical patterns of water resources and consumption be different than the geopolitical patterns of oil resources and consumption?

2. Why does "an obscenely large portion of the world's population lack regular access to fresh water and sanitization?"

3. Refer to your map, and map out the "profound mismatch between the distribution of the human population and the availability of fresh water." Make some critical observations regarding that pattern.

4. How might climate change make the world's future water challenge even more challenging?

5. How might the water challenges of the industrialized world be different than the challenges facing "developing" nations?

ERIK R. PETERSON is senior vice president of the Center for Strategic and International Studies and director of its Global Strategy Institute. **RACHEL A. POSNER** is assistant director of the CSIS Global Water Futures project.

Wet Dreams: Water Consumption in America

CYNTHIA BARNETT

During America's retreat to the suburbs in the 1950s, large home lots and disposable incomes allowed for a new marker of success: the backyard swimming pool. For the rest of the 20th century, residential pools symbolized upward mobility and offered a sense of seclusion not possible at city pools. The following decades redefined our relationship with water itself—from essence of life to emblem of luxury. By the time of the 21st-century housing run-up, even the plain blue pool had lost its luster. Adornments were needed: floating fire pits, glass portholes, and vanishing edges.

The amenity to envy was no longer the diving board. The must-have, now, was the waterfall. No community glorified the trend like Granite Bay, California, nestled on the north shores of Folsom Lake, near Sacramento.

In Granite Bay's best backyards, rocky waterfalls cascade into swimming pools with grottoes and swim-up bars. Thick bushes and trees bearing flowers and fruit adorn the watery wonders, making a place naturally dominated by needlegrass and sedge look more like Fiji. Groomed lawns, a quarter acre and larger, complete the unnatural tableau and help push average water consumption in Granite Bay to among the highest on Earth, nearly 500 gallons a person each day. Even when drought conditions cut federal water deliveries to California farmers, Granite Bay residents continued to consume water as if it were as plentiful as air.

Spectacular squander in the middle of a water crisis is not much of a shock in the United States, where we use about half our daily household water bounty outdoors. What is surprising, however, is to find some of the world's worst waste in the Sacramento metropolitan area, since Greater Sacramento has become a national leader in finding solutions to America's energy and climate challenges. Landing regularly on lists of top green and livable cities, Sacramento also has earned this startling ranking: It squanders more water than anywhere else in California. Residents of the metro region use nearly 300 gallons of water per person every day. By comparison, the equally affluent residents of Perth, Australia, use about 75 gallons per day. Londoners tap about 42 gallons per day. The water-rich Dutch use about 33 gallons daily.

Somehow, America's green craze missed the blue.

Sacramento is by no means unique. Even as our green consciousness evolves, we often manage to ignore water. Across the United States, we give little thought to our water consumption even as we replace incandescent bulbs with LEDs.

How is that?

One part of the answer is the illusion of water abundance. When we twist the tap, we're rewarded with fresh, clean water. Water is also our cheapest necessity. Four-dollar-a-gallon gasoline helped drive consumers to more efficient cars, while our water is so subsidized that many Americans pay less than a tenth of a penny a gallon for clean freshwater delivered right into our homes.

"Water is just too easy to take for granted," says Tom Gohring, executive director of the Sacramento Water Forum, which works to find solutions to the region's water woes. "It's always there."

This is true in Sustainable Sacramento, and it's true in the scorched Southwest. The most conspicuous water consumption in America is often found in those parts of the country where water shortages are most serious. Nationwide, we use an average of 147 gallons each day. In Las Vegas, it's 227 gallons per person—in one of the most water-scarce metro areas of the United States.

Vegas swimming pools make Granite Bay's look like they came from the Kmart garden department. But in both locales the extreme illusion of abundance makes it all but impossible for people who live and play there to notice their personal connection to the nation's water crisis—to understand how wasteful water consumption in one house, in one backyard, multiplied by 310 million Americans, equals trouble for the generations to come.

Profligate water use today will imperil future generations, the same as profligate use of oil. But water is much more important to our future than oil. That's because there are no alternatives to water, no new substitute for life's essential ingredient being cooked from corn, french fry grease, or algae.

Towering above the Colorado River, Hoover Dam stands as a breathtaking marvel of U.S. engineering. Its reservoir, Lake Mead, supplies water to millions of Americans and another million acres of farmland. The dam's iconic symbolism makes a study by the University of California's Scripps Institution of Oceanography that much more unsettling. In a grim paper titled

"When Will Lake Mead Go Dry?" marine physicist Tim Barnett (no relation to this article's author) and climate scientist David Pierce say there's a 50-50 chance it will happen by 2021. The Scripps scientists say they were "stunned at the magnitude of the problem and how fast it was coming at us."

A dried-up Lake Mead is only the most dramatically visible of the collapses that scientists say could play out in the seven states that rely on the Colorado River and its tributaries as ever-increasing water use, ever-growing population, and a changing climate shrink its flow. Scientists say that the 20th century, when America built its grand waterworks and divvied up its rivers, was the wettest in a thousand years. Now, the wet period is over and the Southwest is expected to become dryer. Trees are already showing the strain, dying off and burning at unprecedented rates. People must adjust, conclude Barnett and Pierce, to forestall "a major societal and economic disruption in the desert Southwest."

This dry, dusty American future is not confined to the desert. In the Great Plains, farmers are depleting the enormous High Plains Aquifer, which underlies 225,000 square miles from Wyoming to Texas, far faster than it can recharge. And Florida has so overpumped its once abundant groundwater that the hundred-thousand-square-mile sponge known as the Floridan Aquifer, one of the most productive aquifers in the world, can no longer supply the state's drinking-water needs.

But here's the confounding thing: Practically every scientific study that describes these catastrophes also concludes that it doesn't have to be this way. In the Southeast, the Great Plains, and even in the arid states of the Southwest, it's possible to reverse this parched path.

America needs nothing less than a revolution in how we use water. We must change not only the wasteful ways we consume water in our homes, businesses, farms, and energy plants but also the inefficient ways we move water to and away from them. This revolution will bring about the ethical use of water in every sector. A water ethic is as essential—and as possible—as past awakenings to threats against our environment and ourselves, from the large scale (the way we halted use of DDT and other deadly chemicals) to the family level (the way we got used to setting out recycling bins).

In all, America guzzles about 410 billion gallons of water per day. That's more than the daily flow of the entire Mississippi River. Power plants and agriculture top the list in water consumption, with agricultural irrigation accounting for about 40 percent of all freshwater sucked up in the United States each day.

Many Americans seem resigned to the notion that agriculture and big industries require a ton of water, and there's not much we can do to change that. But this is like throwing up our hands and concluding that because coal plants are the nation's top emitters of greenhouse gases, there's nothing we can do about climate change.

It is time, now, to turn our attention to water. The overtapping of nearly every major river and aquifer in the nation, and the inability of our political institutions to change course, call for our involvement. Citizens were ahead of politicians when it came to green living. The same will be true of water. A water ethic means deliberately different choices and the political backbone to make them: No wasted water in agriculture. Water-efficient power plants. Restoring floodplains rather than

building taller and taller levees. Reusing water and harvesting rain to irrigate our lawns and cool commercial air conditioners. It's a turn from the vast waterworks of the 20th century toward local solutions. It's an appreciation for "local water" in the same way we're embracing local produce.

In that spirit, the blue revolution begins in our own backyards. Coming in third after power plants and agriculture, at about 43 billion gallons a day, are public and private utilities. That's where the majority of us get our household water. Moving, filtering, and treating all of this water takes a remarkable amount of energy. And then the lion's share of this painstakingly purified drinking water is used on grass. Waterfalls and grottoes aside, the distance between Americans and their global neighbors who use less than 50 gallons of water per person each day is about one-third of an acre: the average size of the American lawn. Using satellite analysis in the early 2000s, research scientist Cristina Milesi found that, between our homes, highway medians, golf greens, and grassy sports fields, lawns are America's largest crop, with 63,240 square miles in turfgrass nationwide. That's larger than most individual American states.

To irrigate this "51st state," Milesi estimates that we use as much as 19 trillion gallons of water per year. That's more than it takes to irrigate all the feed grain in the nation. "People don't believe their water use makes a difference, especially because agricultural consumption is so high," Milesi says. "But water is probably the most important issue facing urban areas in the future—and the primary pressure point on urban water use is the lawn."

It's not that we don't have enough water. It's that we don't have enough water to waste by pouring off 19 trillion gallons a year, most of it drinking water. Sure, some of our lawn water, spiked with pesticides and fertilizers, percolates back underground. But much of it becomes so-called stormwater that never makes its way back to streams and rivers; in the coastal United States, hundreds of millions of gallons of freshwater shoot out to sea every day. All of this despite multimillion-dollar public-education programs to convince Americans that they need not water their grass every day—or even every other day—to keep it green.

But grass is not the root of our country's water problems; it's a 63,240-square-mile symptom of the real ailment—our lack of an ethic for water in America. The illusion of abundance gives us a false sense of security that there's enough water for anything, anytime. New subdivision in the desert? We'll find the water. Kids bugging you to take them to the nation's largest water park, with its 1.2 million–gallon wave pool that holds more than 20,000 bathtubs full of water?

Jump right in.

The sand-plain region of Wisconsin was both muse and refuge for Aldo Leopold, whose *A Sand County Almanac* has inspired our evolving ecological awareness ever since it was published posthumously in 1949.

If people could see how closely their children's and grandchildren's well-being is tied to the health of the land, Leopold believed, personal ethics would drive them to cooperate not only on behalf of their families and communities but also for the natural world they inhabit. This land ethic, wrote Leopold, "enlarges the boundaries of the community to include soils, waters, plants, and animals, or collectively: the land." By *land,* he meant the entire web of life, from climate to water, says his biographer, Curt Meine.

I started my search for a water ethic at Leopold's famous "humble shack" near the Wisconsin River, preserved in quiet posterity by the Aldo Leopold Foundation. Visitors to central Wisconsin can walk through the forest and prairie, and along the sandy banks of the river.

Visitors *can,* but not many do. Thousands a day, though, flock to a tourist strip that lies 10 miles from the spot where Leopold wrote. They come to worship water, although not the sort in the river. For here is the largest concentration of water parks on the planet.

The red sandstone bluffs of Wisconsin began drawing nature-loving tourists in the mid-1800s to see "the dells"—from the French *dalles,* or flagstone. The town bisected by the Wisconsin River and the dells changed its name in 1931 to the Wisconsin Dells to capitalize on its scenic draw. In the ensuing decades, one attraction after another opened along the tourist strip. Then, in 1994, Stan Anderson, a resort owner, decided to shore up foul-weather business by building an indoor kids' water park. Children went nuts, and so did their parents.

Fifteen years later, the Wisconsin Dells are no longer the main attraction in the Wisconsin Dells.

Today, the Wisconsin Dells overflow with 20 water parks—including the biggest indoor and outdoor water parks in the world—that slosh around about 20 million gallons of water. The biggest attraction in the biggest park is called—what else?—the Big Kahuna, a 1.2 million–gallon wave pool.

During operating hours, the 20 parks constantly pump groundwater from an aquifer that scientists say is robust. But the wave pool is an homage to America's illusion of water abundance—particularly to the children exposed solely to its chlorinated wonders. Most American kids no longer know about the watershed they live in, where their house water comes from, or where it goes when they flush. If children's love for water is cradled only within the bright-colored resin sides of a thrill ride, never the wondrous red sides of a sandstone bluff, future Americans will have ever less understanding of, and value for, our freshwater resources.

The overwhelming popularity of the Dells proves that humans love water. We begin life in water, and we're drawn to it from the day we're born. Somehow, we have to harness that natural affinity to create a shared water ethic: an aquatic revival of Leopold's land ethic that would help Americans see that our future ecological—and economic—prosperity depends on how well we take care of the water around us.

Sixty years after Leopold's call for a land ethic, most of us take some personal responsibility for the planet. But when it comes to water, we've gone in the opposite direction. Today, we use four times the amount of water, per person, that we did in 1950. The nation's illusion of water abundance blinds us to how our own backyard garden hose connects to the bigger picture.

With a shared water ethic, we would live well, with much less water. Not just less in our own backyards, but also less

across industries. It doesn't make sense for local government to require citizens to lay off the lawn sprinklers, then approve a new subdivision atop the community's most important water-recharge area. The fundamental belief in water as a national treasure to be preserved has to catch on at every level of society.

The American illusion of water abundance follows a long and peculiar tradition. Throughout history, humans flaunted water as a symbol of power, wealth, and control of nature.

In 17th-century France, Louis XIV built some of the greatest water features in the world at the gardens of Versailles. The colossal fountains, pools, and waterfalls were positioned so that the sovereign and his visitors would never lose sight of water during garden tours that lasted from morning until night.

But here's what the royal visitors didn't know: There wasn't enough water at Versailles to keep all those fountain jets soaring, pools overflowing, and waterfalls cascading. A secret palace staff would scurry ahead of the king's touring parties, signaling their whereabouts with an elaborate system of flags and whistles to convey when it was safe to shut down one group of fountains and turn on the next one.

The American illusion of abundance is likewise carefully maintained. We have gotten so good at harnessing water and moving it around cities and regions that Americans, like the visitors to Versailles, have never had to think about how it all works. The constant reengineering of natural systems bolsters the illusion of abundance. Two mighty rivers, the American and the Sacramento, run through the middle of California's capital city. How can it be water stressed? The same can be said of south Florida, surrounded by the Everglades and pummeled regularly by rains that flood the streets. Yet these watersheds on opposite coasts of America have been manipulated to the point of near ruin. The Everglades of Florida and California's Sacramento–San Joaquin Delta were two of the most water-abundant ecosystems on one of the most water-abundant continents. Today, they are among the best arguments for a blue revolution: They're both dying of thirst.

Critical Thinking

1. Why have Americans held the illusion of water abundance for so long?
2. Describe what you would consider the ethical use of water.
3. Make a list of the ways you have personally seen or experienced the maluse of water.
4. If all fresh water is essentially the product of the hydrological cycle (precipitation), who owns the rain—public or private interests? Explain.

CYNTHIA BARNETT *is a longtime journalist who has reported on freshwater issues from the Suwannee River to Singapore. In addition to* Blue Revolution, *she is the author of* Mirage: Florida and the Vanishing Water of the Eastern U.S. *Excerpted from* Blue Revolution: Unmaking America's Water Crisis *by Cynthia Barnett (Beacon, 2011). Reprinted with permission from Beacon Press.*

Water Footprints of Nations: Water Use by People as a Function of Their Consumption Pattern

A. Y. Hoekstra, A. K. Chapagain

Introduction

Databases on water use traditionally show three columns of water use: water withdrawals in the domestic, agricultural and industrial sector respectively (Gleick, 1993; FAO, 2003). A water expert being asked to assess the water demand in a particular country will generally add the water withdrawals for the different sectors of the economy. Although useful information, this does not tell much about the water actually needed by the people in the country in relation to their consumption pattern. The fact is that many goods consumed by the inhabitants of a country are produced in other countries, which means that it can happen that the real water demand of a population is much higher than the national water withdrawals do suggest. The reverse can be the case as well: national water withdrawals are substantial, but a large amount of the products are being exported for consumption elsewhere.

In 2002, the water footprint concept was introduced in order to have a consumption-based indicator of water use that could provide useful information in addition to the traditional production-sector-based indicators of water use (Hoekstra and Hung, 2002). The water footprint of a nation is defined as the total volume of freshwater that is used to produce the goods and services consumed by the people of the nation. Since not all goods consumed in one particular country are produced in that country, the water footprint consists of two parts: use of domestic water resources and use of water outside the borders of the country.

The water footprint has been developed in analogy to the ecological footprint concept as was introduced in the 1990s. The 'ecological footprint' of a population represents the area of productive land and aquatic ecosystems required to produce the resources used, and to assimilate the wastes produced, by a certain population at a specified material standard of living, wherever on earth that land may be located. Whereas the 'ecological footprint' thus quantifies the *area* needed to sustain people's

living, the 'water footprint' indicates the *water* required to sustain a population.

The water footprint concept is closely linked to the virtual water concept. Virtual water is defined as the volume of water required to produce a commodity or service. The concept was introduced by Allan in the early 1990s (Allan, 1993, 1994) when studying the option of importing virtual water (as opposed to real water) as a partial solution to problems of water scarcity in the Middle East. Allan elaborated on the idea of using virtual water import (coming along with food imports) as a tool to release the pressure on the scarcely available domestic water resources. Virtual water import thus becomes an alternative water source, next to endogenous water sources. Imported virtual water has therefore also been called 'exogenous water' (Haddadin, 2003).

When assessing the water footprint of a nation, it is essential to quantify the flows of virtual water leaving and entering the country. If one takes the use of domestic water resources as a starting point for the assessment of a nation's water footprint, one should subtract the virtual water flows that leave the country and add the virtual water flows that enter the country.

The objective of this study is to assess and analyse the water footprints of nations. The study builds on two earlier studies. Hoekstra and Hung (2002, 2005) have quantified the virtual water flows related to the international trade of crop products. Chapagain and Hoekstra (2003) have done a similar study for livestock and livestock products. The concerned time period in these two studies is 1995–1999. The present study takes the period of 1997–2001 and refines the earlier studies by making a number of improvements and extensions.

Method

A nation's water footprint has two components, the internal and the external water footprint. The internal water footprint *(IWFP)* is defined as the use of domestic water resources to produce goods and services consumed by inhabitants of the country. It is the sum of the total water volume used from the domestic water resources in the national economy *minus* the volume of virtual water export to other countries insofar related to export of domestically produced products:

$$IWFP = AWU + IWW + DWW - VWE_{dom} \qquad (1)$$

Here, *AWU* is the agricultural water use, taken equal to the evaporative water demand of the crops; *IWW* and *DWW* are the water withdrawals in the industrial and domestic sectors respectively; and VWE_{dom} is the virtual water export to other countries insofar related to export of domestically produced products. The agricultural water use includes both effective rainfall (the portion of the total precipitation which is retained by the soil and used for crop production) and the part of irrigation water used effectively for crop production. Here we do not include irrigation losses in the term of agricultural water use assuming that they largely return to the resource base and thus can be reused.

The external water footprint of a country *(EWFP)* is defined as the annual volume of water resources used in other countries to produce goods and services consumed by the inhabitants of the country concerned. It is equal to the so-called virtual water import into the country *minus* the volume of virtual water exported to other countries as a result of re-export of imported products.

$$EWFP = VWI - VWE_{re-export} \qquad (2)$$

Both the internal and the external water footprint include the use of *blue water* (ground and surface water) and the use of *green water* (moisture stored in soil strata).

The use of domestic water resources comprises water use in the agricultural, industrial and domestic sectors. For the latter two sectors we have used data from AQUASTAT (FAO, 2003). Though significant fractions of domestic and industrial water withdrawals do not evaporate but return to either the groundwater or surface water system, these return flows are generally polluted, so that they have been included in the water footprint calculations. The total volume of water use in the agricultural sector has been calculated in this study based on the total volume of crop produced and its corresponding virtual water content. For the calculation of the virtual water content of crop and livestock products we have used the methodology as described in Chapagain and Hoekstra (2004). In summary, the virtual water content (m³/ton) of primary crops has been calculated based on crop water requirements and yields. Crop water requirement have been calculated per crop and per country using the methodology developed by FAO (Allen *et al.,* 1998). The virtual water content of crop products is calculated based on product fractions (ton of crop product obtained per ton of primary crop) and value fractions (the market value of one crop product divided by the aggregated market value of all crop products derived from one primary crop). The virtual water content (m³/ton) of live animals has been calculated based on the virtual water content of their feed and the volumes of drinking and service water consumed during their lifetime. We have calculated the virtual water content for eight major animal

categories: beef cattle, dairy cows, swine, sheep, goats, fowls/poultry (meat purpose), laying hens and horses. The calculation of the virtual water content of livestock products is again based on product fractions and value fractions.

Virtual water flows between nations have been calculated by multiplying commodity trade flows by their associated virtual water content:

$$VWF\,[n_e, n_i, c] = CT\,[n_e, n_i, c] \times VWC\,[n_e, c] \qquad (3)$$

in which *VWF* denotes the virtual water flow (m³yr⁻¹) from exporting country n_e to importing country n_i as a result of trade in commodity *c; CT* the commodity trade (ton yr⁻¹) from the exporting to the importing country; and *VWC* the virtual water content (m³ ton⁻¹) of the commodity, which is defined as the volume of water required to produce the commodity in the exporting country. We have taken into account the trade between 243 countries for which international trade data are available in the Personal Computer Trade Analysis System of the International Trade Centre, produced in collaboration with UNCTAD/WTO. It covers trade data from 146 reporting countries disaggregated by product and partner countries (ITC, 2004). We have carried out calculations for 285 crop products and 123 livestock products. The virtual water content of an industrial product can be calculated in a similar way as described earlier for agricultural products. There are however numerous categories of industrial products with a diverse range of production methods and detailed standardised national statistics related to the production and consumption of industrial products are hard to find. As the global volume of water used in the industrial sector is only 716Gm³/yr (≈10% of total global water use), we have—per country—simply calculated an average virtual water content per dollar added value in the industrial sector (m³/US$) as the ratio of the industrial water withdrawal (m³/yr) in a country to the total added value of the industrial sector (US$ /yr), which is a component of the Gross Domestic Product.

Water Needs by Product

The total volume of water used globally for crop production is 6390 Gm³/yr at field level. Rice has the largest share in the total volume water used for global crop production. It consumes about 1359 Gm³/yr, which is about 21% of the total volume of water used for crop production at field level. The second largest water consumer is wheat (12%). . . . Although the total volume of the world rice production is about equal to the wheat production, rice consumes much more water per ton of production. The difference is due to the higher evaporative demand for rice production. As a result, the global average virtual water content of rice (paddy) is 2291 m³/ton and for wheat 1334 m³/ton.

The virtual water content of rice (broken) that a consumer buys in the shop is about 3420 m³/ton. This is larger than the virtual water content of paddy rice as harvested from the field because of the weight loss if paddy rice is processed into broken rice. The virtual water content of some selected crop and livestock products for a number of selected countries are presented in Table 1.

Table 1 Average virtual water content of some selected products for a number of selected countries (m³/ton)

	USA	China	India	Russia	Indonesia	Australia	Brazil	Japan	Mexico	Italy	Netherlands	World average*
Rice (paddy)	1275	1321	2850	2401	2150	1022	3082	1221	2182	1679		2291
Rice (husked)	1656	1716	3702	3118	2793	1327	4003	1586	2834	2180		2975
Rice (broken)	1903	1972	4254	3584	3209	1525	4600	1822	3257	2506		3419
Wheat	849	690	1654	2375		1588	1616	734	1066	2421	619	1334
Maize	489	801	1937	1397	1285	744	1180	1493	1744	530	408	909
Soybeans	1869	2617	4124	3933	2030	2106	1076	2326	3177	1506		1789
Sugar cane	103	117	159		164	141	155	120	171			175
Cotton seed	2535	1419	8264		4453	1887	2777		2127			3644
Cotton lint	5733	3210	18694		10072	4268	6281		4812			8242
Barley	702	848	1966	2359		1425	1373	697	2120	1822	718	1388
Sorghum	782	863	4053	2382		1081	1609		1212	582		2583
Coconuts		749	2255		2701		1590		1954			2545
Millet	2143	1863	3269	2892		1951		3100	4534			4596
Coffee (green)	4864	6290	12180		17665		13972		28119			17373
Coffee (roasted)	5790	7488	14500		21030		16633		33475			20682
Tea (made)		1110	7002	3002	9474		6592	4940				9205
Beef	13193	12560	16482	21028	14818	17112	16961	11019	37762	21167	11681	15497
Pork	3946	2211	4397	6947	3938	5909	4818	4962	6559	6377	3790	4856
Goat meat	3082	3994	5187	5290	4543	3839	4175	2560	10252	4180	2791	4043
Sheep meat	5977	5202	6692	7621	5956	6947	6267	3571	16878	7572	5298	6143
Chicken meat	2389	3652	7736	5763	5549	2914	3913	2977	5013	2198	2222	3918
Eggs	1510	3550	7531	4919	5400	1844	3337	1884	4277	1389	1404	3340
Milk	695	1000	1369	1345	1143	915	1001	812	2382	861	641	990
Milk powder	3234	4648	6368	6253	5317	4255	4654	3774	11077	4005	2982	4602
Cheese	3457	4963	6793	6671	5675	4544	4969	4032	11805	4278	3190	4914
Leather (bovine)	14190	13513	17710	22575	15929	18384	18222	11864	40482	22724	12572	16656

*For the primary crops, world averages have been calculated as the ratio of the global water use for the production of a crop to the global production volume. For processed products, the global averages have been calculated as the ratio of the global virtual water trade volume to the global product trade volume.

In general, livestock products have a higher virtual water content than crop products. This is because a live animal consumes a lot of feed crops, drinking water and service water in its lifetime before it produces some output. We consider here an example of beef produced in an industrial farming system. It takes in average 3 years before it is slaughtered to produce about 200 kg of boneless beef. It consumes nearly 1300 kg of grains (wheat, oats, barley, corn, dry peas, soybean meal and other small grains), 7200 kg of roughages (pasture, dry hay, silage and other roughages), 24 cubic meter of water for drinking and 7 cubic meter of water for servicing. This means that to produce one kilogram of boneless beef, we use about 6.5 kg of grain, 36 kg of roughages, and 155l of water (only for drinking and servicing). Producing the volume of feed requires about 15340l of water in average. With every step of food processing we loose part of the material as a result of selection and inefficiencies. The higher we go up in the product chain, the higher will be the virtual water content of the product. For example, the global average virtual water content of maize, wheat and rice (husked) is 900, 1300 and 3000 m^3/ton respectively, whereas the virtual water content of chicken meat, pork and beef is 3900, 4900 and 15500 m^3/ton respectively. However, the virtual water content of products strongly varies from place to place, depending upon the climate, technology adopted for farming and corresponding yields.

The units used so far to express the virtual water content of various products are in terms of cubic meters of water per ton of the product. A consumer might be more interested to know how much water it consumes per unit of consumption. One cup of coffee requires for instance 140l of water in average, one hamburger 2400l and one cotton T-shirt 2000l (Table 2).

The global average virtual water content of industrial products is 80l per US$. In the USA, industrial products take nearly 100l per US$. In Germany and the Netherlands, average virtual water content of industrial products is about 50l per US$. Industrial products from Japan, Australia and Canada take only 10–15l per US$. In world's largest developing nations, China and India, the average virtual water content of industrial products is 20–25l per US$.

Water Footprints of Nations

The global water footprint is 7450 Gm^3/yr, which is 1240 m^3/cap/yr in average. In absolute terms, India is the country with the largest footprint in the world, with a total footprint of 987 Gm^3/yr. However, while India contributes 17% to the global population, the people in India contribute only 13% to the global water footprint. On a relative basis, it is the people of the USA that have the largest water footprint, with 2480 m^3/yr per capita, followed by the people in south European countries such as Greece, Italy and Spain (2300–2400 m^3/yr per capita). High water footprints can also be found in Malaysia and Thailand. At the other side of the scale, the Chinese people have a relatively low water footprint with an average of 700 m^3/yr per capita. The average per capita water footprints . . . are shown in Table 3 for a few selected countries.

Table 2 Global average virtual water content of some selected products, per unit of product

Product	Virtual water content (litres)
1 glass of beer (250 ml)	75
1 glass of milk (200 ml)	200
1 cup of coffee (125 ml)	140
1 cup of tea (250 ml)	35
1 slice of bread (30 g)	40
1 slice of bread (30 g) with cheese (10 g)	90
1 potato (100 g)	25
1 apple (100 g)	70
1 cotton T-shirt (250 g)	2000
1 sheet of A4-paper (80 g/m^2)	10
1 glass of wine (125 ml)	120
1 glass of apple juice (200 ml)	190
1 glass of orange juice (200 ml)	170
1 bag of potato crisps (200 g)	185
1 egg (40 g)	135
1 hamburger (150 g)	2400
1 tomato (70 g)	13
1 orange (100 g)	50
1 pair of shoes (bovine leather)	8000
1 microchip (2 g)	32

The size of the global water footprint is largely determined by the consumption of food and other agricultural products. . . . The estimated contribution of agriculture to the total water use (6390 Gm^3/yr) is even bigger than suggested by earlier statistics due to the inclusion of green water use (use of soil water). If we include irrigation losses, which globally add up to about 1590 Gm^3/yr (Chapagain and Hoekstra, 2004), the total volume of water used in agriculture becomes 7980 Gm^3/yr. About one third of this amount is blue water withdrawn for irrigation; the remaining two thirds is green water (soil water).

The four major direct factors determining the water footprint of a country are: volume of consumption (related to the gross national income); consumption pattern (e.g. high versus low meat consumption); climate (growth conditions); and agricultural practice (water use efficiency). In rich countries, people generally consume more goods and services, which immediately translates into increased water footprints. But it is not consumption volume alone that determines the water demand of people. The composition of the consumption package is relevant too, because some goods in particular require a lot of water (bovine meat, rice). In many poor countries it is a combination of unfavourable climatic conditions (high evaporative demand) and bad agricultural practice (resulting in low water productivity) that contributes to a high water footprint. Underlying factors that contribute to bad agricultural practice

Table 3 Composition of the water footprint for some selected countries. Period: 1997–2001

Country	Population	Use of domestic water resources					Use of foreign water resources			Water footprint			Water footprint by consumption category			
		Domestic water withdrawal (Gm³/yr)	Crop evapotranspiration* For national consumption (Gm³/yr)	Crop evapotranspiration* For export (Gm³/yr)	Industrial water withdrawal For national consumption (Gm³/yr)	Industrial water withdrawal For export (Gm³/yr)	For national consumption Agricultural goods (Gm³/yr)	For national consumption Industrial goods (Gm³/yr)	For re-export of imported products (Gm³/yr)	Total (Gm³/yr)	Per capita (m³/cap/yr)	Domestic water Internal water footprint (m³/cap/yr)	Agricultural goods Internal water footprint (m³/cap/yr)	Agricultural goods External water footprint (m³/cap/yr)	Industrial goods Internal water footprint (m³/cap/yr)	Industrial goods External water footprint (m³/cap/yr)
Australia	19071705	6.51	14.03	68.67	1.229	0.12	0.78	4.02	4.21	26.56	1393	341	736	41	64	211
Bangladesh	129942975	2.12	109.98	1.38	0.344	0.08	3.71	0.34	0.13	116.49	896	16	846	29	3	3
Brazil	169109675	11.76	195.29	61.01	8.666	1.63	14.76	3.11	5.20	233.59	1381	70	1155	87	51	18
Canada	30649675	8.55	30.22	52.34	11.211	20.36	7.74	5.07	22.62	62.80	2049	279	986	252	366	166
China	1257521250	33.32	711.10	21.55	81.531	45.73	49.99	7.45	5.69	883.39	702	26	565	40	65	6
Egypt	63375735	4.16	45.78	1.55	6.423	0.66	12.49	0.64	0.49	69.50	1097	66	722	197	101	10
France	58775400	6.16	47.84	34.63	15.094	12.80	30.40	10.69	31.07	110.19	1875	105	814	517	257	182
Germany	82169250	5.45	35.64	18.84	18.771	13.15	49.59	17.50	38.48	126.95	1545	66	434	604	228	213
India	1007369125	38.62	913.70	35.29	19.065	6.04	13.75	2.24	1.24	987.38	980	38	907	14	19	2
Indonesia	204920450	5.67	236.22	22.62	0.404	0.06	26.09	1.58	2.74	269.96	1317	28	1153	127	2	8
Italy	57718000	7.97	47.82	12.35	10.133	5.60	59.97	8.69	20.29	134.59	2332	138	829	1039	176	151
Japan	126741225	17.20	20.97	0.40	13.702	2.10	77.84	16.38	4.01	146.09	1153	136	165	614	108	129
Jordan	4813708	0.21	1.45	0.07	0.035	0.00	4.37	0.21	0.22	6.27	1303	44	301	908	7	43
Mexico	97291745	13.55	81.48	12.26	2.998	1.13	35.09	7.05	7.94	140.16	1441	139	837	361	31	72
Netherlands	15865250	0.44	0.50	2.51	2.562	2.20	9.30	6.61	52.84	19.40	1223	28	31	586	161	417
Pakistan	136475525	2.88	152.75	7.57	1.706	1.28	8.55	0.33	0.67	166.22	1218	21	1119	63	12	2
Russia	145878750	14.34	201.26	8.96	13.251	34.83	41.33	0.80	3.94	270.98	1858	98	1380	283	91	5
South Africa	42387403	2.43	27.32	6.05	1.123	0.40	7.18	1.42	2.10	39.47	931	57	644	169	26	33
Thailand	60487800	1.83	120.17	38.49	1.239	0.55	8.73	2.49	3.90	134.46	2223	30	1987	144	20	41
United Kingdom	58669403	2.21	12.79	3.38	6.673	1.46	34.73	16.67	12.83	73.07	1245	38	218	592	114	284
USA	280343325	60.80	334.24	138.96	170.777	44.72	74.91	55.29	45.62	696.01	2483	217	1192	267	609	197
Global total/avg.	5994251631	344	5434	957	476	240	957	240	427	7452	1243	57	907	160	79	40

*Includes both green and blue water use in agriculture

and thus high water footprints are the lack of proper water pricing, the presence of subsidies, the use of water inefficient technology and lack of awareness of simple water saving measures among farmers.

The influence of the various determinants varies from country to country. The water footprint of the USA is high (2480 m^3/cap/yr) partly because of large meat consumption per capita and high consumption of industrial products. The water footprint of Iran is relatively high (1624 m^3/cap/yr) partly because of low yields in crop production and partly because of high evapotranspiration. In the USA the industrial component of the water footprint is 806 m^3/cap/yr whereas in Iran it is only 24 m^3/cap/yr.

The aggregated external water footprints of nations in the world constitute 16% of the total global water footprint. . . . However, the share of the external water footprint strongly varies from country to country. Some African countries, such as Sudan, Mali, Nigeria, Ethiopia, Malawi and Chad have hardly any external water footprint, simply because they have little import. Some European countries on the other hand, e.g. Italy, Germany, the UK and the Netherlands have external water footprints contributing 50–80% to the total water footprint. The agricultural products that contribute most to the external water footprints of nations are: bovine meat, soybean, wheat, cocoa, rice, cotton and maize.

Eight countries—India, China, the USA, the Russian Federation, Indonesia, Nigeria, Brazil and Pakistan—together contribute fifty percent to the total global water footprint. India (13%), China (12%) and the USA (9%) are the largest consumers of the global water resources. . . .

Both the size of the national water footprint and its composition differs between countries. . . . On the one end . . . [is] China with a relatively low water footprint per capita, and on the other end the USA. In the rich countries consumption of industrial goods has a relatively large contribution to the total water footprint if compared with developing countries. The water footprints of the USA, China, India and Japan are presented in more detail in Figure 1. The contribution of the external water footprint to the total water footprint is very large in Japan if compared to the other three countries. The consumption of industrial goods very significantly contributes to the total water footprint of the USA (32%), but not in India (2%).

Conclusion

The global water footprint is 7450 Gm^3/yr, which is in average 1240 m^3/cap/yr. The differences between countries are large: the USA has an average water footprint of 2480 m^3/cap/yr whereas China has an average water footprint of 700 m^3/cap/yr. There are four most important direct factors explaining high water footprints. A first factor is the total volume of consumption, which is generally related to gross national income of a country. This partially explains the high water footprints of for instance the USA, Italy and Switzerland. A second factor behind a high water footprint can be that people have a water-intensive consumption pattern. Particularly high consumption of meat significantly contributes to a high water footprint. This

factor partially explains the high water footprints of countries such as the USA, Canada, France, Spain, Portugal, Italy and Greece. The average meat consumption in the United States is for instance 120kg/yr, more than three times the world-average meat consumption. Next to meat consumption, high consumption of industrial goods significantly contributes to the total water footprints of rich countries. The third factor is climate. In regions with a high evaporative demand, the water requirement per unit of crop production is relatively large. This factor partially explains the high water footprints in countries such as Senegal, Mali, Sudan, Chad, Nigeria and Syria. A fourth factor that can explain high water footprints is water-inefficient agricultural practice, which means that water productivity in terms of output per drop of water is relatively low. This factor partly explains the high water footprints of countries such as Thailand, Cambodia, Turkmenistan, Sudan, Mali and Nigeria. In Thailand for instance, rice yields averaged 2.5 ton/ha in the period 1997–2001, while the global average in the same period was 3.9 ton/ha.

Reducing water footprints can be done in various ways. A first way is to break the seemingly obvious link between economic growth and increased water use, for instance by adopting production techniques that require less water per unit of product. Water productivity in agriculture can be improved for instance by applying advanced techniques of rainwater harvesting and supplementary irrigation. A second way of reducing water footprints is to shift to consumptions patterns that require less water, for instance by reducing meat consumption. However, it has been debated whether this is a feasible road to go, since the world-wide trend has been that meat consumption increases rather than decreases. Probably a broader and subtler approach will be needed, where consumption patterns are influenced by pricing, awareness raising, labelling of products or introduction of other incentives that make people change their consumption behaviour. Water costs are generally not well reflected in the price of products due to the subsidies in the water sector. Besides, the general public is—although often aware of energy requirements—hardly aware of the water requirements in producing their goods and services.

A third method that can be used—not yet broadly recognized as such—is to shift production from areas with low water-productivity to areas with high water productivity, thus increasing global water use efficiency (Chapagain et al., 2005a). For instance, Jordan has successfully externalised its water footprint by importing wheat and rice products from the USA, which has higher water productivity than Jordan.

The water footprint of a nation is an indicator of water use in relation to the consumption volume and pattern of the people. As an aggregated indicator it shows the total water requirement of a nation, a rough measure of the impact of human consumption on the natural water environment. More information about the precise components and characteristics of the total water footprint will be needed, however, before one can make a more balanced assessment of the effects on the natural water systems. For instance, one has to look at what is blue versus green water use, because use of blue water often affects the environment more than green water use. Also it is relevant to consider the

Article 25. Water Footprints of Nations: Water Use by People as a Function of Their Consumption Pattern

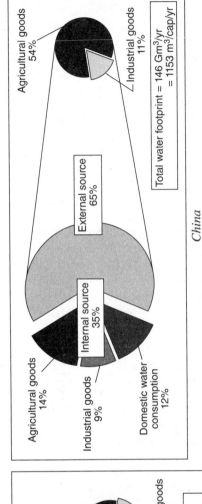

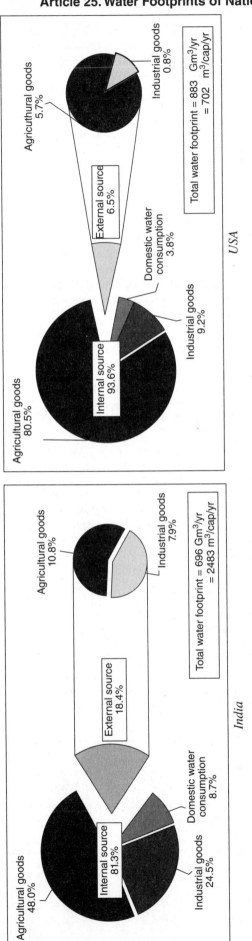

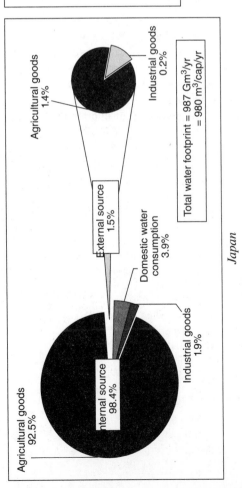

Figure 1   Details of the water footprints of the USA. China, India and Japan. Period: 1997–2001

internal versus the external water footprint. Externalising the water footprint for instance means externalising the environmental impacts. Also one has to realise that some parts of the total water footprint concern use of water for which no alternative use is possible, while other parts relate to water that could have been used for other purposes with higher added value. There is a difference for instance between beef produced in extensively grazed grasslands of Botswana (use of green water without alternative use) and beef produced in an industrial livestock farm in the Netherlands (partially fed with imported irrigated feed crops).

The current study has focused on the quantification of consumptive water use, i.e. the volumes of water from groundwater, surface water and soil water that evaporate. The effect of water pollution was accounted for to a limited extent by including the (polluted) return flows in the domestic and industrial sector. The calculated water footprints thus consist of two components: consumptive water use and wastewater production. The effect of pollution has been underestimated however in the current calculations of the national water footprints, because one cubic metre of wastewater should not count for one, because it generally pollutes much more cubic metres of water after disposal (various authors have suggested a factor of ten to fifty). The impact of water pollution can be better assessed by quantifying the dilution water volumes required to dilute waste flows to such extent that the quality of the water remains below agreed water quality standards. We have shown this in a case study for the water footprints of nations related to cotton consumption (Chapagain et al., 2005b).

International water dependencies are substantial and are likely to increase with continued global trade liberalisation. Today, 16% of global water use is not for producing products for domestic consumption but for making products for export. Considering this substantial percentage and the upward trend, we suggest that future national and regional water policy studies should include an analysis of international or interregional virtual water flows.

References

Allan JA (1993) Fortunately there are substitutes for water otherwise our hydro-political futures would be impossible. In: Priorities for water resources allocation and management, ODA, London, pp 13–26

Allan JA (1994) Overall perspectives on countries and regions. In: Rogers P, Lydon P (eds) Water in the Arab World: perspectives and prognoses. Harvard University Press, Cambridge, Massachusetts, pp 65–100

Allen RG, Pereira LS, Raes D, Smith M (1998) Crop evapotranspiration—Guidelines for computing crop water requirements—FAO Irrigation and Drainage Paper 56, FAO, Rome, Italy, www.fao.org/docrep/X0490E/x0490e00.htm

Chapagain AK, Hoekstra AY (2003) Virtual water flows between nations in relation to trade in livestock and livestock products. Value of Water Research Report Series No. 13, UNESCO-IHE, Delft, The Netherlands, www.waterfootprint.org/Reports/Report13.pdf

Chapagain AK, Hoekstra AY, Savenije HHG (2005a) Saving water through global trade, Value of Water Research Report Series No. 17, UNESCO-IHE, Delft, the Netherlands, www.waterfootprint.org/Reports/Report17.pdf

Chapagain AK, Hoekstra AY, Savenije HHG, Gautam R (2005b) The water footprint of cotton consumption. Value of Water Research Report Series No. 18, UNESCO-IHE, The Netherlands, www.waterfootprint.org/Reports/Report18.pdf

Chapagain AK, Hoekstra AY (2004) Water footprints of nations, Value of Water Research Report Series No. 16, UNESCO-IHE, Delft, The Netherlands, www.waterfootprint.org/Reports/Report16.pdf

FAO (2003) AQUASTAT 2003. Food and Agriculture Organization of the United Nations, Rome, Italy, ftp://ftp.fao.org/agl/aglw/aquastat/aquastat2003.xls

Gleick PH (ed) (1993) Water in crisis: A guide to the world's fresh water resources. Oxford University Press, Oxford, UK

Haddadin MJ (2003) Exogenous water: A conduit to globalization of water resources. In: Hoekstra AY (ed) Virtual water trade: Proceedings of the International Expert Meeting on Virtual Water Trade. Value of Water Research Report Series No. 12, UNESCO-IHE, Delft, The Netherlands, www.waterfootprint.org/Reports/Report12.pdf

Hoekstra AY, Hung PQ (2002) Virtual water trade: A quantification of virtual water flows between nations in relation to international crop trade. Value of Water Research Report Series No. 11, UNESCO-IHE Institute for Water Education, Delft, The Netherlands, www.waterfootprint.org/Reports/Report11.pdf

Hoekstra AY, Hung PQ (2005) Globalisation of water resources: International virtual water flows in relation to crop trade. Global Environmental Change 15(1):45–56

ITC (2004) PC-TAS version 1997–2001 in HS or SITC, CD-ROM. International Trade Centre, Geneva

Critical Thinking

1. Define "water footprint." To what degree would a region's environmental characteristics impact its water footprint?

2. Referring to Tables 1, 2, and 3, describe ways that a country's geopolitical-economy might "shape" the water data (footprints) provided in those tables.

3. What is virtual water? Do all countries have access to virtual water?

4. What county has the highest national water footprint? The lowest? How might this change over the next 20 years?

5. Refer to Figure 1 and write a brief, concise paragraph articulating what each of the four charts is illustrating.

The Big Melt

BROOK LARMER

Glaciers in the high heart of Asia feed its greatest rivers, lifelines for two billion people. Now the ice and snow are diminishing.

The gods must be furious.

It's the only explanation that makes sense to Jia Son, a Tibetan farmer surveying the catastrophe unfolding above his village in China's mountainous Yunnan Province. "We've upset the natural order," the devout, 52-year-old Buddhist says. "And now the gods are punishing us."

On a warm summer afternoon, Jia Son has hiked a mile and a half up the gorge that Ming-yong Glacier has carved into sacred Mount Kawagebo, looming 22,113 feet high in the clouds above. There's no sign of ice, just a river roiling with silt-laden melt. For more than a century, ever since its tongue lapped at the edge of Mingyong village, the glacier has retreated like a dying serpent recoiling into its lair. Its pace has accelerated over the past decade, to more than a football field every year—a distinctly unglacial rate for an ancient ice mass.

"This all used to be ice ten years ago," Jia Son says, as he scrambles across the scree and brush. He points out a yak trail etched into the slope some 200 feet above the valley bottom. "The glacier sometimes used to cover that trail, so we had to lead our animals over the ice to get to the upper meadows."

Around a bend in the river, the glacier's snout finally comes into view: It's a deathly shade of black, permeated with pulverized rock and dirt. The water from this ice, once so pure it served in rituals as a symbol of Buddha himself, is now too loaded with sediment for the villagers to drink. For nearly a mile the glacier's once smooth surface is ragged and cratered like the skin of a leper. There are glimpses of blue-green ice within the fissures, but the cracks themselves signal trouble. "The beast is sick and wasting away," Jia Son says. "If our sacred glacier cannot survive, how can we?"

It is a question that echoes around the globe, but nowhere more urgently than across the vast swath of Asia that draws its water from the "roof of the world." This geologic colossus—the highest and largest plateau on the planet, ringed by its tallest mountains—covers an area greater than western Europe, at an average altitude of more than two miles. With nearly 37,000 glaciers on the Chinese side alone, the Tibetan Plateau and its surrounding arc of mountains contain the largest volume of ice outside the polar regions. This ice gives birth to Asia's largest and most legendary rivers, from the Yangtze and the Yellow to the Mekong and the Ganges—rivers that over the course of history have nurtured civilizations, inspired religions, and sustained ecosystems. Today they are lifelines for some of Asia's most densely settled areas, from the arid plains of Pakistan to the thirsty metropolises of northern China 3,000 miles away. All told, some two billion people in more than a dozen countries—nearly a third of the world's population—depend on rivers fed by the snow and ice of the plateau region.

But a crisis is brewing on the roof of the world, and it rests on a curious paradox: For all its seeming might and immutability, this geologic expanse is more vulnerable to climate change than almost anywhere else on Earth. The Tibetan Plateau as a whole is heating up twice as fast as the global average of 1.3°F over the past century—and in some places even faster. These warming rates, unprecedented for at least two millennia, are merciless on the glaciers, whose rare confluence of high altitudes and low latitudes make them especially sensitive to shifts in climate.

For thousands of years the glaciers have formed what Lonnie Thompson, a glaciologist at Ohio State University, calls "Asia's freshwater bank account"—an immense storehouse whose buildup of new ice and snow (deposits) has historically offset its annual runoff (withdrawals). Glacial melt plays its most vital role before and after the rainy season, when it supplies a greater portion of the flow in every river from the Yangtze (which irrigates more than half of China's rice) to the Ganges and the Indus (key to the agricultural heartlands of India and Pakistan). But over the past half century, the balance has been lost, perhaps irrevocably. Of the 680 glaciers Chinese scientists monitor closely on the Tibetan Plateau, 95 percent are shedding more ice than they're adding, with the heaviest losses on its southern and eastern edges. "These glaciers are not simply retreating," Thompson says. "They're losing mass from the surface down."

The ice cover in this portion of the plateau has shrunk more than 6 percent since the 1970s—and the damage is still greater in Tajikistan and northern India, with 35 percent and 20 percent declines respectively over the past five decades. The rate of melting is not uniform, and a number of glaciers in the Karakoram Range on the western edge of the plateau are actually advancing. This anomaly may result from increases in snowfall in the higher latitude—and therefore colder—Karakorams, where snow and ice are less vulnerable to small temperature increases. The gaps in scientific knowledge are still great, and in the Tibetan Plateau they are deepened by the region's remoteness and political sensitivity—as well as by the inherent complexities of climate science.

Though scientists argue about the rate and cause of glacial retreat, most don't deny that it's happening. And they believe the worst may be yet to come. The more dark areas that are exposed by melting, the more sunlight is absorbed than reflected, causing temperatures to rise faster. (Some climatologists believe this warming feedback loop could intensify the Asian monsoon, triggering more violent storms and flooding in places such as Bangladesh and Myanmar.) If current trends hold, Chinese scientists believe that 40 percent of the plateau's glaciers could disappear by 2050. "Full-scale glacier shrinkage is inevitable," says Yao Tandong, a glaciologist at China's Institute of Tibetan Plateau Research. "And it will lead to ecological catastrophe."

The potential impacts extend far beyond the glaciers. On the Tibetan Plateau, especially its dry northern flank, people are already affected by a warmer climate. The grasslands and wetlands are deteriorating, and the permafrost that feeds them with spring and summer melt is retreating to higher elevations. Thousands of lakes have dried up. Desert now covers about one-sixth of the plateau, and in places sand dunes lap across the highlands like waves in a yellow sea. The herders who once thrived here are running out of options.

Along the plateau's southern edge, by contrast, many communities are coping with too much water. In alpine villages like Mingyong, the glacial melt has swelled rivers, with welcome side effects: expanded croplands and longer growing seasons. But such benefits often hide deeper costs. In Mingyong, surging meltwater has carried away topsoil; elsewhere, excess runoff has been blamed for more frequent flooding and landslides. In the mountains from Pakistan to Bhutan, thousands of glacial lakes have formed, many potentially unstable. Among the more dangerous is Imja Tsho, at 16,400 feet on the trail to Nepal's Island Peak. Fifty years ago the lake didn't exist; today, swollen by melt, it is a mile long and 300 feet deep. If it ever burst through its loose wall of moraine, it would drown the Sherpa villages in the valley below.

This situation—too much water, too little water—captures, in miniature, the trajectory of the overall crisis. Even if melting glaciers provide an abundance of water in the short run, they portend a frightening endgame: the eventual depletion of Asia's greatest rivers. Nobody can predict exactly when the glacier retreat will translate into a sharp drop in runoff. Whether it happens in 10, 30, or 50 years depends on local conditions, but the collateral damage across the region could be devastating. Along with acute water and electricity shortages, experts predict a plunge in food production, widespread migration in the face of ecological changes, even conflicts between Asian powers.

The nomads' tent is a pinprick of white against a canvas of green and brown. There is no other sign of human existence on the 14,000-foot-high prairie that seems to extend to the end of the world. As a vehicle rattles toward the tent, two young men emerge, their long black hair horizontal in the wind. Ba O

and his brother Tsering are part of an unbroken line of Tibetan nomads who for at least a thousand years have led their herds to summer grazing grounds near the headwaters of the Yangtze and Yellow Rivers.

Inside the tent, Ba O's wife tosses patties of dried yak dung onto the fire while her four-year-old son plays with a spool of sheep's wool. The family matriarch, Lu Ji, churns yak milk into cheese, rocking back and forth in a hypnotic rhythm. Behind her are two weathered Tibetan chests topped with a small Buddhist shrine: a red prayer wheel, a couple of smudged Tibetan texts, and several yak butter candles whose flames are never allowed to go out. "This is the way we've always done things," Ba O says. "And we don't want that to change."

But it may be too late. The grasslands are dying out, as decades of warming temperatures—exacerbated by overgrazing—turn prairie into desert. Watering holes are drying up, and now, instead of traveling a short distance to find summer grazing for their herds, Ba O and his family must trek more than 30 miles across the high plateau. Even there the grass is meager. "It used to grow so high you could lose a sheep in it," Ba O says. "Now it doesn't reach above their hooves." The family's herd has dwindled from 500 animals to 120. The next step seems inevitable: selling their remaining livestock and moving into a government resettlement camp.

Across Asia the response to climate-induced threats has mostly been slow and piecemeal, as if governments would prefer to leave it up to the industrialized countries that pumped the greenhouse gases into the atmosphere in the first place. There are exceptions. In Ladakh, a bone-dry region in northern India and Pakistan that relies entirely on melting ice and snow, a retired civil engineer named Chewang Norphel has built "artificial glaciers"—simple stone embankments that trap and freeze glacial melt in the fall for use in the early spring growing season. Nepal is developing a remote monitoring system to gauge when glacial lakes are in danger of bursting, as well as the technology to drain them. Even in places facing destructive monsoonal flooding, such as Bangladesh, "floating schools" in the delta enable kids to continue their education—on boats.

But nothing compares to the campaign in China, which has less water than Canada but 40 times more people. In the vast desert in the Xinjiang region, just north of the Tibetan Plateau, China aims to build 59 reservoirs to capture and save glacial runoff. Across Tibet, artillery batteries have been installed to launch rain-inducing silver iodide into the clouds. In Qinghai the government is blocking off degraded grasslands in hopes they can be nurtured back to health. In areas where grasslands have already turned to scrub desert, bales of wire fencing are rolled out over the last remnants of plant life to prevent them from blowing away.

Along the road near the town of Madoi are two rows of newly built houses. This is a resettlement village for Tibetan nomads, part of a massive and controversial program to relieve pressure on the grasslands near the sources of Chinas three major rivers—the Yangtze, Yellow, and Mekong—where nearly half of Qinghai Province's 530,000 nomads have traditionally lived. Tens of thousands of nomads here have had to give up their way of life, and many more—including, perhaps, Ba O—may follow.

The subsidized housing is solid, and residents receive a small annual stipend. Even so, Jixi Lamu, a 33-year-old woman in a traditional embroidered dress, says her family is stuck in limbo, dependent on government handouts. "We've spent the $400 we had left from selling off our animals," she says. "There was no future with our herds, but there's no future here either." Her husband is away looking for menial work. Inside the one-room house, her mother sits on the bed, fingering her prayer beads. A Buddhist shrine stands on the other side of the room, but the candles have burned out.

It is not yet noon in Delhi, just 180 miles south of the Himalayan glaciers. But in the narrow corridors of Nehru Camp, a slum in this city of 16 million, the blast furnace of the north Indian summer has already sent temperatures soaring past 105 degrees Fahrenheit. Chaya, the 25-year-old wife of a fortune-teller, has spent seven hours joining the mad scramble for water that, even today, defines life in this heaving metropolis—and offers a taste of what the depletion of Tibet's water and ice portends.

Chaya's day began long before sunrise, when she and her five children fanned out in the darkness, armed with plastic jugs of every size. After daybreak, the rumor of a tap with running water sent her stumbling in a panic through the slum's narrow corridors. Now, with her containers still empty and the sun blazing overhead, she has returned home for a moment's rest. Asked if she's eaten anything today, she laughs: "We haven't even had any tea yet."

Suddenly cries erupt—a water truck has been spotted. Chaya leaps up and joins the human torrent in the street. A dozen boys swarm onto a blue tanker, jamming hoses in and siphoning the water out. Below, shouting women jostle for position with their containers. In six minutes the tanker is empty. Chaya arrived too late and must move on to chase the next rumor of water.

Delhi's water demand already exceeds supply by more than 300 million gallons a day, a shortfall worsened by inequitable distribution and a leaky infrastructure that loses an estimated 40 percent of the water. More than two-thirds of the city's water is pulled from the Yamuna and the Ganges, rivers fed by Himalayan ice. If that ice disappears, the future will almost certainly be worse. "We are facing an unsustainable situation," says Diwan Singh, a Delhi environmental activist. "Soon—not in thirty years but in five to ten—there will be an exodus because of the lack of water."

The tension already seethes. In the clogged alleyway around one of Nehru Camp's last functioning taps, which run for one hour a day, a man punches a woman who cut in line, leaving a purple welt on her face. "We wake up every morning fighting over water," says Kamal Bhate, a local astrologer watching the melee. This one dissolves into shouting and finger-pointing, but the brawls can be deadly. In a nearby slum a teenage boy was recently beaten to death for cutting in line.

As the rivers dwindle, the conflicts could spread. India, China, and Pakistan all face pressure to boost food production to keep up with their huge and growing populations. But climate change and diminishing water supplies could reduce cereal yields in South Asia by 5 percent within three decades. "We're going to see rising tensions over shared water resources,

including political disputes between farmers, between farmers and cities, and between human and ecological demands for water," says Peter Gleick, a water expert and president of the Pacific Institute in Oakland, California. "And I believe more of these tensions will lead to violence."

The real challenge will be to prevent water conflicts from spilling across borders. There is already a growing sense of alarm in Central Asia over the prospect that poor but glacier-heavy nations (Tajikistan, Kyrgyzstan) may one day restrict the flow of water to their parched but oil-rich neighbors (Uzbekistan, Kazakhstan, Turkmenistan). In the future, peace between Pakistan and India may hinge as much on water as on nuclear weapons, for the two countries must share the glacier-dependent Indus.

The biggest question mark hangs over China, which controls the sources of the region's major rivers. Its damming of the Mekong has sparked anger downstream in Indochina. If Beijing follows through on tentative plans to divert the Brahmaputra, it could provoke its rival, India, in the very region where the two countries fought a war in 1962.

For the people in Nehru Camp, geopolitical concerns are lost in the frenzied pursuit of water. In the afternoon, a tap outside the slum is suddenly turned on, and Chaya, smiling triumphantly, hauls back a full, ten-gallon jug on top of her head. The water is dirty and bitter, and there are no means to boil it. But now, at last, she can give her children their first meal of the day: a piece of bread and a few spoonfuls of lentil stew. "They should be studying, but we keep shooing them away to find water," Chaya says. "We have no choice, because who knows if we'll find enough water tomorrow."

Fatalism may be a natural response to forces that seem beyond our control. But Jia Son, the Tibetan farmer watching Mingyong Glacier shrink, believes that every action counts—good or bad, large or small. Pausing on the mountain trail, he makes a guilty confession. The melting ice, he says, may be his fault.

When Jia Son first noticed the rising temperatures—an unfamiliar trickle of sweat down his back about a decade ago—he figured it was a gift from the gods. Winter soon lost some of its brutal sting. The glacier began releasing its water earlier in the summer, and for the first time in memory villagers had the luxury of two harvests a year.

Then came the Chinese tourists, a flood of city dwellers willing to pay locals to take them up to see the glacier. The Han tourists don't always respect Buddhist traditions; in their gleeful hollers to provoke an icefall, they seem unaware of the calamity that has befallen the glacier. Still, they have turned a poor village into one of the region's wealthiest. "Life is much easier now," says Jia Son, whose simple farmhouse, like all in the village, has a television and government-subsidized satellite dish. "But maybe our greed has made Kawagebo angry."

He is referring to the temperamental deity above his village. One of the holiest mountains in Tibetan Buddhism, Kawagebo has never been conquered, and locals believe its summit—and its glacier—should remain untouched. When a Sino-Japanese expedition tried to scale the peak in 1991, an avalanche near the top of the glacier killed all 17 climbers. Jia Son remains convinced the deaths were not an accident but an act of divine retribution. Could Mingyong's retreat be another sign of Kawagebo's displeasure?

Jia Son is taking no chances. Every year he embarks on a 15-day pilgrimage around Kawagebo to show his deepening Buddhist devotion. He no longer hunts animals or cuts down trees. As part of a government program, he has also given up a parcel of land to be reforested. His family still participates in the village's tourism cooperative, but Jia Son makes a point of telling visitors about the glacier's spiritual significance. "Nothing will get better," he says, "until we get rid of our materialistic thinking."

It's a simple pledge, perhaps, one that hardly seems enough to save the glaciers of the Tibetan Plateau—and stave off the water crisis that seems sure to follow. But here, in the shadow of one of the world's fastest retreating glaciers, this lone farmer has begun, in his own small way, to restore the balance.

Critical Thinking

1. Sketch a concept map of the linkages between Asia's population centers, location of glaciers, and their agricultural regions. Locate potential areas for conflict over water resources.

2. If water resources (via glaciers) became too scarce, how will Asia feed its people?

3. What other environmental impacts in the region may also result from glacial loss?

4. How does Asia's potential water scarcity issue relate to the issues discussed in Article 24, "The World's Water Challenge?"

Eating Fossil Fuels

DALE ALLEN PFEIFER

October 3, 2003, 1200 PDT, (FTW)—Human beings (like all other animals) draw their energy from the food they eat. Until the last century, all of the food energy available on this planet was derived from the sun through photosynthesis. Either you ate plants or you ate animals that fed on plants, but the energy in your food was ultimately derived from the sun.

It would have been absurd to think that we would one day run out of sunshine. No, sunshine was an abundant, renewable resource, and the process of photosynthesis fed all life on this planet. It also set a limit on the amount of food that could be generated at any one time, and therefore placed a limit upon population growth. Solar energy has a limited rate of flow into this planet. To increase your food production, you had to increase the acreage under cultivation, and displace your competitors. There was no other way to increase the amount of energy available for food production. Human population grew by displacing everything else and appropriating more and more of the available solar energy.

The need to expand agricultural production was one of the motive causes behind most of the wars in recorded history, along with expansion of the energy base (and agricultural production is truly an essential portion of the energy base). And when Europeans could no longer expand cultivation, they began the task of conquering the world. Explorers were followed by conquistadors and traders and settlers. The declared reasons for expansion may have been trade, avarice, empire or simply curiosity, but at its base, it was all about the expansion of agricultural productivity. Wherever explorers and conquistadors traveled, they may have carried off loot, but they left plantations. And settlers toiled to clear land and establish their own homestead. This conquest and expansion went on until there was no place left for further expansion. Certainly, to this day, landowners and farmers fight to claim still more land for agricultural productivity, but they are fighting over crumbs. Today, virtually all of the productive land on this planet is being exploited by agriculture. What remains unused is too steep, too wet, too dry or lacking in soil nutrients.[1]

Just when agricultural output could expand no more by increasing acreage, new innovations made possible a more thorough exploitation of the acreage already available. The process of "pest" displacement and appropriation for agriculture accelerated with the industrial revolution as the mechanization of agriculture hastened the clearing and tilling of land and augmented the amount of farmland which could be tended by one person. With every increase in food production, the human population grew apace.

At present, nearly 40% of all land-based photosynthetic capability has been appropriated by human beings.[2] In the United States we divert more than half of the energy captured by photosynthesis.[3] We have taken over all the prime real estate on this planet. The rest of nature is forced to make due with what is left. Plainly, this is one of the major factors in species extinctions and in ecosystem stress.

The Green Revolution

In the 1950s and 1960s, agriculture underwent a drastic transformation commonly referred to as the Green Revolution. The Green Revolution resulted in the industrialization of agriculture. Part of the advance resulted from new hybrid food plants, leading to more productive food crops. Between 1950 and 1984, as the Green Revolution transformed agriculture around the globe, world grain production increased by 250%.[4] That is a tremendous increase in the amount of food energy available for human consumption. This additional energy did not come from an increase in incipient sunlight, nor did it result from introducing agriculture to new vistas of land. The energy for the Green Revolution was provided by fossil fuels in the form of fertilizers (natural gas), pesticides (oil), and hydrocarbon fueled irrigation.

The Green Revolution increased the energy flow to agriculture by an average of 50 times the energy input of traditional agriculture.[5] In the most extreme cases, energy consumption by agriculture has increased 100 fold or more.[6]

In the United States, 400 gallons of oil equivalents are expended annually to feed each American (as of data provided in 1994).[7] Agricultural energy consumption is broken down as follows:

- 31% for the manufacture of inorganic fertilizer
- 19% for the operation of field machinery
- 16% for transportation
- 13% for irrigation
- 08% for raising livestock (not including livestock feed)
- 05% for crop drying
- 05% for pesticide production
- 08% miscellaneous[8]

Energy costs for packaging, refrigeration, transportation to retail outlets, and household cooking are not considered in these figures.

To give the reader an idea of the energy intensiveness of modern agriculture, production of one kilogram of nitrogen for fertilizer requires the energy equivalent of from 1.4 to 1.8 liters of diesel fuel. This is not considering the natural gas feedstock.[9] According to The Fertilizer Institute (www.tfi.org), in the year from June 30 2001 until June 30 2002 the United States used 12,009,300 short tons of nitrogen fertilizer.[10] Using the low figure of 1.4 liters diesel equivalent per kilogram of nitrogen, this equates to the energy content of 15.3 billion liters of diesel fuel, or 96.2 million barrels.

Of course, this is only a rough comparison to aid comprehension of the energy requirements for modern agriculture.

In a very real sense, we are literally eating fossil fuels. However, due to the laws of thermodynamics, there is not a direct correspondence between energy inflow and outflow in agriculture. Along the way, there is a marked energy loss. Between 1945 and 1994, energy input to agriculture increased 4-fold while crop yields only increased 3-fold.[11] Since then, energy input has continued to increase without a corresponding increase in crop yield. We have reached the point of marginal returns. Yet, due to soil degradation, increased demands of pest management and increasing energy costs for irrigation (all of which is examined below), modern agriculture must continue increasing its energy expenditures simply to maintain current crop yields. The Green Revolution is becoming bankrupt.

Fossil Fuel Costs

Solar energy is a renewable resource limited only by the inflow rate from the sun to the earth. Fossil fuels, on the other hand, are a stock-type resource that can be exploited at a nearly limitless rate. However, on a human timescale, fossil fuels are non-renewable. They represent a planetary energy deposit which we can draw from at any rate we wish, but which will eventually be exhausted without renewal. The Green Revolution tapped into this energy deposit and used it to increase agricultural production.

Total fossil fuel use in the United States has increased 20-fold in the last 4 decades. In the US, we consume 20 to 30 times more fossil fuel energy per capita than people in developing nations. Agriculture directly accounts for 17% of all the energy used in this country.[12] As of 1990, we were using approximately 1,000 liters (6.41 barrels) of oil to produce food on one hectare of land.[13]

In 1994, David Pimentel and Mario Giampietro estimated the output/input ratio of agriculture to be around 1.4.[14] For 0.7 Kilogram-Calories (kcal) of fossil energy consumed, U.S. agriculture produced 1 kcal of food. The input figure for this ratio was based on FAO (Food and Agriculture Organization of the UN) statistics, which consider only fertilizers (without including fertilizer feedstock), irrigation, pesticides (without including pesticide feedstock), and machinery and fuel for field operations. Other agricultural energy inputs not considered were energy and machinery for drying crops, transportation for inputs and outputs to and from the farm, electricity, and construction and maintenance of farm buildings and infrastructures. Adding in estimates for these energy costs brought the input/output energy ratio down to 1.[15] Yet this does not include the energy expense of packaging, delivery to retail outlets, refrigeration or household cooking.

In a subsequent study completed later that same year (1994), Giampietro and Pimentel managed to derive a more accurate ratio of the net fossil fuel energy ratio of agriculture.[16] In this study, the authors defined two separate forms of energy input: Endosomatic energy and Exosomatic energy. Endosomatic energy is generated through the metabolic transformation of food energy into muscle energy in the human body. Exosomatic energy is generated by transforming energy outside of the human body, such as burning gasoline in a tractor. This assessment allowed the authors to look at fossil fuel input alone and in ratio to other inputs.

Prior to the industrial revolution, virtually 100% of both endosomatic and exosomatic energy was solar driven. Fossil fuels now represent 90% of the exosomatic energy used in the United States and other developed countries.[17] The typical exo/endo ratio of pre-industrial, solar powered societies is about 4 to 1. The ratio has changed tenfold in developed countries, climbing to 40 to 1. And in the United States it is more than 90 to 1.[18] The nature of the way we use endosomatic energy has changed as well.

The vast majority of endosomatic energy is no longer expended to deliver power for direct economic processes. Now the majority of endosomatic energy is utilized to generate the flow of information directing the flow of exosomatic energy driving machines. Considering the 90/1 exo/endo ratio in the United States, each endosomatic kcal of energy expended in the US induces the circulation of 90 kcal of exosomatic energy. As an example, a small gasoline engine can convert the 38,000 kcal in one gallon of gasoline into 8.8 KWh (Kilowatt hours), which equates to about 3 weeks of work for one human being.[19]

In their refined study, Giampietro and Pimentel found that 10 kcal of exosomatic energy are required to produce 1 kcal of food delivered to the consumer in the U.S. food system. This includes packaging and all delivery expenses, but excludes household cooking).[20] *The U.S. food system consumes ten times more energy than it produces in food energy.* This disparity is made possible by nonrenewable fossil fuel stocks.

Assuming a figure of 2,500 kcal per capita for the daily diet in the United States, the 10/1 ratio translates into a cost of 35,000 kcal of exosomatic energy per capita each day. However, considering that the average return on one hour of endosomatic labor in the U.S. is about 100,000 kcal of exosomatic energy, the flow of exosomatic energy required to supply the daily diet is achieved in only 20 minutes of labor in our current system. Unfortunately, if you remove fossil fuels from the equation, the daily diet will require 111 hours of endosomatic labor per capita; that is, *the current U.S. daily diet would require nearly three weeks of labor per capita to produce.*

Quite plainly, as fossil fuel production begins to decline within the next decade, there will be less energy available for the production of food.

Soil, Cropland and Water

Modern intensive agriculture is unsustainable. Technologically-enhanced agriculture has augmented soil erosion, polluted and overdrawn groundwater and surface water, and even (largely due to increased pesticide use) caused serious public health and environmental problems. Soil erosion, overtaxed cropland and water resource overdraft in turn lead to even greater use of fossil fuels and hydrocarbon products. More hydrocarbon-based fertilizers must be applied, along with more pesticides; irrigation water requires more energy to pump; and fossil fuels are used to process polluted water.

It takes 500 years to replace 1 inch of topsoil.[21] In a natural environment, topsoil is built up by decaying plant matter and weathering rock, and it is protected from erosion by growing plants. In soil made susceptible by agriculture, erosion is reducing productivity up to 65% each year.[22] Former prairie lands, which constitute the bread basket of the United States, have lost one half of their topsoil after farming for about 100 years. This soil is eroding 30 times faster than the natural formation rate.[23] Food crops are much hungrier than the natural grasses that once covered the Great Plains. As a result, the remaining topsoil is increasingly depleted of nutrients. Soil erosion and mineral depletion removes about $20 billion worth of plant nutrients from U.S. agricultural soils every year.[24] Much of the soil in the Great Plains is little more than a sponge into which we must pour hydrocarbon-based fertilizers in order to produce crops.

Every year in the U.S., more than 2 million acres of cropland are lost to erosion, salinization and water logging. On top of this, urbanization, road building, and industry claim another 1 million acres annually from farmland.[24] Approximately three-quarters of the land area in the United States is devoted to agriculture and commercial forestry.[25] The expanding human population is putting increasing pressure on land availability. Incidentally, only a small portion of U.S. land area remains available for the solar energy technologies necessary to support a solar energy-based economy. The land area for harvesting biomass is likewise limited. For this reason, the development of solar energy or biomass must be at the expense of agriculture.

Modern agriculture also places a strain on our water resources. Agriculture consumes fully 85% of all U.S. freshwater resources.[26] Overdraft is occurring from many surface water resources, especially in the west and south. The typical example is the Colorado River, which is diverted to a trickle by the time it reaches the Pacific. Yet surface water only supplies 60% of the water used in irrigation. The remainder, and in some places the majority of water for irrigation, comes from ground water aquifers. Ground water is recharged slowly by the percolation of rainwater through the earth's crust. Less than 0.1% of the stored ground water mined annually is replaced by rainfall.[27] The great Ogallala aquifer that supplies agriculture, industry and home use in much of the southern and central plains states has an annual overdraft up to 160% above its recharge rate. The Ogallala aquifer will become unproductive in a matter of decades.[28]

We can illustrate the demand that modern agriculture places on water resources by looking at a farmland producing corn. A corn crop that produces 118 bushels/acre/year requires more than 500,000 gallons/acre of water during the growing season. The production of 1 pound of maize requires 1,400 pounds (or 175 gallons) of water.[29] Unless something is done to lower these consumption rates, modern agriculture will help to propel the United States into a water crisis.

In the last two decades, the use of hydrocarbon-based pesticides in the U.S. has increased 33-fold, yet each year we lose more crops to pests.[30] This is the result of the abandonment of traditional crop rotation practices. Nearly 50% of U.S. corn land is grown continuously as a monoculture.[31] This results in an increase in corn pests, which in turn requires the use of more pesticides. Pesticide use on corn crops had increased 1,000-fold even before the introduction of genetically engineered, pesticide resistant corn. However, corn losses have still risen 4-fold.[32]

Modern intensive agriculture is unsustainable. It is damaging the land, draining water supplies and polluting the environment. And all of this requires more and more fossil fuel input to pump irrigation water, to replace nutrients, to provide pest protection, to remediate the environment and simply to hold crop production at a constant. Yet this necessary fossil fuel input is going to crash headlong into declining fossil fuel production.

US Consumption

In the United States, each person consumes an average of 2,175 pounds of food per person per year. This provides the U.S. consumer with an average daily energy intake of 3,600 Calories. The world average is 2,700 Calories per day.[33] Fully 19% of the U.S. caloric intake comes from fast food. Fast food accounts for 34% of the total food consumption for the average U.S. citizen. The average citizen dines out for one meal out of four.[34]

One third of the caloric intake of the average American comes from animal sources (including dairy products), totaling 800 pounds per person per year. This diet means that U.S. citizens derive 40% of their calories from fat—nearly half of their diet.[35]

Americans are also grand consumers of water. As of one decade ago, Americans were consuming 1,450 gallons/day/capita (g/d/c), with the largest amount expended on agriculture. Allowing for projected population increase, consumption by 2050 is projected at 700 g/d/c, which hydrologists consider to be minimal for human needs.[36] This is without taking into consideration declining fossil fuel production.

To provide all of this food requires the application of 0.6 million metric tons of pesticides in North America per year. This is over one fifth of the total annual world pesticide use, estimated at 2.5 million tons.[37] Worldwide, more nitrogen fertilizer is used per year than can be supplied through natural sources. Likewise, water is pumped out of underground aquifers at a much higher rate than it is recharged. And stocks of important minerals, such as phosphorus and potassium, are quickly approaching exhaustion.[38]

Total U.S. energy consumption is more than three times the amount of solar energy harvested as crop and forest products. The United States consumes 40% more energy annually than the total amount of solar energy captured yearly by all U.S. plant biomass. Per capita use of fossil energy in North America is five times the world average.[39]

Our prosperity is built on the principal of exhausting the world's resources as quickly as possible, without any thought to our neighbors, all the other life on this planet, or our children.

Population & Sustainability

Considering a growth rate of 1.1% per year, the U.S. population is projected to double by 2050. As the population expands, an estimated one acre of land will be lost for every person added to the U.S. population. Currently, there are 1.8 acres of farmland available to grow food for each U.S. citizen. By 2050, this will decrease to 0.6 acres. 1.2 acres per person is required in order to maintain current dietary standards.[40]

Presently, only two nations on the planet are major exporters of grain: the United States and Canada.[41] By 2025, it is expected that the U.S. will cease to be a food exporter due to domestic demand. The impact on the U.S. economy could be devastating, as food exports earn $40 billion for the U.S. annually. More importantly, millions of people around the world could starve to death without U.S. food exports.[42]

Domestically, 34.6 million people are living in poverty as of 2002 census data.[43] And this number is continuing to grow at an alarming rate. Too many of these people do not have a sufficient diet. As the situation worsens, this number will increase and the United States will witness growing numbers of starvation fatalities.

There are some things that we can do to at least alleviate this tragedy. It is suggested that streamlining agriculture to get rid of losses, waste and mismanagement might cut the energy inputs for food production by up to one-half.[35] In place of fossil fuel-based fertilizers, we could utilize livestock manures that are now wasted. It is estimated that livestock manures contain 5 times the amount of fertilizer currently used each year.[36] Perhaps most effective would be to eliminate meat from our diet altogether.[37]

Mario Giampietro and David Pimentel postulate that a sustainable food system is possible only if four conditions are met:

1. Environmentally sound agricultural technologies must be implemented.
2. Renewable energy technologies must be put into place.
3. Major increases in energy efficiency must reduce exosomatic energy consumption per capita.
4. Population size and consumption must be compatible with maintaining the stability of environmental processes.[38]

Providing that the first three conditions are met, with a reduction to less than half of the exosomatic energy consumption per capita, the authors place the maximum population for a sustainable economy at 200 million.[39] Several other studies have produced figures within this ballpark (**Energy and Population**, Werbos, Paul J. www.dieoff.com/page63.htm; **Impact of Population**

Growth on Food Supplies and Environment, Pimentel, David, et al. www.dieoff.com/page57.htm).

Given that the current U.S. population is in excess of 292 million,[40] that would mean a reduction of 92 million. *To achieve a sustainable economy and avert disaster, the United States must reduce its population by at least one-third.* The black plague during the 14th Century claimed approximately one-third of the European population (and more than half of the Asian and Indian populations), plunging the continent into a darkness from which it took them nearly two centuries to emerge.[41]

None of this research considers the impact of declining fossil fuel production. The authors of all of these studies believe that the mentioned agricultural crisis will only begin to impact us after 2020, and will not become critical until 2050. The current peaking of global oil production (and subsequent decline of production), along with the peak of North American natural gas production will very likely precipitate this agricultural crisis much sooner than expected. Quite possibly, a U.S. population reduction of one-third will not be effective for sustainability; the necessary reduction might be in excess of one-half. And, for sustainability, global population will have to be reduced from the current 6.32 billion people[42] to 2 billion—a reduction of 68% or over two-thirds. The end of this decade could see spiraling food prices without relief. And the coming decade could see massive starvation on a global level such as never experienced before by the human race.

Three Choices

Considering the utter necessity of population reduction, there are three obvious choices awaiting us.

We can-as a society-become aware of our dilemma and consciously make the choice not to add more people to our population. This would be the most welcome of our three options, to choose consciously and with free will to responsibly lower our population. However, this flies in the face of our biological imperative to procreate. It is further complicated by the ability of modern medicine to extend our longevity, and by the refusal of the Religious Right to consider issues of population management. And then, there is a strong business lobby to maintain a high immigration rate in order to hold down the cost of labor. Though this is probably our best choice, it is the option least likely to be chosen.

Failing to responsibly lower our population, we can force population cuts through government regulations. Is there any need to mention how distasteful this option would be? How many of us would choose to live in a world of forced sterilization and population quotas enforced under penalty of law? How easily might this lead to a culling of the population utilizing principles of eugenics?

This leaves the third choice, which itself presents an unspeakable picture of suffering and death. Should we fail to acknowledge this coming crisis and determine to deal with it, we will be faced with a die-off from which civilization may very possibly never revive. We will very likely lose more than the numbers necessary for sustainability. Under a die-off scenario, conditions will deteriorate so badly that the surviving human population would be a negligible fraction of the present population. And those survivors would suffer from the trauma of living through the death of their civilization, their neighbors, their friends and their families. Those survivors will have seen their world crushed into nothing.

The questions we must ask ourselves now are, how can we allow this to happen, and what can we do to prevent it? Does our present lifestyle mean so much to us that we would subject ourselves and our children to this fast approaching tragedy simply for a few more years of conspicuous consumption?

Author's Note

This is possibly the most important article I have written to date. It is certainly the most frightening, and the conclusion is the bleakest I have ever penned. This article is likely to greatly disturb the reader; it has certainly disturbed me. However, it is important for our future that this paper should be read, acknowledged and discussed.

I am by nature positive and optimistic. In spite of this article, I continue to believe that we can find a positive solution to the multiple crises bearing down upon us. Though this article may provoke a flood of hate mail, it is simply a factual report of data and the obvious conclusions that follow from it.

Endnotes

1. **Availability of agricultural land for crop and livestock production,** Buringh, P. Food and Natural Resources, Pimentel. D. and Hall. C.W. (eds), Academic Press, 1989.

2. **Human appropriation of the products of photosynthesis,** Vitousek, P.M. et al. Bioscience 36, 1986. www.science.duq .edu/esm/unit2-3

3. **Land, Energy and Water: the constraints governing Ideal US Population Size,** Pimental, David and Pimentel, Marcia. **Focus,** Spring 1991. **NPG Forum,** 1990. www.dieoff.com/ page136.htm

4. **Constraints on the Expansion of Global Food Supply,** Kindell, Henry H. and Pimentel, David. Ambio Vol. 23 No. 3, May 1994. The Royal Swedish Academy of Sciences. www .dieoff.com/page36htm

5. **The Tightening Conflict: Population, Energy Use, and the Ecology of Agriculture,** Giampietro, Mario and Pimentel, David, 1994. www.dieoff.com/page69.htm

6. Op. Cit. See note 4.

7. **Food, Land, Population and the U.S. Economy,** Pimentel, David and Giampietro, Mario. Carrying Capacity Network, 11/21/1994. www.dieoff.com/page55.htm

8. **Comparison of energy inputs for inorganic fertilizer and manure based corn production,** McLaughlin, N.B., et al. Canadian Agricultural Engineering, Vol. 42, No. 1, 2000.

9. Ibid.

10. **US Fertilizer Use Statistics.** www.tfi.org/Statistics/USfertuse2 .asp

11. **Food, Land, Population and the U.S. Economy, Executive Summary,** Pimentel, David and Giampietro, Mario. Carrying Capacity Network, 11/21/1994. www.dieoff.com/ page40.htm

12. Ibid.

13. Op. Cit. See note 3.

14. Op. Cit. See note 7.

15. Ibid.

16. Op. Cit. See note 5.

17. Ibid.

18. Ibid.

19. Ibid.

20. Ibid.

21. Op. Cit. See note 11.

22. Ibid.

23. Ibid.

24. Ibid.

25. Ibid.

26. Op Cit. See note 3.

27. Op Cit. See note 11.

28. Ibid.

29. Ibid.

30. Ibid.

31. Op. Cit. See note 3.

32. Op. Cit. See note 5.

33. Op. Cit. See note 3.

34. Op. Cit. See note 11.

35. **Food Consumption and Access,** Lynn Brantley, et al. Capital Area Food Bank, 6/1/2001. www.clagettfarm.org/ purchasing.html

36. Op. Cit. See note 11.

37. Ibid.

38. Op. Cit. See note 5.

39. Ibid.

40. Ibid.

41. Op. Cit. See note 11.

42. Op. Cit. See note 4.

43. Op. Cit. See note 11.

44. **Poverty 2002.** The U.S. Census Bureau. www.census.gov/hhes/ poverty/poverty02/pov02hi.html

45. Op. Cit. See note 3.

46. Ibid.

47. **Diet for a Small Planet,** Lappé, Frances Moore. Ballantine Books, 1971-revised 1991. www.dietforasmallplanet.com/

48. Op. Cit. See note 5.

49. Ibid.

50. **U.S. and World Population Clocks.** U.S. Census Bureau. www.census.gov/main/www/popclock.html

51. **A Distant Mirror,** Tuckman Barbara. Ballantine Books, 1978.

52. Op. Cit. See note 40.

Critical Thinking

1. What does the author mean by "eating fossil fuels"?

2. Outline the arguments used in the article to support the author's thesis.

3. Explain how the idea of the "competitive exclusion principle" in Article 13 can be applied to the human history of agriculture.

4. Assess the efficacy of the author's three choices in light of the articles presented in Unit I regarding the nature of our consumption.

The Myth of Mountaintop Removal Mining

Big Coal says it's a tough choice: we can have prosperity and jobs or a pristine environment, but not both. That's a Big Lie

BETH WELLINGTON

C NN correspondent Soledad O'Brien's recent piece on mountaintop removal (MTR) in the Appalachian mountains has the troubling title, "Steady job or healthy environment: what [sic] would you choose?"

How about we choose both?

In any case, MTR does not, despite industry claims, deliver employment to offset its environmental damage. It's simply a win-win for Big Coal and its political supporters, and a lose-lose for ordinary people who live in mining areas. Whatever the industry would have you believe, basing an economy on coal is not a sustainable development plan. A study by the Appalachian Regional Commission noted the effects of mining on employment in Central Appalachia:

> As employment in Central Appalachia's mining sector has declined over time . . . many counties that were already typically experiencing relatively poor and tenuous economic circumstances . . . have been unable to successfully adapt to changing economic conditions.

Michael Hendryx and Melissa M Ahem found similar results when they investigated the region: "The heaviest coal mining areas of Appalachia had the poorest socio-economic conditions."

In addition to the negative impact on employment, mountaintop removal has terrible effects on the land. Rob Goodwin of Coal River Mountain Watch recently said of the land around Southern Appalachia:

> Southern Appalachia is unique. Because there were no glaciers here, the topsoil is some of the oldest in the world and that's why there are ramps, ginseng and molly moochers [morels], among other valuable species. What you are doing here on this mine site is destroying the 10,000-year-old species that, regardless of what you do, will not grow back.

The health toll is also steep, several academic studies have indicated. This week, West Virginia's junior senator, Joe Manchin, was bashing the EPA at a constituent breakfast in Huntington. The Senate now has before it the House plan, the so-called Clean Water Cooperative Federalism Act of 2011 (HR 2018). which would restrict the EPA's ability to veto permits issued by the Army Corps of Engineers. Janet Keating, executive director of Ohio Valley Environmental Coalition, attended and referred Manchin to what he'd said on CNN, that there is no clear evidence of human health impacts from MTR. She then handed him copies of the 18 studies showing or suggesting health impacts. Manchin told Keating that at the time the CNN show was taped, these studies were not available.

Huh? I had copies of those studies. Surely a US Senator and former governor has as much access to published information as I do?

The health and economic problems caused by coal may explain why we're not buying the attacks on the EPA. A majority of voters in four Appalachian states want their water protected and disapprove of mountaintop mining. The same day Manchin was in Huntington, Lake Research Partners and Bellwether Research & Consulting released the results (pdf) of a poll commissioned by Appalachian Mountain Advocates. Earthjustice and the Sierra Club. Of 1,315 people interviewed in Kentucky, West Virginia, Tennessee and Virginia, "Three-quarters support fully enforcing—and even increasing protections in—the Clean Water Act to safeguard streams, rivers and lakes in their states from mountaintop removal coal mining. . . . Just 8% of voters oppose it. Support for this proposal is far-reaching, encompassing solid majorities of Democrats (86%), independents (76%), Republicans (71%) and Tea Party supporters (67%)."

The reaction to the poll from Jason Hayes, the communications director for the American Coal Council?

> They're doing a numbers job. They need to frighten people. They need more membership dollars. . . . It's all very frightening if you don't understand what's going on.

What's going in is that this is an industry that spends money for fancy websites to "dispel myths"—for instance, by telling you that "reclamation" returns the mountainsides to their original state. For a more typical picture of reclamation, you might want to check out the PBS film "Razing Appalachia". And speaking of coal dollars, they appear to be benefiting our local politicians, including Manchin. As Manuel Quinones and Elana Schor pointed out:

> Senator Joe Manchin (Democrat, West Virginia) is more than just a supporter of his state's influential coal producers—he's a full-fledged industry insider. On his financial disclosures for 2009 and 2010, Manchin reported significant earnings from Enersystems Inc, a coal brokerage that he helped run before his political star rose. In the 19 months before winning his Senate seat in a hard-fought special election, Manchin reported operating income of $1,363,916 from Enersystems. His next disclosure showed $417,255 in Enersystems income.

Of course, Manchin says his investments are in a blind trust, but do you think he doesn't know that what's good for the coal industry is good for Joe Manchin? As filmmaker Mari-Lynn Evans has said of the CNN programme referred to by Keating:

> Mountain top removal mining is not an issue of jobs v the environment. It is an issue of corporate profits and corrupt politicians v the health and safety of human beings living under MTR sites in Appalachia. WVU scientists estimate over 11,000 people die in Appalachia each year because of coal. MTR mining provides less than 4,000 direct MTR jobs in West Virginia. Does that mean, for every MTR job, we must accept that those jobs will cost each of us the lives of two or three of our friends and loved ones? This is jobs v genocide. If you don't understand that, then you don't understand the story

And the rest of us are paying, too, if not to the same extent. Air and water pollution travel past the immediate region. Social costs to the environment and health are not payed by the coal industry, and thus artificially lower the cost of coal energy and encourage its consumption. As economists Todd L Cherry of Appalacian State University and Jason Shogrenb of the Univisty of Wyoming pointed out in a 2002 study . . . (before the devastation of mountaintop removal had reached its current levels):

> Coal is by far the most under-priced energy resource: the price per ton of coal was about $30, but the external costs

are nearly $160. Also including climate change risks, the external costs would be about $190 per ton

And prices are further lowered by vast subsidies. . . . And think of the opportunity costs [forgone] to develop cleaner energy. Interestingly, another full-fledged insider, Dick Kelly, who is retiring as chief executive officer of Exel Energy, told journalist Don Selby:

> We've got to get off fossil fuels. . . . The quicker the better. All [that some members of Congress] are worried about is the next two or six years when they run for re-election. They just keep kicking the can down the road. . . . I don't know how they can deny the science. I really don't. . . . I think one of the misconceptions is that many people believe that wind is just outrageously expensive. Truth is, wind power competes very well with natural gas. The technology is getting better. We are getting a lot more kilowatts out of our windmills now. Even solar has come down 50% in the last two years. . . . I'd be OK if there were never any more coal.

How long will our politicans favour the coal industry with subsidies and lax safety and health regulations? Consider the cost of this choice—in lives, health and damage to the air we breath, the water we drink and the land that provides us with nourishment and recreation. Wouldn't it be better for them to enact polices that support efficiency, conservation and alternative energy sources with a lower environmental impact, such as wind, solar and geothermal?

That way, we might actually get steady jobs *and* a healthy environment.

Critical Thinking

1. "Big Coal says we can have prosperity and jobs or a pristine environment, but not both." Why does the author say "that's a big lie"?

2. What would the wealthiest 1% of Americans think about this article?

3. What would the world's poorest billion people think about this issue of mountaintop removal mining?

4. How could we apply full cost accounting to the issue of mountaintop removal mining?

The Efficiency Dilemma

If our machines use less energy, will we just use them more?

DAVID OWEN

In April, the federal government adopted standards for automobiles requiring manufacturers to improve the average fuel economy of their new-car fleets thirty percent by 2016. The *Times,* in an editorial titled "Everybody Wins," said the change would produce "a trifecta of benefits." Those benefits were enumerated last year by Steven Chu, the Secretary of Energy: a reduction in total oil consumption of 1.8 billion barrels; the elimination of nine hundred and fifty million metric tons of greenhouse-gas emissions; and savings, for the average American driver, of three thousand dollars.

Chu, who shared the Nobel Prize in Physics in 1997, has been an evangelist for energy efficiency, and not just for vehicles. I spoke with him in July, shortly after he had conducted an international conference called the Clean Energy Ministerial, at which efficiency was among the main topics. "I feel very passionate about this," he told me. "We in the Department of Energy are trying to get the information out that efficiency really does save money and doesn't necessarily mean that you're going to have to make deep sacrifices."

Energy efficiency has been called "the fifth fuel" (after coal, petroleum, nuclear power, and renewables); it is seen as a cost-free tool for accelerating the transition to a green-energy economy. In 2007, the United Nations Foundation said that efficiency improvements constituted "the largest, the most evenly geographically distributed, and least expensive energy resource." Last year, the management-consulting firm McKinsey & Company concluded that a national efficiency program could eliminate "up to 1.1 gigatons of greenhouse gases annually." The environmentalist Amory Lovins, whose thinking has influenced Chu's, has referred to the replacement of incandescent light bulbs with compact fluorescents as "not a free lunch, but a lunch you're paid to eat," since a fluorescent bulb will usually save enough electricity to more than offset its higher purchase price. Tantalizingly, much of the technology required to increase efficiency is well understood. The World Economic Forum, in a report called "Towards a More Energy Efficient World," observed that "the average refrigerator sold in the United States today uses three-quarters less energy than the 1975 average, even though it is 20 percent larger and costs 60 percent less"—an improvement that Chu cited in his conversation with me.

But the issue may be less straightforward than it seems. The thirty-five-year period during which new refrigerators have plunged in electricity use is also a period during which the global market for refrigeration has burgeoned and the world's total energy consumption and carbon output, including the parts directly attributable to keeping things cold, have climbed. Similarly, the first fuel-economy regulations for U.S. cars—which were enacted in 1975, in response to the Arab oil embargo—were followed not by a steady decline in total U.S. motor-fuel consumption but by a long-term rise, as well as by increases in horsepower, curb weight, vehicle miles travelled (up a hundred percent since 1980), and car ownership (America has about fifty million more registered vehicles than licensed drivers). A growing group of economists and others have argued that such correlations aren't coincidental. Instead, they have said, efforts to improve energy efficiency can more than negate any environmental gains—an idea that was first proposed a hundred and fifty years ago, and which came to be known as the Jevons paradox.

Great Britain in the middle of the nineteenth century was the world's leading military, industrial, and mercantile power. In 1865, a twenty-nine-year-old Englishman named William Stanley Jevons published a book, "The Coal Question," in which he argued that the bonanza couldn't last. Britain's affluence, he wrote, depended on its endowment of coal, which the country was rapidly depleting. He added that such an outcome could not be delayed through increased "economy" in the use of coal—what we refer to today as energy efficiency. He concluded, in italics, *"It is wholly a confusion of ideas to suppose that the economical use of fuel is equivalent to a diminished consumption. The very contrary is the truth."*

He offered the example of the British iron industry. If some technological advance made it possible for a blast furnace to produce iron with less coal, he wrote, then profits would rise, new investment in iron production would be attracted, and the price of iron would fall, thereby stimulating additional demand. Eventually, he concluded, "the greater number of furnaces will more than make up for the diminished consumption of each." Other examples of this effect abound. In a paper published in 1998, the Yale economist William D. Nordhaus estimated the cost of lighting throughout human history. An ancient Babylonian, he calculated, needed to work more than forty-one hours to acquire enough lamp oil to provide a thousand lumen-hours of light—the equivalent of a seventy-five-watt incandescent bulb burning for about an hour. Thirty-five hundred years later, a contemporary of Thomas Jefferson's could buy the same amount of illumination, in the form of tallow candles, by working for about five hours and twenty minutes. By 1992, an average American, with access to compact fluorescents, could do the same in less than half a second. Increasing the energy efficiency of illumination is nothing new; improved lighting has been "a lunch you're paid to eat" ever since humans upgraded from cave fires (fifty-eight hours of labor for our early Stone Age ancestors). Yet our efficiency gains haven't reduced the energy we expend on illumination or shrunk our energy consumption over all. On the contrary, we now generate light so extravagantly that darkness itself is spoken of as an endangered natural resource.

Jevons was born in Liverpool in 1835. He spent two years at University College, in London, then went to Australia, where he had been offered a job as an assayer at a new mint, in Sydney. He left after five years, completed his education in England, became a part-time college instructor, and published a well-received book on gold markets. "The Coal Question" made him a minor celebrity; it was admired by John Stuart Mill and William Gladstone, and it inspired the government to investigate his findings. In 1871, he published "The Theory of Political Economy," a book that's still considered one of the founding texts of mathematical economics. He drowned a decade later, at the age of forty-six, while swimming in the English Channel. In 1905, John Maynard Keynes, who was then twenty-two and a graduate student at Cambridge University, wrote to Lytton Strachey that he had discovered a "thrilling" book: Jevons's "Investigations in Currency and Finance." Keynes wrote of Jevons, "I am convinced that he was one of *the* minds of the century."

Jevons might be little discussed today, except by historians of economics, if it weren't for the scholarship of another English economist, Len Brookes. During the nineteen-seventies oil crisis, Brookes argued that devising ways to produce goods with less oil—an obvious response to higher prices—would merely accommodate the new prices, causing energy consumption to be higher than it would have been if no effort to increase efficiency had been made; only later did he discover that Jevons had anticipated him by more than a century. I spoke with Brookes recently. He told me, "Jevons is very simple. When we talk about increasing energy efficiency, what we're really talking about is increasing the productivity of energy. And, if you increase the productivity of anything, you have the effect of reducing its implicit price, because you get more return for the same money—which means the demand goes up."

Nowadays, this effect is usually referred to as "rebound"—or, in cases where increased consumption more than cancels out any energy savings, as "backfire." In a 1992 paper, Harry D. Saunders, an American researcher, provided a concise statement of the basic idea: "With fixed real energy price, energy efficiency gains will increase energy consumption above where it would be without these gains."

In 2000, the journal *Energy Policy* devoted an entire issue to rebound. It was edited by Lee Schipper, who is now a senior research engineer at Stanford University's Precourt Energy Efficiency Center. In an editorial, Schipper wrote that the question was not whether rebound exists but, rather, "how much the effect appears, how rapidly, in which sectors, and in what manifestations." The majority of the *Energy Policy* contributors concluded that there wasn't a lot to worry about. Schipper, in his editorial, wrote that the articles, taken together, suggested that "rebounds are significant but do not threaten to rob society of most of the benefits of energy efficiency improvements."

I spoke with Schipper recently, and he told me that the Jevons paradox has limited applicability today. "The key to understanding Jevons," he said, "is that processes, products, and activities where energy is a very high part of the cost—in this country, a few metals, a few chemicals, air travel—are the only ones whose variable cost is very sensitive to energy. That's it." Jevons wasn't wrong about nineteenth-century British iron smelting, he said; but the young and rapidly growing industrial world that Jevons lived in no longer exists.

Most economists and efficiency experts have come to similar conclusions. For example, some of them say that when you increase the fuel efficiency of cars you lose no more than about ten percent of the fuel savings to increased use. And if you look at the whole economy, Schipper said, rebound effects are comparably trivial. "People like Brookes would say—they don't quite know how to say it, but they seem to want to say the extra growth is more than the saved energy, so it's like a backfire. The problem is, that's never been observed on a national level."

But troublesome questions have lingered, and the existence of large-scale rebound effects is not so easy to dismiss. In 2004, a committee of the House of Lords invited a number of experts to help it grapple with a conundrum: the United Kingdom, like a number of other countries, had spent heavily to increase energy efficiency in an attempt to reduce its greenhouse emissions. Yet energy consumption and carbon output in Britain—as in the rest of the world—had continued to rise. Why?

Most economic analyses of rebound focus narrowly on particular uses or categories of uses: if people buy a more efficient clothes dryer, say, what will happen to the energy they use as they dry clothes? (At least one such study has concluded that, for appliances in general, rebound is nonexistent.) Brookes dismisses such "bottom-up" studies, because they ignore or understate the real consumption effects, in economies as a whole.

A good way to see this is to think about refrigerators, the very appliances that the World Economic Forum and Steven Chu cited as efficiency role models for reductions in energy use. The first refrigerator I remember is the one my parents owned when I was little. They acquired it when they bought their first house, in 1954, a year before I was born. It had a tiny, uninsulated freezer compartment, which seldom contained much more than a few aluminum ice trays and a burrow-like mantle of frost. (Frost-free freezers stay frost-free by periodically heating their cooling elements—a trick that wasn't widely in use yet.) In the sixties, my parents bought a much improved model—which presumably was more efficient, since the door closed tight, by means of a rubberized magnetic seal rather than a mechanical latch. But our power consumption didn't fall, because the old refrigerator didn't go out of service; it moved into our basement, where it remained plugged in for a further twenty-five years—mostly as a warehouse for beverages and leftovers—and where it was soon joined by a stand-alone freezer. Also, in the eighties, my father added an icemaker to his bar, to supplement the one in the kitchen fridge.

This escalation of cooling capacity has occurred all over suburban America. The recently remodelled kitchen of a friend of mine contains an enormous side-by-side refrigerator, an enormous side-by-side freezer, and a drawer-like under-counter mini-fridge for beverages. And the trend has not been confined to households. As the ability to efficiently and inexpensively chill things has grown, so have opportunities to buy chilled things—a potent positive-feedback loop. Gas stations now often have almost as much refrigerated shelf space as the grocery stores of my early childhood; even mediocre hotel rooms usually come with their own small fridge (which, typically, either is empty or—if it's a minibar—contains mainly things that don't need to be kept cold), in addition to an icemaker and a refrigerated vending machine down the hall.

The steadily declining cost of refrigeration has made eating much more interesting. It has also made almost all elements of food production more cost-effective and energy-efficient: milk lasts longer if you don't have to keep it in a pail in your well. But there are environmental downsides, beyond the obvious one that most of the electricity that powers the world's refrigerators is generated by burning fossil fuels. James McWilliams, who is the author of the recent book "Just Food," told me, "Refrigeration and packaging convey to the consumer a sense that what we buy will last longer than it does. Thus, we buy enough stuff to fill our capacious Sub-Zeros and, before we know it, a third of it is past its due date and we toss it." (The item that New Yorkers most often throw away unused, according to the anthropologist-in-residence at the city's Department of Sanitation, is vegetables.) Jonathan Bloom, who runs the Website wastedfood.com and is the author of the new book "American Wasteland," told me that, since the mid-nineteen-seventies, per-capita food waste in the United States has increased by half, so that we now throw away forty percent of all the edible food we produce. And when we throw away food we don't just throw away nutrients; we also throw away the energy we used in keeping it cold as we lost interest in it, as well as the energy that went into growing, harvesting, processing, and transporting it, along with its proportional share of our staggering national consumption of fertilizer, pesticides, irrigation water, packaging, and landfill capacity. According to a 2009 study, more than a quarter of U.S. freshwater use goes into producing food that is later discarded.

Efficiency improvements push down costs at every level—from the mining of raw materials to the fabrication and transportation of finished goods to the frequency and intensity of actual use—and reduced costs stimulate increased consumption. (Coincidentally or not, the growth of American

refrigerator volume has been roughly paralleled by the growth of American body-mass index.) Efficiency-related increases in one category, furthermore, spill into others. Refrigerators are the fraternal twins of air-conditioners, which use the same energy-hungry compressor technology to force heat to do something that nature doesn't want it to. When I was a child, cold air was a far greater luxury than cold groceries. My parents' first house—like eighty-eight percent of all American homes in 1960—didn't have air-conditioning when they bought it, although they broke down and got a window unit during a heat wave, when my mom was pregnant with me. Their second house had central air-conditioning, but running it seemed so expensive to my father that, for years, he could seldom be persuaded to turn it on, even at the height of a Kansas City summer, when the air was so humid that it felt like a swimmable liquid. Then he replaced our ancient Carrier unit with a modern one, which consumed less electricity, and our house, like most American houses, evolved rapidly from being essentially un-air-conditioned to being air-conditioned all summer long.

Modern air-conditioners, like modern refrigerators, are vastly more energy efficient than their mid-twentieth-century predecessors—in both cases, partly because of tighter standards established by the Department of Energy. But that efficiency has driven down their cost of operation, and manufacturing efficiencies and market growth have driven down the cost of production, to such an extent that the ownership percentage of 1960 has now flipped: by 2005, according to the Energy Information Administration, eighty-four percent of all U.S. homes had air-conditioning, and most of it was central. Stan Cox, who is the author of the recent book "Losing Our Cool," told me that, between 1993 and 2005, "the energy efficiency of residential air-conditioning equipment improved twenty-eight percent, but energy consumption for A.C. by the average air-conditioned household rose thirty-seven percent." One consequence, Cox observes, is that, in the United States, we now use roughly as much electricity to cool buildings as we did for all purposes in 1955.

As "Losing Our Cool" clearly shows, similar rebound effects permeate the economy. The same technological gains that have propelled the growth of U.S. residential and commercial cooling have helped turn automobile air-conditioners, which barely existed in the nineteen-fifties, into standard equipment on even the least luxurious vehicles. (According to the National Renewable Energy Laboratory, running a mid-sized car's air-conditioning increases fuel consumption by more than twenty percent.) And access to cooled air is self-reinforcing: to someone who works in an air-conditioned office, an un-air-conditioned house quickly becomes intolerable, and vice versa. A resident of Las Vegas once described cars to me as "devices for transporting air-conditioning between buildings."

In less than half a century, increased efficiency and declining prices have helped to push access to air-conditioning almost all the way to the bottom of the U.S. income scale—and now those same forces are accelerating its spread all over the world. According to Cox, between 1997 and 2007 the use of air-conditioners tripled in China (where a third of the world's units are now manufactured, and where many air-conditioner purchases have been subsidized by the government). In India, air-conditioning is projected to increase almost tenfold between 2005 and 2020; according to a 2009 study, it accounted for forty percent of the electricity consumed in metropolitan Mumbai.

All such increases in energy-consuming activity can be considered manifestations of the Jevons paradox. Teasing out the precise contribution of a particular efficiency improvement isn't just difficult, however; it may be impossible, because the endlessly ramifying network of interconnections is too complex to yield readily to empirical, mathematics-based analysis. Most modern studies of energy rebound are "bottom-up" by necessity: it's only at the micro end of the economics spectrum that the number of mathematical variables can be kept manageable. But looking for rebound only in individual consumer goods, or in closely cropped economic snapshots, is as futile and misleading as trying to analyze the global climate with a single thermometer.

Schipper told me, "In the end, the impact of rebound is small, in my view, for one very key reason: energy is a small share of the economy. If sixty percent of our economy were paying for energy, then anything that moved it down by ten percent would liberate a huge amount of resources. Instead, it's between six and eight percent for primary energy, depending on exactly what country you're in." ("Primary energy" is the energy in oil, coal, wind, and other natural resources before it's been converted into electricity or into refined or synthetic fuels.) Schipper believes that cheap energy is an environmental problem, but he also believes that, because we can extract vastly more economic benefit from a ton of coal than nineteenth-century Britons did, efficiency gains now have much less power to stimulate consumption. This concept is closely related to one called "decoupling," which suggests that the growing efficiency of machines has weakened the link between energy use and economic activity, and also to the idea of "decarbonization," which holds that, for similar reasons, every dollar we spend represents a shrinking quantity of greenhouse gas.

These sound like environmentally valuable trends—yet they seem to imply that the world's energy and carbon challenges are gradually solving themselves, since decoupling and decarbonization, like increases in efficiency, are nothing new. One problem with decoupling, as the concept is often applied, is that it doesn't account for energy use and carbon emissions that have not been eliminated but merely exported out of the region under study (say, from California to a factory in China). And there's a more fundamental problem, described by the Danish researcher Jørgen S. Nørgård, who has called energy decoupling "largely a statistical delusion." To say that energy's economic role is shrinking is a little like saying, "I have sixteen great-great-grandparents, eight great-grandparents, four grandparents, and two parents—the world's population must be imploding." Energy production may account for only a small percentage of our economy, but its falling share of G.D.P. has made it more important, not less, since every kilowatt we generate supports an ever larger proportion of our well-being. The logic misstep is apparent if you imagine eliminating primary energy from the world. If you do that, you don't end up losing "between six and eight percent" of current economic activity, as Schipper's formulation might suggest; you lose almost everything we think of as modern life.

Blake Alcott, an ecological economist, has made a similar case in support of the existence of large-scale Jevons effects. Recently, he told me, "If it is true that greater efficiency in using a resource means less consumption of it—as efficiency environmentalists say—then less efficiency would logically mean more consumption. But this yields a reductio ad absurdum: engines and smelters in James Watt's time, around 1800, were far less efficient than today's, but is it really imaginable that, had technology been frozen at that efficiency level, a greater population would now be using vastly more fossil fuel than we in fact do?" Contrary to the argument made by "decouplers," we aren't gradually reducing our dependence on energy; rather, we are finding ever more ingenious ways to leverage B.T.U.s. Between 1984 and 2005, American electricity production grew by about sixty-six percent—and it did so despite steady, economy-wide gains in energy efficiency. The increase was partly the result of population growth; but per-capita energy consumption rose, too, and it did so even though energy use per dollar of G.D.P. fell by roughly half. Besides, population growth itself can be a Jevons effect: the more efficient we become, the more people we can sustain; the more people we sustain, the more energy we consume.

The Model T was manufactured between 1908 and 1927. According to the Ford Motor Company, its fuel economy ranged between thirteen and twenty-one miles per gallon. There are vehicles on the road today that do worse than that; have we really made so little progress in more than a hundred years? But focussing on miles per gallon is the wrong way to assess the environmental impact of cars. Far more revealing is to consider the productivity of driving. Today, in contrast to the early nineteen-hundreds, any American with a license can cheaply travel almost anywhere, in almost any weather, in extraordinary comfort; can drive for thousands of miles with no maintenance other than refuelling; can easily find gas, food, lodging, and just about anything else within a short distance of almost any road; and can order and eat meals without undoing a seat belt or turning off the ceiling-mounted DVD player.

A modern driver, in other words, gets vastly more benefit from a gallon of gasoline—makes far more economical use of fuel—than any Model T owner ever did. Yet motorists' energy consumption has grown by mind-boggling amounts, and, as the productivity of driving has increased and the cost of getting around has fallen, the global market for cars has surged. (Two of the biggest road-building efforts in the history of the world are currently under way in India and China.) And developing small, inexpensive vehicles that get a hundred miles to the gallon would only exacerbate that trend. The problem with efficiency gains is that we inevitably reinvest them in additional consumption. Paving roads reduces rolling friction, thereby boosting miles per gallon, but it also makes distant destinations seem closer, thereby enabling people to live in sprawling, energy-gobbling subdivisions far from where they work and shop.

Chu has said that drivers who buy more efficient cars can expect to save thousands of dollars in fuel costs; but, unless those drivers shred the money and add it to a compost heap, the environment is unlikely to come out ahead, as those dollars will inevitably be spent on goods or activities that involve fuel consumption—say, on increased access to the Internet, which is one of the fastest-growing energy drains in the world. (Cox writes that, by 2014, the U.S. computer network alone will each year require an amount of energy equivalent to the total electricity consumption of Australia.) The problem is exactly what Jevons said it was: the economical use of fuel is not equivalent to a diminished consumption. Schipper told me that economy-wide Jevons effects have "never been observed," but you can find them almost anywhere you look: they are the history of civilization.

J evons died too soon to see the modern uses of oil and natural gas, and he obviously knew nothing of nuclear power. But he did explain why "alternative" energy sources, such as wind, hydropower, and biofuels (in his day, mainly firewood and whale oil), could not compete with coal: coal had replaced them, on account of its vastly greater portability, utility, and productivity. Early British steam engines were sometimes used to pump water to turn water wheels; we do the equivalent when we burn coal to make our toothbrushes move back and forth.

Decreasing reliance on fossil fuels is a pressing global need. The question is whether improving efficiency, rather than reducing total consumption, can possibly bring about the desired result. Steven Chu told me that one of the appealing features of the efficiency discussions at the Clean Energy Ministerial was that they were never contentious. "It was the opposite," he said. "No one was debating about who's responsible, and there was no finger-pointing or trying to lay blame." This seems encouraging in one way but dismaying in another. Given the known level of global disagreement about energy and climate matters, shouldn't there have been *some* angry table-banging? Advocating efficiency involves virtually no political risk—unlike measures that do call for sacrifice, such as capping emissions or putting a price on carbon or increasing energy taxes or investing heavily in utility-scale renewable-energy facilities or confronting the deeply divisive issue of global energy equity. Improving efficiency is easy to endorse: we've been doing it, globally, for centuries. It's how we created the problems we're now trying to solve.

Efficiency proponents often express incredulity at the idea that squeezing more consumption from less fuel could somehow carry an environmental cost. Amory Lovins once wrote that, if Jevons's argument is correct, "we should mandate inefficient equipment to save energy." As Lovins intended, this seems laughably illogical—but is it? If the only motor vehicle available today were a 1920 Model T, how many miles do you think you'd drive each year, and how far do you think you'd live from where you work? No one's going to "mandate inefficient equipment," but, unless we're willing to do the equivalent—say, by mandating costlier energy—increased efficiency, as Jevons predicted, can only make our predicament worse.

A t the end of "The Coal Question," Jevons concluded that Britain faced a choice between "brief greatness and longer continued mediocrity." His preference was for mediocrity, by which he meant something like "sustainability." Our world is different from his, but most of the central arguments of his book still apply. Steve Sorrell, who is a senior fellow at Sussex University and a co-editor of a recent comprehensive book on rebound, called "Energy Efficiency and Sustainable Consumption," told me, "I think the point may be that Jevons has yet to be disproved. It is rather hard to demonstrate the validity of his proposition, but certainly the historical evidence to date is wholly consistent with what he was arguing." That might be something to think about as we climb into our plug-in hybrids and continue our journey, with ever-increasing efficiency, down the road paved with good intentions.

Critical Thinking

1. Describe the "efficiency dilemma."

2. Outline the facts and data used to support the idea of "Jevons' Paradox."

3. Do you see a connection between increasing energy efficiency and continued overconsumption? Is so, describe what you see.

4. How might increasing energy efficiency and the material goods and services it creates actually hinder our quest for environmental sustainability?

Jevons' Paradox and the Perils of Efficient Energy Use

GREG LINDSAY

It's a given among Peak Oilers and New Urbanists alike that the imminent and permanent return of high oil prices will send convulsions through the suburban American landscape. But it's one thing when professional Jeremiahs like James Howard Kunstler preach this to the converted week after week [1], and something else [2] when the Urban Land Institute and PricewaterhouseCoopers advise commercial real estate investors to "shy away from fringe places in the exurbs and places with long car commutes or where getting a quart of milk takes a 15-minute drive." Oil shocks will do what urban planners can't seem to and the government won't (through sharply higher gas taxes or putting a price on carbon): force people to live at greater densities.

In books like *$20 Per Gallon* [3] and *Why Your World Is About To Get A Whole Lot Smaller* [4]—both published last year, in the wake of 2008's real estate bubble-burst—the end of cheap oil is presented as a good thing, a chance to press the reset button on civilization and live more locally and sustainably. Kunstler goes further in 2005's *The Long Emergency* and in subsequent blog posts and novels, painting peak oil as a cleansing fire that will burn away exurban places like Pasco County, Florida. Last Wednesday, I drove for hours through the ground zero of Florida's foreclosure crisis [5], a scrolling landscape of strip malls, auto dealerships and billboards promising motorists that their stock market losses had been someone else's fault (and that you should sue them). The apocalypse would be a small price to pay for no more of this.

How else to explain the hostility directed at Amory Lovins by Kunstler and others [6]? Lovins identified the hard and soft paths of fossil fuels versus conservation and renewables thirty-four years ago [7], and has since written books like *Winning the Oil Endgame* and *Small is Beautiful,* in which he called for a massively distributed, solar-powered "microgrid [8]". But Lovins earned ridicule for his still-unrealized vision of a "hypercar [9]" made of composites and electric drive trains three-to-five times more efficient than existing models. The hypercar, Kunstler wrote, "Would have only promoted the unhelpful idea that Americans can continue to lead urban lives in the rural setting." (To add insult to injury, Lovins' Rocky Mountain Institute is accessible only by car.)

Why unhelpful? In a phrase: Jevons' Paradox. Nearly a century before the geologist M. King Hubbert began calculating peak oil, the economist William Stanley Jevons [10] discovered, to his horror, peak coal. In *The Coal Question,* published in 1865, Jevons raised the questions which haunt sustainability advocates to this day: "Are we wise in allowing the commerce of this country to rise beyond the point at which we can long maintain it?" He estimated Britain's coal production would reach a peak in less than a hundred years, with calamitous economic and Malthusian consequences. The engine of coal's demise would be the same invention that was created to conserve it: the steam engine. But it made burning coal so efficient, that instead of conserving coal, it drove the price down until everyone was burning it. This is Jevons' Paradox: the more efficiently you use a resource, the more of it you will use. Put another way: the better the machine—or fuel—the broader its adoption.

A corollary is the Piggy Principle [11]: instead of saving the energy conserved through efficiency, we find new ways to spend it, leading to greater consumption than before. No wonder Kunstler is alarmed that a hyper-efficient hypercar would lead to hyper-sprawl—it's only been the pattern throughout all of human history [12]. Maybe the worst thing that could happen to new urbanism would be an incredibly efficient new car (or fuel) that allows Americans (and, increasingly, the Chinese) to carry on as before, as an oil glut allowed us to do between 1979 and 2001. Crisis is on their side.

Jevons' peak coal reckoning was postponed by a new fuel source discovered a few years earlier in the Pennsylvania hills: oil. Today, there is another liquid fuel source on the horizon, provided it can scale: next-generation biofuels. Peak Oilers take it as an article of faith that biofuels won't work (and for now they have both physics and economics in their corner). But reading books like the ones mentioned above (or watching films like *The End of Suburbia* and *Collapse*) one gets the feeling they're actively rooting against them as well. "A crisis is a terrible thing to waste," Paul Romer once said. Especially if you waste one by solving it and forgetting it ever happened.

Energy, transportation and urbanism are inextricably entwined, but as far as I can tell, no one has asked the founders of biofuel startups what kind of world they envision if they succeed. The assumption is more of the same. Only more of it. Last Thursday, I was in Washington D.C. for a briefing sponsored by the Biotech Industry Organization (a lobbying group) to update lawmakers on their progress. Executives from Solazyme, Algenol, HP Biopetroleum, Gevo, and Coskata took turns explaining how sunlight/sugars/miscanthus/waste products would be converted by algae/microorganisms/yeast into oil/ethanol/isobutanol. Each laid out plans to leave the lab behind and begin scaling up production to millions of gallons per year.

The urban consequences of a potentially endless, cheap(er), greener source of liquid fuel weren't on the agenda that morning, but I did bring it up at breakfast with Jonathan Wolfson, the CEO and co-founder of Solazyme. His company hopes to produce oil using algae (albeit by brewing it in fermenters instead of growing it in ponds). Solazyme touts resulting biodiesel is the only one of its kind that can be poured into a diesel gas tank unblended.

Peak oil's big crunch "isn't creative destruction," he said. "It's *destruction* destruction." Sketching two circles on his place setting, one large and one small, he explained the large represented the roughly 150 billion gallons of gasoline Americans consumed last year, and the small one 50 billion gallons of diesel. His goal wasn't to produce or replace gasoline, he said. Electric cars would do that—ideally electric-biodiesel hybrids.

He was unfamiliar with Jevons' Paradox, but grasped the dilemma immediately. "Whether it's because of that, or whether it's because the world is projected to grow to 9 billion people by 2050, one way or another you have the same end problem, which is that we need more moving forward, not less," he said. "We need more energy, and we need more food, not less of either." He hoped algal oils would offer a solution to both—Solazyme has already begun selling algae-based nutritional supplements, and announced last week that it's working with Unilever to explore replacing petrochemicals in its soaps.

"I'm very receptive to urban planning," he continued. "I live in San Francisco, and I live in a place where I can walk to the market and walk to a movie, and I don't need to drive around the city and very often I don't. What I'm not receptive to is when you take that argument to the Luddite level. There's no really important technology that's been developed that couldn't be used in a negative way. But that's not a valid argument against new technology. Biofuels can either be very good or very bad, depending on where your biomass comes from, but they're like everything else. When you say 'Okay, because of the worst case, we don't want this technology. So we don't want to look at alternatives to liquid petroleum, because we know petroleum is going to run out, and what we really believe is that everyone should be walking anyway.' Well guess what? That's just not realistic."

Two days earlier, Accenture published a survey [13] of 9,000 individuals in 22 countries about their attitudes on energy: 90% were concerned by rising energy costs, and 76% by the prospect of shortages; 83% were concerned by climate change, and 89% thought it was important to reduce their country's reliance on fossil fuels. But barely a third thought they should do so by using less energy; the remainder believed their governments should find new sources, stat.

"We cannot address climate change or energy security unless we both create new sources of clean energy and reduce consumer demand," said the report's primary author. "But our survey shows that consumers do not think lower energy use is a priority."

Critical Thinking

1. In reference to Article 30, what does Article 31 add to your understanding of Jevons' Paradox?

2. What is the Piggy Principle? What does it suggest about planetary consumption?

3. How does the Piggy Principle relate to the articles in Unit I—The Nature of Our Consumption: *Ecophagy?*

4. Why might a hyperefficient hypercar, or "hyperproduct" of any kind (e.g. hyperfactories, hyperplanes, hyperhouses) actually run contrary to the ideals of a sustainable world?

Links:

1. http://kunstler.com/blog/
2. www.commoncurrent.com/notes/2O10/03/urban-resilience-for-dummiesp-p.html
3. http://seedmagazine.com/content/article/the_coming_oil-free_utopia/
4. www.jeffrubinssmallerworld.com/
5. www.newyorker.com/reporting/2009/02/09/090209fa_fact_packer
6. www.amazon.com/Green-Metropolis-Smaller-Driving-Sustainability/dp/1594488827
7. www.theatlantic.com/magazine/archive/2009/07/the-elusive-green-economy/7554/
8. www.fastcompany.com/magazine/137l/beyond-the-grid.html
9. move.rmi.org/markets-in-motion/case-studies/automotive/hypercar.html
10. en.wikipedia.org/wiki/William_Stanley_Jevons
11. www.theoildrum.com/node/6245?utm_source=feedburner&utm_medium=feed&utm_campaign=Feed%3A+theoildrum+%28The+Oil+Drum%29
12. www.cesaremarchetti.org/archive/electronic/basic_instincts.pdf
13. newsroom.accenture.com/news/consumers+reject+lower+energy+use+as+the+an

Rich Countries Launch Great Land Grab to Safeguard Food Supply

JULIAN BORGER

States and companies target developing nations
Small farmers at risk from industrial-scale deals

Rich governments and corporations are triggering alarm for the poor as they buy up the rights to millions of hectares of agricultural land in developing countries in an effort to secure their own long-term food supplies.

The head of the UN Food and Agriculture Organisation, Jacques Diouf, has warned that the controversial rise in land deals could create a form of "neo-colonialism", with poor states producing food for the rich at the expense of their own hungry people.

Rising food prices have already set off a second "scramble for Africa". This week, the South Korean firm Daewoo Logistics announced plans to buy a 99-year lease on a million hectares in Madagascar. Its aim is to grow 5m tonnes of corn a year by 2023, and produce palm oil from a further lease of 120,000 hectares (296,000 acres), relying on a largely South African workforce. Production would be mainly earmarked for South Korea, which wants to lessen dependence on imports.

"These deals can be purely commercial ventures on one level, but sitting behind it is often a food security imperative backed by a government," said Carl Atkin, a consultant at Bidwells Agribusiness, a Cambridge firm helping to arrange some of the big international land deals.

Madagascar's government said that an environmental impact assessment would have to be carried out before the Daewoo deal could be approved, but it welcomed the investment. The massive lease is the largest so far in an accelerating number of land deals that have been arranged since the surge in food prices late last year.

"In the context of arable land sales, this is unprecedented," Atkin said. "We're used to seeing 100,000-hectare sales. This is more than 10 times as much."

At a food security summit in Rome, in June, there was agreement to channel more investment and development aid to African farmers to help them respond to higher prices by producing more. But governments and corporations in some cash-rich but land-poor states, mostly in the Middle East, have opted not to wait for world markets to respond and are trying to guarantee their own long-term access to food by buying up land in poorer countries.

According to diplomats, the Saudi Binladin Group is planning an investment in Indonesia to grow basmati rice, while tens of thousands of hectares in Pakistan have been sold to Abu Dhabi investors.

Arab investors, including the Abu Dhabi Fund for Development, have also bought direct stakes in Sudanese agriculture. The president of the UEA, Khalifa bin Zayed, has said his country was considering large-scale agricultural projects in Kazakhstan to ensure a stable food supply.

Even China, which has plenty of land but is now getting short of water as it pursues breakneck industrialisation, has begun to explore land deals in south-east Asia. Laos, meanwhile, has signed away between 2m-3m hectares, or 15% of its viable farmland. Libya has secured 250,000 hectares of Ukrainian farmland, and Egypt is believed to want similar access. Kuwait and Qatar have been chasing deals for prime tracts of Cambodia rice fields.

Eager buyers generally have been welcomed by sellers in developing world governments desperate for capital in a recession. Madagascar's land reform minister said revenue would go to infrastructure and development in flood-prone areas.

Sudan is trying to attract investors for almost 900,000 hectares of its land, and the Ethiopian prime minister, Meles Zenawi, has been courting would-be Saudi investors.

"If this was a negotiation between equals, it could be a good thing. It could bring investment, stable prices and predictability to the market," said Duncan Green, Oxfam's head of research. "But the problem is, [in] this scramble for soil I don't see any place for the small farmers."

Alex Evans, at the Centre on International Cooperation, at New York University, said: "The small farmers are losing out already. People without solid title are likely to be turfed off the land."

Details of land deals have been kept secret so it is unknown whether they have built-in safeguards for local populations.

Steve Wiggins, a rural development expert at the Overseas Development Institute, said: "There are very few economies of scale in most agriculture above the level of family farm because managing [the] labour is extremely difficult." Investors might also have to contend with hostility. "If I was a political-risk adviser to [investors] I'd say 'you are taking a very big risk'. Land is an extremely sensitive thing. This could go horribly wrong if you don't learn the lessons of history."

Critical Thinking

1. Why is it not a good idea for rich countries to buy up agricultural land in poor countries?

2. Aside from the obvious increase in the hunger issue in poor countries, how can this "land grab" contribute to other global problems?

3. If, however, the United States is buying up some of this land to ensure our food security, why should we say "no"?

Global Urbanization: Can Ecologists Identify a Sustainable Way Forward?

Robert I. McDonald

The year 2007 was the first year in which more than half of humanity lived in cities. Over the next 25 years, the world will see the addition of nearly one million km^2 of urban area, occurring in tens of thousands of cities around the globe. The form these new neighborhoods take will affect our planet's ecology profoundly. Here, I highlight the connection between urban form and ecosystem service generation and consumption, I also discuss how urban form controls energy use, and hence oil security and climate change. I argue that only by directly addressing the implications of urban growth as a research subject will ecologists meet their responsibility to provide a foundation for a sustainable biosphere, a mandate of the Ecological Society of America.

—*Front Ecol Environ* 2008; 6(2):

In a nutshell:

- Urban growth in both the developed and developing world will be the main obstacle to achieving sustainable development
- The form of new urban areas will determine per capita resource use, now and for decades to come
- Urbanization in developing countries could dramatically increase oil consumption, with implications for global warming and energy security
- Conducting multiple before-and-after assessments of urban sustainability projects may be the greatest contribution ecologists can make to sustainable development

The Coming Urbanization

Sometime in 2007, humanity crossed a momentous milestone: for the first time ever, the majority of humans are living in cities (UNPD 2005a). In the next 25 years, 1.7 billion new people will move into urban areas (UNPD 2005b), and new settlements in the developing world will spread to cover an area the size of California (Angel *et al.* 2005). Most of these settlements will be in the developing world (UN-HABITAT 2006), where new-found urban lifestyles and increased affluence could lead to dramatically increased resource use. This resource use, especially of oil and other fossil fuels, will have implications for global warming and the security of nations. Humanity is building the equivalent of a city the size of Vancouver—about 600 000 people—twice every week (Peñalosa 2006. . .). What does the form of these new cities mean for the goal of sustainable development? And what role can ecologists play in helping to realize that goal?

The ongoing process of urbanization occurring in the world's poorer countries is the result of a predictable demographic shift: as a country's development proceeds, an increasing proportion of its population lives in urban areas. The increase in urban population is partly due to natural growth—the greater number of births than deaths among people already living in cities—and partly due to migration from rural areas to urban areas (NRC 2003). Lack of employment or safety sometimes pushes migrants from rural areas, while the increased economic and personal opportunities in cities can also attract migrants (eg Ma 2002). The precise mix of drivers for rural-urban migration seems to vary by region, and is still the subject of much debate (eg Roy *et al* 1992; deHaan 1997).

The best quantitative source of global data on urbanization comes from the UN Population Division, although some concerns remain about the varying definitions of "urban" used by different member countries (Cohen 2004). These data show that urbanization has already largely ceased in the US (80.1% urban), Europe (72.2% urban), Latin America and the Caribbean (77.4% urban), and South America (81.6% urban). However, Africa (38.3% urban) and countries in Southeast Asia, such as China (40.4% urban) and India (28.7% urban), are just beginning their shift, and will undergo rapid urbanization over the next several decades. Global urban population is predicted to increase from 3.2 billion to 4.9 billion over the next 25 years (UNPD 2005b). For comparison, the population of Europe in 2000 was 729 million (US Census Bureau, Population Division nd) and average household size was around 2.5 people per household (UNECE 2005), which means that there were approximately 291 million housing units. A 1.7 billion increase in urban dwellers, assuming five persons per household (typical for developing countries), would mean that more housing units will be built in the next 25 years than currently exist in all of Europe. . . .

Recent work by the World Bank allows us to translate urban population growth to urban area growth. In a survey of more than 120 cities from around the world, Angel *et al.* (2005) found that the biggest single variable controlling the per capita growth in area of a city was its income. On average, each new resident in a city adds 355 m^2 of developed area in a rich country, 125 m^2 in a middle-income country, and 85 m^2 in a poor country. As with any estimate of urban area, the values obtained are dependent on the definition of "urban" used in the particular study. However, if we accept this definition of "urban", and assume the current shift toward lower urban population density continues at the same rate, a reasonable estimate of total new urban area by 2030 is around 900 000 km^2. This will be split fairly evenly between the developed and developing nations, although its causes are quite different in each case: in the former, the driver is predominantly low-density settlement patterns (Jackson 1985), while in the latter, the driver is predominantly urban population growth (UN-HABITAT 2006). It is also important to note that around 46% of all urban population growth will occur in relatively small cities, with populations of fewer than 500 000 people (UNPD 2005b). Urbanization is therefore a fundamentally dispersed process, occurring in tens of thousands of cities.

Total urban area will remain a relatively small fraction of the Earth's terrestrial surface, from 0.3% today to 1% in 2030 (Angel *et al.* 2005). However, the ecological footprint of these cities will encompass an area many hundreds of times larger (eg Rees 1999; McGranahan and Satterthwaite 2003). Humanity will be entering a new world, a world of the city, by the city, and for the city. Urbanization will be the main obstacle to, and opportunity for, sustainable development.

Urban Form Matters

Humanity is building the grand cities of the 21st century, and their design, good or bad, is fundamentally a human decision. Even in capitalist economies, where urban development decisions are made by millions of different landowners, the pattern of development is largely a result of specific human plans, a joint consequence of topography (eg Muller 2001), zoning law (eg Munroe *et al.* 2005), and the geography of highways, rail lines, and mass transit (Handy 2005). This overall pattern of development can be referred to generally as urban form, and quantified using landscape metrics (cf Alberti 2005) of land-use intensity (eg people ha^{-1}), land-use heterogeneity (eg the number of different functional land uses), and land-use connectivity (eg degree of aggregation of land uses).

There is abundant evidence in the literature to show that urban form affects resource use (eg Folke *et al.* 1997; Pickett *et al.* 2001), although the exact relationship between the two seems to vary substantially, depending on the resource in question and the socioeconomic context of the city. In general, there is a great need for high-quality, municipal-level data that can be used to build mechanistic models of how resource use responds to changes in urban form (Pataki *et al* 2006). The form of the new cities in the developing world is doubly important, because of the tendency for patterns of past urban development to remain locked in the future. Past patterns tend to persist because major infrastructure (eg a road corridor) creates a physical imprint that lasts for decades. Moreover, past development patterns tend to condition the shape of new development. Therefore, the form of urban development today will control resource use for generations to come. Below, I focus on one particular facet: the link between urban form and energy use. . . .

One crucial way in which urban form affects energy use is through its control of, and mutual relationship with, the transportation sector (Banister *et al.* 1997). In developing countries, migration from the countryside to the city is typically associated with an increase in income, which in turn generates an increase in per-capita environmental consumption rates. If we control for the level of economic development, we can examine more closely how, at a given income level, alternative urban forms result in different resource utilization rates. For example, the average resident of Tokyo drives one-sixth as many kilometers per year as the average resident of Sacramento, despite having similar incomes (Kenworthy and Laube 1999). Personal automobile use is usually correlated with the density of an urban region, with use increasing dramatically at densities below 30 people per ha (Cameron *et al.* 2004). Finally, more energy-efficient public transportation systems require a density above a similar minimum threshold to be successful (Badoe and Miller 2000; Bento *et al.* 2005). While the empirical correlation between density and energy efficiency of transportation is clear, the exact mechanisms underlying this correlation are still uncertain (eg Anderson *et al.* 1996; Alberti 1999; Mindali *et al.* 2004).

Another way in which urban form affects energy use is through the heating and cooling of buildings (Steemers 2003). Heating a single-family, detached home in Oslo usually consumes 10 000 KW hr^{-1} per person per year, whereas heating an apartment in a larger building usually costs 7000 KW hr^{-1} per person per year, (Holden and Norland 2005). Generally, heating and cooling a single-family, detached home is less efficient than for row houses or multi-unit buildings. The quality of building construction also has an important influence on per-capita energy use, based on numerous factors, including building orientation, insulation, and choice of heating/cooling system. For example, new homes in Oslo consume 3000 fewer kW hr^{-1} annually than older homes in Oslo, correcting for house size, presumably because of improvements in some of the design components (Holden and Norland 2005).

Ecological ramifications

Energy use in cities generally results in CO_2 emissions in one form or another. Essentially all motorized personal transport burns gasoline or diesel; US oil use alone accounts for 10% of global emissions (EIA 2006). Most heating and cooling of homes is either directly powered by fossil fuels or by electricity from power plants that burn fossil fuels. Given this link between urban form, energy use, and CO_2 emissions, it is clear that the urban form of cities in the developing world has serious implications for global warming. Although the Kyoto Protocol does not set binding emissions reduction targets for developing countries, it creates one of the first working examples of a payment-for-ecosystem-services scheme, via the Clean Development Mechanism (CDM). The CDM allows carbon trading between First World countries (ie Annex 1 countries) and the developing world. If countries exceed their quota of greenhouse-gas emissions, they can purchase carbon credits from a developing country that has implemented some project to reduce their emissions. For example, Dhakal (2003) shows that a modest increase in electric vehicle use in Kathmandu, converting all three-wheeled passenger travel to electric motors and 20% of the bus-travel to trolley buses run on overhead electric lines, would save 20 400 tons of CO_2 annually. As this upgrade could claim credit for several years' worth of decreases in emissions, the value of the project on the global carbon market could be in excess of US$500 000, which would cover a sizeable portion of the cost of the project.

Urbanization in the developing world, along with rising incomes, may dramatically increase the number of potential oil consumers (Riley 2002), raising concerns about global oil security. Most discussions about oil security focus on the potential supply of oil, which is key for oil price and availability over the short term (Cleveland and Kaufmann 2003). It seems reasonable, however, to examine not just supply-side issues but demand-side issues as well, and the aggregate effect on oil prices (cf Awerbuch and Sauter 2006). While oil-industry forecasts are notoriously uncertain, there is general agreement that, over the medium to long term, demand will grow faster than supply for the simple reason that, while desire for oil may increase without bound, oil supply is limited, in part by geologic constraints (for a pessimistic view on this issue, see Bentley [2002]). If the price of oil rises in response (cf Skeer and Wang 2007), some new sources of oil, such as oil shales, will become economically viable, and demand will abate somewhat; however, these adjustments will probably not alter the general long-term trend toward higher oil prices (IEA 2006). It is important to realize that the rise in global energy consumption is not preordained, but depends on the energy efficiency of rapidly growing cities, particularly their density and use of public transit (Kenworthy and Laube 1999). It is not far-fetched, therefore, to argue that the oil demand created by rapidly urbanizing countries is of great importance in terms of security, because it threatens to drive oil prices upward and to increase the likelihood of conflicts over oil supply.

Urbanization will also profoundly affect ecosystem services and the underlying biodiversity that maintains them (McGranahan *et al.* 2006). For example, the spatial concentration of humanity in cities results in the surrounding sources of freshwater being partially or fully appropriated for urban use. Over time, cities in many parts of the world may have to search farther afield to find sufficient freshwater (cf MA 2005), often constructing major hydrological diversions in the process (eg Righter 2005). Similarly, the concentration of people in cities inevitably leads to increases in the concentration of the pollution they generate as well (eg Van der Zee *et al.* 1998), surpassing the ability of natural ecosystems to successfully absorb pollutants without becoming degraded (Pouyat *et al.* 1995). The political power of cities also often means that residents' desire for aesthetic and recreational pursuits substantially affects nearby natural resource management practices (eg Rothman 2004). Finally, the expansion of urban area has impacts on biodiversity that, while localized to a small portion of the Earth's surface, can cumulatively be quite considerable (McDonald *et al.* in review).

Payment for Sustainable Urbanization

The form of urban development is fundamentally a local decision, both for democratic and pragmatic reasons. Local citizens have the right to shape their city's development and the local knowledge necessary to make wise decisions. Many policy responses to urbanization, such as zoning and tax regimes, must be decided at this local level. However, the developed nations have a role to play in fostering sustainable development. First, they have a moral responsibility to help alleviate extreme poverty and environmental degradation. Second, there is a subset of sustainable development issues in which developed country aid appears to be a win-win scenario. Economists might call this an example of Coasian bargaining, in which a negative environmental externality is averted by a payment from another group affected by the externality. A classic example is when a downstream user of water pays an upstream polluter to stop polluting. Conservation NGOs have taken to calling it "payment for ecosystem service" (PES). The rapidly urbanizing countries, which have a legal and moral right to develop, as defined in the UN Declaration on the Right to Development, are creating substantial environmental "externalities". The developed countries could produce environmental benefits for themselves and reduce this "externality" by paying to promote more responsible urbanization in the rapidly urbanizing countries.

PES schemes have become increasingly common, with one review documenting more than 287 examples of payment for forest ecosystem services (Landell-Mills and Porras 2002). Lessons can be learned from these past markets (Salzman 2005) and applied to future PES schemes involving urbanization and the environment. First, it is important to clearly define what ecosystem services are important, how much service provision is needed, and who will ultimately pay for it. Given the vast difference in wealth, it is likely that developed countries will pay developing countries for more sustainable urban development (ie the beneficiary pays rather than the polluter). Second, the spatial scale of a particular ecosystem service, as well as the governance structure, will determine if PES is feasible and desirable (Chan et al. 2006). For ecosystem services that must be generated close to those receiving the service, it may be fairly easy to establish PES, as fewer institutions will be involved in the administration of the scheme, However, these local PES schemes may not have much potential in cities with few financial resources. For example, day-to-day recreation opportunities can best be provided by local governments responding to their citizens' desires, but in poorer countries, adequate funds for the purchase of parklands may be lacking. On the other hand, for ecosystem services that may be generated far from those receiving them, it may be difficult to set up a PES scheme due to the large number of institutions involved, as well as the consensual nature of international environmental treaties. However, PES schemes at this larger spatial scale may generate a far more active market between developed and undeveloped countries. For example, an increase in sequestration of CO_2 is of global value, in that everyone is affected by global warming, and it may therefore be relatively well-funded as a PES scheme. However, a working international market like the Clean Development Mechanism has been very difficult to construct due to the many groups involved in the negotiations.

The dispersed nature of future urbanization means that technologies or strategies aimed at increasing urban sustainability must have several characteristics. First, they must be scalable, that is, feasible to implement in thousands of cities. Second, they must be flexible, able to be adapted to myriad local circumstances. Third, they must be fairly inexpensive, within the budget of most developing countries. This may seem a tall order, but numerous projects seeking sustainable solutions are currently underway.

Treating Sustainable Development Projects as Experiments

One way for ecologists and environmental scientists to make a major contribution to sustainable urban development would be to view particular sustainability projects as experiments. This, while perhaps difficult for ecologists used to studying the workings of "natural" ecosystems, is quite similar in spirit to the new paradigm of evidence-based development work (Banerjee 2007), evidence-based conservation (Sutherland et al. 2004), and adaptive experimentation (Cook et al. 2004). Before any sustainability project is undertaken, assessments should be made of the current ecosystem services at the project site and at a set of similar control sites. After implementation, the effect on ecosystem services at the project site can be compared with the change in ecosystem services at control sites. In this way, sustainability science can move beyond a plethora of case studies of specific places and topics to a more solid scientific understanding of the measured benefits of sustainability projects.

This evidence-based philosophy will require ecologists and conservation biologists to change their focus from the elaboration of theories internal to their field to a more inherently collaborative venture. These sustainability projects will require a team of ecologists, economists, policy makers, and development advocates. While pleas for interdisciplinary collaboration are now commonplace in ecology, much further integration is needed. Only then will the scientific community begin to play its crucial role, describing how practical changes in urban forms can make future urbanization compatible with the dream of sustainable development. Sustainability is currently a policy goal and a platitude—it needs to be transformed into an evidence-based and empirically grounded science, with close links to ecology and the study of ecosystem services.

References

Alberti M. 1999, Urban patterns and environmental performance: what do we know? *J Plan Educ Res* **19**: 151–63.

Alberti M. 2005. The effects of urban patterns on ecosystem function, *Int Regional Sci Rev* **28**: 168–92.

Anderson WP, Kanaroglou PS, and Miller EJ. 1996. Urban form, energy, and the environment: a review of issues, evidence, and policy. *Urban Stud* **33**: 7–35.

Angel S, Sheppard S, Civco D, *et al.* 2005. The dynamics of global urban expansion. Washington, DC: Transport and Urban Development Department, World Bank.

Awerbuch S and Sauter R. 2006. Exploiting the oil-GDP effect to support renewables deployment. *Energ Policy* **34**: 2805–19.

Badoe DA and Miller EJ. 2000. Transportation-land-use interaction: empirical findings in North America, and their implications for modeling. *Transport Res D-Tr E* **5**: 235–63.

Banerjee A. 2007. Making aid work. Cambridge, MA: MIT Press.

Banister D, Watson S, and Wood C. 1997. Sustainable cities: transport, energy, and urban form. *Environ Plana B* **24**: 125—43.

Bentley RW. 2002. Global oil and gas depletion: an overview. *Energ policy* **30**: 189–205.

Bento AM, Cropper ML, Mobarak AM, and Vinha K. 2005. The effects of urban spatial structure on travel demand in the United States. *Rev Econ Stat* **87**. 466–78.

Cameron I, Lyons T, and Kenworthy J. 2004. Trends in vehicle kilometers of travel in world cities, 1960–1990: underlying drivers and policy responses. *Transport Policy* **11**: 287–98.

Chan K, Shaw M, Cameron D, *et al.* 2006. Conservation planning for ecosystem services. *PLOS Biol* **4**: 2138–52.

Cleveland CJ and Kaufmann RK. 2003. Oil supply and oil politics: deja vu all over again. *Energ Policy* **31**: 485–89.

Cohen B. 2004. Urban growth in developing countries: a review of current trends and a caution regarding existing forecasts. *World Dev* **32**: 23–51.

Cook WM, Casagrande D, Hope D, *et al.* 2004. Learning to roll with the punches: adaptive experimentation in human-dominated systems. *Front Ecol Environ* **2**: 467–74.

deHaan A. 1997. Rural-urban migration and poverty: the case of India. *IDS Bull-l Dev Stud* **28**: 35–47.

Dhakal S. 2003. Implications of transportation policies on energy and environment in Kathmandu Valley, Nepal, *Energ Policy* **31;** 1493–1507.

EIA (Energy Information Administration). 2006. Emissions of green house gases in the United States 2005. Washington, DC: Energy Information Administration, US Department of Energy.

Folke C, Jansson A, Larsson J, and Costanza R. 1997. Ecosystem appropriation by cities. Ambio **26**: 167–72.

Handy S. 2005. Smart growth and the transportation-land use connection; what does the research tell us? *Int Regional Sci Rev* **28**: 146–67.

Holden E and Norland IT. 2005. Three challenges for the compact city as a sustainable urban form: household consumption of energy and transport in eight residential areas in the greater Oslo region. *Urban Stud* **42**: 2145–66.

IEA (International Energy Agency). 2006. World energy outlook 2006. Paris, France; International Energy Agency.

Jackson K. 1985. Crabgrass frontier. New York, NY: Oxford University Press.

Kenworthy J and Laube F. 1999. Patterns of automobile dependence in cities: an international overview of key physical and economic dimensions with some implications for urban policy. *Transport Res A-Pol* **33**: 691–723.

Landell-Mills N and Porras I. 2002. Silver bullet or fool's gold? A global review of markets for forest environmental services and their impact on the poor. London, UK: International Institute for Environment and Development.

Ma L. 2002. Urban transformation in China, 1949–2000: a review and research agenda. *Environ Plana A* **34;** 1545–69.

McDonald RI, Kareiva P, and Forman R. The implications of urban growth for global protected areas and biodiversity conservation. *Nature.* In review.

McGranahan G, Marcotullio P, Bai X, *et al.* 2006. Urban systems. In: Hassan R, Scholes R, and Ash N (Eds). Ecosystems and human well-being: current state and trends. Washington, DC: Island Press.

McGranahan G and Satterthwaite D. 2003. Urban centers; an assessment of sustainability. *Annu Rev Environ Resour* **28**: 243–74.

MA (Millenium Ecosystem Assessment). 2005. Ecosystems and human well-being: wetlands and water synthesis. Washington, DC: World Resources Institute.

Mindali O, Raveh A, and Salomon I. 2004. Urban density and energy consumption: a new look at old statistics. *Transport Res A-Pol* **38**: 143–62.

Muller EK. 2001. Industrial suburbs and the growth of metropolitan Pittsburgh, 1870–1920. *J Hist Geogr* **27;** 58–73.

Munroe DK, Croissant C, and York AM. 2005. Land-use policy and landscape fragmentation in an urbanizing region: assessing the impact of zoning. *Appl Geogr* **25**: 121–41.

NRC (National Research Council). 2003. Cities transformed: demographic change and its implication in the developing world. Washington, DC: National Academies Press.

Pataki DE, Alig RJ, Fung AS, *et al.* 2006. Urban ecosystems and the North American carbon cycle. *Global Change Biol* **12**: 2092–2102.

Peñalosa E. 2006. Plenary speech at the World Urban Forum. In: Report of the third session of the World Urban Forum; 2006 June 22; Vancouver, Canada. New York, NY: UN-Habitat.

Pickett STA, Cadenasso ML, Grove JM, *et al.* 2001. Urban ecological systems: linking terrestrial ecological, physical, and socioeconomic components of metropolitan areas. *Annu Rev Ecol Syst* **32;** 127–57.

Pouyat RV, McDonnell MJ, and Pickett STA. 1995. Soil characteristics of oak stands along an urban-rural land-use gradient. *J Environ Qual* **24**: 516–26.

Rees WE. 1999. The built environment and the ecosphere: a global perspective. *Build Res Inf* **27;** 206–20.

Righter R. 2005. The battle over Hetch Hetchy; America's most controversial dam and the birth of modern environmentalism.

New York, NY: Oxford University Press.

Riley K. 2002. Motor vehicles in China: the impact of demographic and economic changes. *Popui Environ* **23**: 479–94.

Rothman H. 2004. The new urban park: Golden Gate National Recreation Area and civic environmentalism. Lawrence, KS: University Press of Kansas.

Roy KC, Tisdell C, and Alauddin M. 1992. Rural-urban migration and poverty in south Asia, *J Contemp A* **22**: 57–72.

Salzman J. 2005, Creating markets for ecosystem services: notes from the field. *New York U Law Rev* **80**: 870–961.

Skeer J and Wang YJ. 2007. China on the move: oil price explosion? *Energ Policy* **35**: 678–91.

Steemers K. 2003. Energy and the city: density, buildings, and transport. *Energ Buildings* **35**: 3–14.

Sutherland WJ, Pullin AS, Dolman PM, and Knight TM. 2004. The need for evidence-based conservation. Trends *Ecol Evol* **19**: 305–08.

UN-HABITAT (UN Human Settlements Programme). 2006. State of the world's cities. New York, NY: UN Human Settlements Programme.

UNECE (United Nations Economic Commission for Europe). 2005. UNECE Trends 2005 Thematic Database. www.unece.org/stats/trends2005/Welcome.html. Viewed 15 Feb 2007.

UNPD (UN Population Division). 2005a. Population challenges and development goals. New York, NY: UN Population Division.

UNPD (UN Population Division). 2005b. World urbanization prospects: the 2005 revision. New York, NY: UN Population Division.

US Census Bureau, Population Division. US Census Bureau International Database. www.census.gov/ipc/www/idb/index.html. Viewed 15 Feb 2007.

Van der Zee SC, Hoek G, Harssema H, and Brunekreef B. 1998. Characterization of particulate air pollution in urban and non-urban areas in the Netherlands. *Atmos Environ* **32**: 3717–29.

Critical Thinking

1. This article says that urban growth in both the developed and developing world will be the main obstacle to achieving sustainable development. Why?

2. What does the author mean by "urban form matters"? Try to explain this in terms of "consuming the land."

3. According to the author, briefly describe the connections between urban form and ecosystem service generation and consumption.

4. How can looking at sustainability projects as "experiments" help achieve the goals of sustainable urban development?

Acknowledgements—RIM was supported by a David H Smith Conservation Fellowship. P Kareiva and RTT Forman provided helpful comments and encouragement on a draft of this manuscript. The Graduate School of Design at Harvard University, and C Steinitz and N Kirkwood in particular, provided space to work.

Development at the Urban Fringe and Beyond: Impacts on Agriculture and Rural Land

RALPH E. HEIMLICH AND WILLIAM D. ANDERSON

I. Overview

In the early 1970's, bipartisan legislation was introduced in Congress to establish a national land-use policy, but failed after extensive debate. In the decades that followed, urban area in the United States has more than doubled. Public concerns about ill-controlled growth once again have raised the issue of the Federal role in land-use policy.

Purpose of This Report

Although land-use issues have traditionally been the prerogative of State and local government, policymakers at the Federal level are increasingly urged to respond to concerns about development and growth, particularly with regard to their impacts on agriculture and rural land uses. While anecdotes are legion, and much has been written by commentators, advocates, and experts, there are surprisingly few places to find a comprehensive picture of land-use changes in urbanizing areas, relative to the rural landscape. This report responds to that need in two ways.

This overview provides a summary of our findings about the forces driving development, its character and impacts on agriculture and rural communities, the means available to channel and control growth, and the pros and cons of potential Federal roles.

The following chapters provide the details, presented in a documented, objective way that make the case for the arguments presented here. A consensus culled from the literature supports some of the points, while original analyses presented in this report have not been published elsewhere.

What Is Sprawl?

This report is about urban development at the edges of cities and in rural areas, sometimes called "urban sprawl." With no widely accepted definition of sprawl (U.S. GAO, 1999; Staley, 1999), attempts to define it range from the expansive to the prescriptive.

Most definitions have some common elements, including:

- Low-density development that is dispersed and uses a lot of land;
- Geographic separation of essential places such as work, homes, schools, and shopping; and
- Almost complete dependence on automobiles for travel.

Without an agreed definition, any growth in suburban areas may be accused of "sprawling."

Short of a return to a form of urban living not seen since before World War II, it is not clear how growth can be accommodated at suburban densities without incurring the worst features of "sprawl." Because "sprawl" is not easily defined, this report is couched in the more neutral terms "development" or "growth," without making implicit judgments about the quality or outcomes of that development or growth. See *Trends In Land Use: Two Kinds of Growth* p. 9.

How to Think about Development

Concerns about development around urban areas are not new, but have arisen periodically during most of the last century, and certainly since automobile ownership became widespread after World War II. Amid the environmental concerns during the 1970's, bipartisan legislation was introduced in Congress to establish a national land-use policy. Recognizing the primacy of State authority over land use, the legislation sought to provide Federal grants to States to strengthen their ability to plan for development and channel growth. After 5 years of debate, the legislation was passed in the Senate, but narrowly defeated in the House on June 11, 1974. What lessons have been learned about urban development and the Federal role in managing it in the 26 years since then?

There are two kinds of growth, but both affect the amount and productivity of agricultural land and create other problems—Our existing urban areas continue to grow into the countryside, and more isolated large-lot housing development is occurring, generally beyond the urban fringe.

At the urban fringe—The urban "fringe" is that part of metropolitan counties that is not settled densely enough to be called "urban." Low-density development (2 or fewer houses per acre) of new houses, roads, and commercial buildings causes urban areas to grow farther out into the countryside, and increases the density of settlement in formerly rural areas. The extent of urbanized areas and urban places, as defined by the Bureau of Census, more than doubled over the last 40 years from 25.5 million acres in 1960 to 55.9 million acres in 1990, and most likely reached about 65 million acres by 2000.

Beyond the urban fringe—Another kind of development often occurs farther out in the rural countryside, beyond the edge of existing urban areas and often in adjacent nonmetropolitan counties. Development of scattered single-family houses removes land from agricultural production and changes the nature of open space, but is not "urban." Large lots dominate this process, and growth in large-lot development has accelerated with business cycles since 1970. Nearly 80 percent of the acreage used for new housing construction in 1994–97—about 2 million acres—is outside urban areas. Almost all of this land (94 percent) is in lots of 1 acre or larger, with 57 percent on lots 10 acres or larger. About 16 percent was located in existing urban areas and 5 percent was on farms. See *Two Kinds of Growth*, p. 12.

Growth in developed areas is increasing, but at rates only slightly higher than in the past—Urbanized areas and urban places increased at about the same 1 million acres per year between 1960 and 1990. Developed land, including residential and other development that is not dense enough to meet urban definitions, increased from 78.4 million acres in 1982 to 92.4 million acres in 1992, and was estimated to be about 107 million acres in 2000. The rate of increase in developed land grew from 1.4 million acres per year to about 1.8 million acres. See *Two Kinds of Growth*, p. 12.

The processes of land-use change are well understood and flow predictably from population growth, household formation, and economic development—Changes in land use are the end result of many forces that drive millions of separate choices made by home-owners, farmers, businesses, and government. The ultimate drivers are population growth and household formation. Economic growth increases income and wealth, and preferences for housing and lifestyles, enabled by new transportation and communications technologies, spur new housing development and new land-use patterns. Metropolitan areas grow organically, following well-known stages of growth.

Almost alone among developed nations, the United States continues to add population from high fertility rates, high immigration, and longer life expectancy, increasing 1 percent per year, or another 150 million people by 2050. Average household size has dropped to 2.6 persons, creating about 1 million new households, the unit of demand for new housing, each year in the 1990's.

Increased income and wealth increased the number of new houses constructed each year by 1.5 million units, faster than the rate of household formation. Two-thirds of these houses are single-family dwellings. While average lot sizes have been dropping near cities as owners turn to townhouses and condominiums, a parallel growth in large-lot (greater than 1 acre) housing has occurred beyond the urban fringe.

Metropolitan expansion since 1950 has occurred because rural people moved off the farms, and residents of the densely populated central cities dispersed to surrounding suburbs. Urbanized areas (excluding towns of 2,500 or more) increased from 106 to 369 and expanded to five times their size. Population density in urbanized areas dropped by more than 50 percent, from 8.4 to 4 people per acre, over the last 50 years. Growth is spilling out of metropolitan areas, as population disperses to rural parts of metropolitan counties and previously rural nonmetropolitan counties.

Enabling this dispersion are investments in new infrastructure such as roads, sewers, and water supplies. New information and communication technologies, such as the Internet and cellular telephone networks, facilitate population in rural areas, and free employment to follow. New retail, office, warehouse, and other commercial development follows in the wake of new housing development, to serve the new population and to employ the relocated labor force. See *Driving Forces*, p. 15.

There are benefits of low-density development that attract people—Living beyond the edge of the city is a lifestyle much sought after by the American people. While 55 percent of Americans living in medium to large cities preferred that location, 45 percent wanted to live in a rural or small town setting 30 or more miles from the city (Brown et al., 1997). Of those living in rural or small towns more than 30 miles from large cities, 35 percent wanted to live closer to the city. The urban fringe is thus under development pressure from both directions. The most obvious benefit is that growth in rural areas has allowed many people, including those who cannot afford city real estate, to buy single-family homes because land costs are cheaper on the fringe than in the core.

The automobile imposes private and social costs in exchange for the comfort, flexibility, low door-to-door travel time, freight-carrying capacity (for shopping trips), cheap long-distance travel, and aesthetic benefits of extensive, automobile-dependent development. Air quality improvements may also result from decentralizing population and employment, because emissions are dispersed over larger rural airsheds and are reduced by higher speeds. Automobile pollution is more strongly related to the number of trips than to the length of each trip, with a major part of auto pollution deriving from cold starts.

Not everyone wants to live the rural lifestyle. The "new urbanism" school of urban design is redesigning conventional suburban developments as small towns and finding a market (Chen, 2000; Duany et al., 2000). In 1992, 55 percent of those surveyed living in large cities (over 50,000) preferred that type of community (Brown et al., 1997). See *Demand for Low-Density Development*, p. 17.

Development imposes direct costs on the communities experiencing it, as well as indirect costs in terms of the rural lands sacrificed to it—A number of studies show that less dense, unplanned development requires higher private and public capital and operating costs than more compact, denser planned development. Eighty-five studies gauging the cost of community services around the country have shown that residential development requires $1.24 in expenditures for public services for every dollar it generates in tax revenues, on average. By contrast, farmland or open space generates only 38 cents in costs for each dollar in taxes paid. See *Impacts on Taxpayers,* p. 28.

Finally, development can disrupt existing social, community, environmental and ecological patterns, imposing a variety of costs on people, wildlife, water, air, and soil quality. Agricultural production has its own negative environmental impacts, but these are generally less severe than those from urban development. See *Impacts on Landscape, Open Space, and Sense of Community,* p. 31.

However, does moving out into the "country" ultimately destroy all the good things that prompt that move? In the words of the National Governor's Association, "In the context of traditional growth patterns, the desire to live the 'American Dream' and purchase a single-family home on a large lot in a formerly open space can produce a negative outcome for society as a whole" (Hirschorn, p. 55).

Continued demand for low-density development despite negative consequences for residents can be understood as a market failure—Consumers, businesses, and communities fail to anticipate the results of development because they often lack information on potential or approved development proposals for surrounding land. When communities fail to plan and zone, there is no institutional framework within which development can proceed, and little information to help housing buyers anticipate their future landscape setting.

Spillovers from development include the loss of rural amenities, open space, and environmental goods when previously existing farms and rural land uses are developed. Negative spillovers from increased housing consumption in developing areas can include traffic congestion, crowding, and destruction of visual amenities. If the landscape features that contribute to rural amenity were marketed in developments, housing prices would be higher.

Real estate markets are based on many small decisions which, when taken without an overall context, produce results that can neither be envisioned by nor anticipated by consumers and developers. Cumulative impacts from this myriad of decisions can be large, but are not reflected in market prices until disamenities become large. Inaccurate judgments about future landscapes are locked in because development is irreversible. See *An Economic Interpretation of the Demand for Low-Density Development,* p. 36.

Urban growth and development is not a threat to national food and fiber production, but may reduce production of some high-value or specialty crops—Despite doubling since 1960, urban area still made up less than 3 percent of U.S. land area in 1990 (excluding Alaska). Developed area, including rural roads and transportation, made up less than 5 percent in 1992. Development affects local agricultural economies and can cause other environmental and resource problems in local areas, but the increase in urban area in the United States poses no threat to U.S. food and fiber production. Some crops in some areas are particularly vulnerable to development. For example,

61 percent of U.S. vegetable production is located in metropolitan areas, but vegetable production takes up less than 1 percent of U.S. cropland. See *Consequences for Farming,* p. 38.

Agriculture can adapt to development, but does so by changing the products and services offered—Low-density, fragmented settlement patterns leave room for agriculture to continue. Farms in metropolitan areas are an increasingly important segment of U.S. agriculture. They make up 33 percent of all farms, 16 percent of cropland, and produce a third of the value of U.S. agricultural output. However, to adapt to rising land values and increasing contact with new residents, farmers may have to change their operations to emphasize higher value products, more intensive production, enterprises that fit better in an urbanizing environment, and a more urban marketing orientation.

Development can be profitable for farmers who can see and take advantage of opportunities in the new situation. Forces of urbanization allow a variety of farm types to coexist. Farms in metropolitan areas are generally smaller, but produce more per acre, have more diverse enterprises, and are more focused on high-value production than nonmetropolitan farms. Metropolitan agriculture is characterized by recreational farmers who follow both farm and non-farm pursuits; a smaller group of adaptive farmers who have accommodated their farm operation to the urban environment; and a residual group of traditional farms that are trying to survive in the face of urbanization. Both of the latter types are generally working farms. See *Consequences for Farming,* p. 38.

Benefits of conserving rural land are difficult to estimate, and vary widely depending on the circumstances—Because there are no markets for some characteristics of land, such as scenic amenity, there are no observable prices apart from the land's value for development. Lacking prices, it is difficult to develop economic benefit measures for policymaking.

Rural lands in a working landscape provide economic benefits as resources for agricultural production, as sources of employment, and through property and income taxes. Working landscapes are defined as farm, ranch, and forest lands actively used in agricultural or forestry production. While agricultural production can create environmental problems of its own, properly managed farmlands provide nonmarket benefits from improving water and air quality, protecting natural bio-diversity, and preserving wetlands relative to development. They create aesthetically pleasing landscapes and can provide social and recreational opportunities. The rural landscape reflects and conserves rural culture and traditions, and maintains traditions of civic leadership and responsibility in voluntary rural institutions, such as fire companies and village boards. See *Impacts on Landscape, Open Space, and Sense of Community,* p. 31.

Based on information and assumptions about the number of acres likely subject to development in the future, and on limited studies of residents' willingness to pay to conserve farmland and open space, we estimate that households would be willing to pay $1.4–$26.6 billion per year to conserve rural lands. In addition, another $0.7–$1.1 billion in sediment and water quality damages would be avoided if the land were prevented from being developed. Conserving land for agriculture helps preserve farming as a part of the rural economy, and is often seen as a bulwark against the worst effects of development. See *Benefits of Farmland and Open Space*, p. 44.

Local governments generally do not develop adequate capacity to plan for and manage growth until it is too late to effectively channel development—Because urban growth processes are well understood, strategically directing development to the most favorable areas well in advance of urban pressures offers the greatest hope for controlling growth. Planning and zoning have generally been upheld by the courts as valid out. If planning is not in place as development begins to occur, property owners' expectations about higher land values can exacerbate property rights conflicts and complicate subsequent growth-control efforts. Local governments often fail to appreciate impending growth facing them, and generally lack capacity to develop adequate responses before growth overwhelms them.

Better planning and zoning is central to the ability to respond to growth. A U.S. General Accounting Office survey found that 75 percent of the communities that were concerned with "sprawl" were highly involved in planning for and managing growth (U.S. GAO, 2000, p. 99).

However many cities and counties may be falling short of what is needed to control and manage growth effectively. A recent survey of Alabama's mayors and county commissioners found that only a minority of the responding officials (18 percent of the mayors and 19 percent of the commissioners) believed they currently had the necessary staff and resources to plan and manage growth effectively. High-growth communities were only somewhat more likely to have the capacity to manage growth than were other communities.

Most of the smaller rural towns do not have a full-time planner. To meet their planning needs, these communities may be served by a circuit riding planner, or several towns and a county may combine their efforts to set up one planning office to serve their joint needs. Even at the county level, rural planners often must spend part of their time doing other duties. See *Local Responses to Growth*, p. 50.

State governments can do more to deal with growth strategically—Our Constitution reserves control of land use to the States, which usually have delegated the responsibility to local governments. Increasingly, States are realizing that local governments cannot adequately address growth pressures that transcend local boundaries. Some States have adopted "smart growth" strategies that actively direct transportation, infrastructure, and other resources to channel growth into appropriate areas.

The term "smart growth" is a catch-all phrase used to describe a group of land-use planning techniques that influence the pattern and density of new development. In general, smart growth strategies represent a movement away from State-imposed requirements for local compliance with State planning goals. Because smart growth strategies tend to use financial incentives to encourage voluntary adoption, they are generally supported by a broad spectrum of interest groups. These strategies also garner support because they direct, rather than inhibit, growth and development. There's no 'one size fits all': the specific smart-growth strategies that have been adopted

vary by location but often share common elements. Smart-growth principles favor investing resources in center cities and older suburbs, supporting mass transit and pedestrian-friendly development, and encouraging mixed-use development while conserving open space, rural amenities, and environmentally sensitive resources (Hirschhorn 2000). These strategies also typically remove financial incentives provided by State funding to develop outside designated growth areas. In essence, smart growth encourages development in designated areas without prohibiting development outside them. See *Slow Growth, No Growth, and Smart Growth*, p. 55.

Existing monetary incentives for conserving rural land are not as effective as they could be—Use-value assessment, enacted in every State, is one of the most widespread public policies aimed at conserving rural land. Under use-value assessment, the owner is taxed based on what the land could earn in agriculture, rather than the higher developed value. We estimated the cost of tax reductions under use-value assessment nationally at $1.1 billion per year.

However, most students of use-value assessment acknowledge that it is not effective at preventing development. Use-value assessment spreads resources over all qualifying rural land, providing a small incentive to conserve land to all landowners. The size of the tax reduction is insufficient to keep land with the highest development potential from conversion, while tax expenditures to less developable land produce little result. Redirecting tax expenditures on use-value assessment could increase the resources available for incentives to conserve the most developable land, but could make some land currently getting the tax subsidy more vulnerable to urbanization and would face stiff opposition from property owners currently enjoying the tax reduction. See *Monetary Incentives for Conserving Farm and Forest Land*, p. 57.

The cost of effective incentives would be large, but if resources were redirected, almost one-third of the cropland with the greatest development potential could be protected—Purchasing the development rights to rural land effectively protects it from being developed. The landowner retains ownership and can continue to farm the land, but the deed restriction continues indefinitely. The implicit economic value of the easement is the difference between the unrestricted or market value of the parcel and its restricted or agricultural value.

Nineteen States have State-level PDR (purchase of development rights) programs using public funds to compensate landowners for the easements on otherwise private farm or forest land. In addition, at least 34 county programs in 11 States operate separate programs. The American Farmland Trust estimates that, nationwide, PDR programs have cumulatively protected 819,490 acres of farmland with an expenditure of $1.2 billion.

We estimate the cost to purchase development rights on cropland most likely subject to urban pressure over the next 30 to 50 years at $88–$130 billion. If tax expenditures currently devoted to use-value assessment were redirected to purchase of development rights, almost one-third of the cropland with greatest potential for development could be protected.

Targeting funds to land under less development pressure could protect the same amount of land at lower cost. For example, development rights on the 25 million acres under medium urban pressure are estimated to cost $25 billion, less than one-third the cost of the 33 million acres under heaviest development pressure. Selecting land with lowest current development pressure would reduce costs to $18 billion.

Even if funds were available to purchase development rights, it may not be desirable to do so. The development pressure exerted on this land will not disappear if this cropland is protected. While some growth might be accommodated in existing urban areas, demand for other rural land would intensify, and growth could fragment even more as development moves out farther into the rural countryside. Purchasing development rights is also no guarantee that the land will be used for working agricultural enterprises. The perpetual deed restrictions could prevent future desirable adjustments in land-use patterns. See *Monetary Incentives for Conserving Farm and Forest Land*, p. 57.

There are neither clear requirements for nor restrictions on Federal roles in managing growth—Historically, authority over land-use decisions has been reserved to the States, who have delegated these powers to local governments. However, the evolution of environmental policy shows an expanding Federal involvement in site-specific, local circumstances that recur across the Nation. The Federal Government has no constitutional mandate to take action on urban growth and development issues, but it can define an appropriate role for itself. See *Potential Federal Roles*, p. 65.

Federal activity in the potential roles identified below is described and pros and cons of expanding each role are enumerated.

Potential Federal Roles

Helping Increase State and Local Planning Capacity— The Federal Government has had a long history of programs to improve the planning capabilities of State and local governments. Perhaps the most notable of these efforts was the HUD 701 planning grant program, established in 1954 (40 USC 461). As late as 1975, the HUD 701 program spent $100 million per year paying as much as two-thirds of the costs of an "ongoing comprehensive planning process" required of all grant recipients. However, the budget was cut to $75 million in 1976 and was gradually phased down until eliminated in the early 1980's.

Within the U.S. Department of Agriculture, the Rural Development Act of 1972 established the Section A-111 Rural Development Planning Grants, also funded into the 1980's. In 1996, the farm bill established new authority for the Rural Business Opportunity Grant program (RBOG), which received $3.5 million in FY2000 appropriations. RBOG provides money to nonprofits, public bodies, Indian tribes, and cooperatives for planning and technical assistance to assist economic development in rural areas. FY 2001 appropriations legislation increased the funding for RBOG to $8 million. Several other smaller USDA grant programs could potentially assist local communities with planning, but they are not specifically directed at planning to guide growth and development and are not integrated into a coordinated program.

Pros—Funding requirements for such programs would be relatively small, and could potentially leverage significant impacts. Impacts from limited funding for such programs could be increased by targeting them to the areas most likely affected by growth in the medium term. Limiting program activities to those most directly relevant to guiding new growth and development would also increase the impact of the program.

Cons—Failures in past programs were attributed to wide use of consultants who provided little service for the money spent, and who did little to add permanently to local government planning capacity. Emphasis on "paper plans" did little to actually direct growth. Targeting funds to areas immediately affected by development wasted resources on efforts that were already too late, while spreading funding widely included areas with little development pressure in reasonable time-frames.

Coordinating Local, Regional, and State Efforts—Urban growth processes often create multi-jurisdictional impacts. Federal coordination and integration have been exercised in other areas of environmental concern, such as water quality, water quantity, and air quality. In addition, the U.S. Office of Management and Budget Circular A-95 review process formerly guided Federal agencies for cooperation with State and local governments in the evaluation, review, and coordination of Federal assistance programs and projects. A-95 review is no longer mandated by the Federal Government, although the process is still voluntarily practiced by some States. USDA has had a long history of area-wide coordination, dating back to efforts like the Great Plains Agricultural Council, the Resource Conservation and Development Council (RC&D), the Small Watershed Program (PL-566), and various river basin planning processes. While these have generally been focused on agricultural, resource, or rural development concerns, their extension to urban development and growth control issues would be reasonable.

Pros—Past Federal funding for transportation, water, and sewer construction and other major infrastructure projects has been identified as a major driver in growth and development. Explicitly monitoring and reviewing potential impacts on urbanization from such investments could, at a minimum, defuse these accusations. Federal funding could serve as a rationale for efforts to coordinate State and local growth control activities, especially where these cross jurisdictional boundaries. Such efforts would cost very little, but would leverage existing expenditures.

Cons—Without convincing resolution to reduce or deny funding to State and local governments that do not cooperate, attempts at coordination could prove futile and frustrating. Congressional attempts to obtain additional funding for local constituents can be at odds with Executive branch notions of coordination and integration.

Coordinating Federal Development Activities and Growth Management Goals—Lines between areas needing development assistance and those suffering from problems of growth and development are geographic ones, and are often exceedingly fine, and shift over time. The Federal Government has had a long history of programs to foster development, and less experience at helping control it. The superficial dichotomy disappears when considered in the context of directing growth and development to appropriate places and under an appropriate timetable, which serves both sets of interests.

Pros—A wide array of rural development and economic development activities in the Departments of Agriculture and Commerce, abetted by less direct activities in the Departments of Housing and Urban Development, Transportation, and Defense, date at least to the War on Poverty and related efforts of the 1960's. The existing institutional structure of these programs could be redirected to growth control and management, but would require new visions by leadership. Some existing resources could be leveraged.

Cons—These programs have become entrenched and rather balkanized and may be difficult to integrate into an effort of sufficient weight to effectively deal with the problem. While pro- and anti-growth interests would hopefully recognize common ground in wellplanned and appropriate development, extremes on both sides may be difficult to persuade, and both sides may be suspicious of Federal help.

Funding Monetary Conservation Incentives—The Federal Government has often been enlisted as an ally with deep pockets, and analogous programs for soil and water conservation, wildlife habitat acquisition, and other land resource issues have existed since the 1930's. USDA's Farmland Protection Program was authorized in the 1996 Farm Act for up to $35 million in matching funds for State programs over 6 years. The initial funding was $33.5 million and it was spent to protect 127,000 acres in over 19 States. The goal of the program is to protect between 170,000 and 340,000 acres of farmland. An additional $10 million was appropriated in FY2000. Direct Federal acquisition of easements is included in USDA's Conservation Reserve Program and Wetland Reserve Program, as well as in several of the U.S. Fish and Wildlife Service's habitat programs.

Pros—Limited Federal funding for farmland protection easements could act as seed money for programs in States with no current program, or as a bonus for States doing a particularly effective job. Utilizing existing State programs may be cost-effective because it both avoids creating a new bureaucracy within the Federal Government and provides an incentive to States that have not yet developed a program to do so. By carefully specifying rules for matching State funding, such a program could avoid discouraging State effort, and could maximize the incentive for new programs.

Cons—As outlined above, the amount of land and resources subject to development is large and State programs are relatively small, posing questions about the effectiveness of a small Federal program and larger questions about the ultimate size needed to make an impact. While the marginal benefits of a small program at this point are likely to be greater than the costs, the wisdom of a larger program becomes problematic. Questions about the displacement of growth and the longrun fate of protected land become more significant as the amount of land protected increases.

Conserving Rural Amenities as Part of Greater Agricultural and Trade Policy Goals—Conserving the amenities provided by rural land is no longer a matter of merely domestic concern. Proposals to direct agri-environmental assistance are widespread in the European Union and other Organization for Economic Cooperation and Development (OECD) countries. Such efforts meet the "green box" requirements for acceptable agricultural policies under agricultural trade reforms in the Uruguay Round of the General Agreement on Tariffs and Trade (GATT). Some proponents of greater Federal involvement in rural land conservation believe that a larger share of Federal funding for agriculture could be directed toward land conservation through agri-environmental payments designed to preserve more of the multiple functions of agriculture in an urbanizing context. While not required by trade agreements to date, such proposals are allowed by them and may garner support from constituents in urbanizing areas, the urban fringe, and among agricultural communities.

Pros—Frameworks for agri-environmental payments have already been proposed in the form of the Conservation Security Act of 2000 (S.3260/H.R. 5511), introduced by Senator Harkin and Congressman Minge, and in the Clinton Administration's proposal for a Conservation Security Program in October 2000. While not explicitly addressing farmland protection, eligible land in urbanizing areas could be included. This kind of program helps align U.S. agricultural support programs with legitimate purposes recognized in trade liberalization agreements.

Cons—The farmland conservation issues in Europe and the United States are fundamentally different. While European efforts are largely aimed at keeping economically marginal farmland from abandonment, U.S. concerns are with preventing otherwise viable farms from being developed. The latter is a far more expensive proposition. Channeling large amounts of assistance to farms in urbanizing areas risks losses if incentives are not sufficiently large to prevent development, and may be pyhrric if protected farms cannot viably continue in operation, despite protection. On balance, preventing the environmental problems from losing farms in urbanizing areas may not yield benefits as large as correcting environmental problems from farming in more rural areas.

Organization of the Remainder of the Report

The remainder of the report provides a more in-depth, documented discussion of this overview. . . .

II. Trends in Land Use: Two Kinds of Growth

In the early 1970's, bipartisan legislation was introduced in Congress to establish a national land-use policy. The proposals, recognizing the primacy of State authority over land use, would have provided Federal grants to States to better manage growth and development. The bills were debated for 5 years and passed by the Senate, but died on a narrow vote in the House on June 11, 1974.

In the decades that followed, urban area in the United States has more than doubled. Some of this growth has been at low densities, with little planning, and has fragmented the rural landscape, prompting communities, States, and the Federal Government to examine more closely unplanned development and its consequences, including the loss of productive farmland. Public concerns about the consequences of ill-controlled growth once again have raised the issue of the Federal role in land-use policy.

Anecdotes of uncontrolled growth across the Nation abound:

- From 1950 to 1990, St. Louis experienced a 355-percent growth in developed land even though population increased by just 35 percent (Missouri Coalition for the Environment).

- Between 1970 and 1990, Kansas City's population grew by 29 percent while developed land increased by 110 percent (Missouri Coalition for the Environment).

- Between 1990 and 1996, the Denver metropolitan region increased by 66 percent. If each county in the Denver metro area grew based on its current comprehensive plan, Denver's urbanized area would swell to 1,150 square miles, an area larger than California's major cities combined (Sierra Club, 1998).

- The Chicago metropolitan area now covers over 3,800 square miles. Over the last decade, the population of the area grew by only 4 percent, but land occupied by housing increased by 46 percent and commercial land uses by 74 percent (U.S. OTA, 1995).

- From 1950 to 1980, population in the Chesapeake Bay watershed increased by 50 percent, while land used for commercial and residential activity climbed 180 percent (EPA, 1993).

- Philadelphia's population increased 2.8 percent between 1970 and 1990, but its developed area increased by 32 percent (U.S. OTA, 1995).

While anecdotes are legion, and much has been written by commentators, advocates, and experts, there are surprisingly few places to find a comprehensive picture of land-use changes in urbanizing areas, relative to the rural landscape. This report responds to that need.

What Is Sprawl?

This report is about urban development at the edges of cities and in rural areas, often referred to as "urban sprawl." There is no widely accepted definition of sprawl (U.S. GAO, 1999; Staley, 1999). Definitions range from the expansive . . .

"When you cannot tell where the country ends and a community begins, that is sprawl. Small towns sprawl, suburbs sprawl, big cities sprawl, and metropolitan areas stretch into giant megalopolises—formless webs of urban development like Swiss cheeses with more holes than cheese."

—U.S. House, 1980

"Cities have become impossible to describe. Their centers are not as central as they used to be, their edges ambiguous, they have no beginnings and apparently no end. Neither words, numbers, nor pictures can adequately comprehend their complex forms and social structures. . . .It's almost as if Frank Lloyd Wright's 1932 tract against the metropolis, *The Disappearing City,* has been vindicated, and the diffusionary proposal of Broad-acre City has become the de facto ideology of urbanism."

—Ingersoll, 1992

to the prescriptive . . .

". . .a spreading, low-density, automobile dependent development pattern of housing, shopping centers, and business parks that wastes land needlessly."

—Pennsylvania 21st Century Environment Commission cited in Staley, 1999

Burchell et al. (1998) devote the first chapter of their report, "The Costs of Sprawl—Revisited," to defining the elusive term. Commonly cited are several features that are captured in urban economist John F. McDonald's characterization:

- Low-density development that is dispersed and uses a lot of land;

- Geographic separation of essential places such as work, homes, schools, and shopping; and

- Almost complete dependence on automobiles for travel.

Myers and Kitsuse (1997) point out that "the very lack of agreed definition about what constitutes density, sprawl or compactness prevents any authoritative measurement." Any growth in suburban areas may be accused of "sprawling." Planned developments at relatively high densities can be accused of accelerating sprawl. As Ewing (1997) points out,

> . . sprawl is a matter of degree. The line between scattered development, a type of sprawl, and multicentered development, a type of compact development by most people's reckoning, is a fine one.
>
> . . Equally elusive is the line between leapfrog development and economically efficient 'discontinuous development', or between commercial strips and 'activity corridors'.

Ewing also suggests that his notion of compact development—which is multicentered, has moderate average densities, and is

Metropolitan, Urban, and Rural Geography

Statistics describing trends in land use are based on one or another geographic entities defined by the U.S. Bureau of the Census (see U.S. Census, Geographic Areas Reference Manual), the USDA National Resources Inventory (NRI), or the American Housing Survey (AHS).

Census of Population . . .

Metropolitan/Nonmetropolitan Area—a core area containing a large population nucleus, together with adjacent communities that have a high degree of economic and social integration with that core. Metro areas are defined in terms of entire counties (except in New England, where towns are used). Metropolitan areas contain a mix of land uses, ranging from deserts, forests, and farms, to suburban landscapes, and include the densest urban core. In 1990, there were 274 metropolitan areas, containing 198.2 million people (80 percent of the total U.S. population) and covering 20 percent of U.S. land area.

Urban/Rural—Census defines urban as comprising all territory, population, and housing units located in urbanized areas (UAs), defined in terms of census tracts, and in places of 2,500 or more inhabitants outside of UAs. In 1990, 187 million people (75 percent of the total) lived in 8,510 places of 2,500 or more covering 2.5 percent of U.S. land area.

Urbanized Areas (UAs) are continuously built-up areas with a population of 50,000 or more, comprised of one or more places—central place(s)—and the adjacent densely settled surrounding area consisting of other places and territory not in defined places.

Urban Places Outside of UAs are any incorporated place or Census-designated place (CDP) with at least 2,500 inhabitants.

Rural Places and Territory not classified as urban are classified as rural. For instance, a rural place is any incorporated place or CDP with fewer than 2,500 inhabitants that is located outside of a UA. A place is either entirely urban or entirely rural.

Urban Fringe consists of rural areas in metropolitan counties. The part of the urban fringe nearest to existing UAs and urban places is likely to grow the fastest and eventually be absorbed when densities rise to urban levels.

Places—Census defines a place as a concentration of population, with a name and local recognition, that is not part of any other place. A place either is legally incorporated under the laws of its respective State or a statistical equivalent that the Census Bureau treats as a Census-designated place (CDP). Not everyone resides in a place; in 1990, approximately 66 million people (26 percent) in the United States lived outside of any place, either in small settlements, in the open countryside, or in the densely settled fringe of large cities in areas that were built-up, but not identifiable as places. Most Census places (19,289 out of a total of 23,435 in 1990) are incorporated.

National Resources Inventory (NRI)

Developed land in the National Resources Inventory consists of urban and built-up areas and land devoted to rural transportation.

Urban and built-up areas consist of residential, industrial, commercial, and institutional land; construction and public administrative sites; railroad yards, cemeteries, airports, golf courses, sanitary landfills, sewage plants, water control structures, small parks, and transportation facilities within urban areas.

Large urban and built-up areas include developed tracts of 10 acres and more.

Small built-up areas include developed tracts of 0.25 to 10 acres, which do not meet the definition of urban area, but are completely surrounded by urban and built-up land.

Rural transportation land includes highways, roads, railroads and rights-of-way outside of urban and built-up areas.

American Housing Survey (AHS)

The American Housing Survey, conducted every 2 years by the Bureau of the Census represents all housing units for the entire Nation, including housing lots on farms. The AHS started the current series in 1980.

Residential area is land devoted to residential housing lots, both urban and rural, based on respondents' estimates of lot size for their house. Sample-based responses are expanded to area totals.

Comparison

Due to differences in data collection techniques and definitions, the NRI estimates of "large urban and built-up areas" are usually higher than the Census "urban area" estimates for nearly all States. The Census urban area series runs from 1950, while the NRI started providing a consistent series in 1982. Prior to the 1982 NRI, Census urban area was the only reliable national source of urban area data available.

The American Housing Survey residential area is the sum of acres in lots used for housing units. While the data have limitations and are not available by State, the series does allow compilation of two important estimates. First, an estimate of the residential component of urban land shows how much land is used for housing in urban areas versus land used for all other urban purposes, such as commercial and industrial sites, institutional uses, urban parks, and all other non-housing urban uses. Second and more important, an estimate is made of land used for residences in rural areas. Recently there appears to be a growing trend toward an increasing demand for more and larger housing lots outside of urban areas. The AHS residential area does not include non-residential areas shown in the Census and NRI, but does include a large area of rural residential land not found in either the Census or the NRI.

continuous except for permanent open spaces or vacant lands to be developed in the near future—is not all that different from Gordon and Richardson's (1997) definition of sprawl.

Short of a return to a form of urban living not seen since before World War II, it is not clear how growth can be accommodated at suburban densities without being accused of being "sprawl."

Some people oppose any change in established land uses and react just as negatively to well-planned, reasonably dense and compact development as others do to "sprawl." Because "sprawl" is so hard to define, we use it only when citing others and set it off in quotation marks. We couch our discussion in the more neutral terms "development" or "growth," without making implicit judgments about the quality or outcomes of that development or growth.

Two Kinds of Growth

Government officials, housing consumers, farmers, and other interest groups appear to be concerned about two kinds of growth. First is the continuing accretion of urban development at the fringes of existing urban areas in rural parts of metropolitan counties. A second kind of growth is the proliferation of more isolated large-lot housing development (1 acre or more) well beyond the urban fringe and into adjacent nonmetropolitan counties. Growth at the edge of existing developed areas gradually shades out into more and more fragmented developments, farther out in the countryside, so there is no clear geographic dividing line between the two kinds of growth. While related, these two forms of growth have qualitatively different causes and have different consequences, especially for agriculture and the environment.

Trends at the Urban Fringe

Even low-density development (2 or fewer houses per acre) of new houses, roads, and commercial buildings at the fringe of existing urban areas can cause greater traffic congestion, loss of open space, loss of agricultural land, and impacts on the natural environment.

The amount of land in urban and developed land uses is measured in different ways, all of which have specific denotations (see box "Metropolitan, Urban, and Rural Geography" . . .). The concept of "urbanized area," defined by the Bureau of Census, includes the densely settled areas within and adjacent to cities with 50,000 people or more, while "urbanized places" include populations of 2,500 people or more that are outside of urbanized areas. Urbanized areas alone increased from 15.9 million acres in 1960 to 39 million acres in 1990, increasing 2.5 times. Total Census urban area (urbanized areas and urban places) more than doubled over the last 40 years from 25.5 million acres in 1960 to 55.9 million acres in 1990. These two categories of urbanization likely reached about 65 million acres by 2000 (. . . Daugherty, 1992).

"Urban and built-up areas" counted in USDA's National Resources Inventory (NRI) include those measured by the Census Bureau, as well as developed areas as small as 10 acres outside urban areas, encompassing some large-lot

development. NRI urban and built-up area increased from 51.9 million acres in 1982 to 76.5 million acres in 1997, and likely rose to about 79 million acres by 2000 (Table 1 . . .). "Developed land" defined by NRI adds the area in rural roads and other transportation developments. By this definition, developed area increased from 73.2 million acres in 1982 to 98.3 million acres in 1997, and likely reached 107 million acres by 2000.

Census-defined urban area has grown by about a million acres per year since 1960, an increase of about 4 percent per year. The rate of increase dropped from 3.5 percent per year in the 1960's and 1970's to 1.8 percent per year in the 1980's. NRI urban and built-up area increased faster than Census urban area in the 1980's, rising 2.9 percent. Much of the increase in NRI urban and built-up area is in less dense, extensive large-lot development beyond the urban fringe and in nonmetropolitan counties. This kind of development will not meet the population density criteria for Census-defined urban area for many years.

Despite doubling since 1960, urban areas still made up less than 3 percent of U.S. land area (excluding Alaska) in 1990. . . . Developed areas, including rural roads and transportation, made up less than 5 percent in 1992. Both kinds of growth (on the metro fringe and large-lot development) take land irreversibly out of commercial agricultural production that might otherwise be available for use. Growth causes social and environmental problems in local areas, but the increase in urban area in the United States poses no threat to U.S. food and fiber production capacity (Vesterby et al., 1994; USDA, 2000).

Trends Beyond the Urban Fringe

Another kind of development occurs beyond the existing urban fringe, often far out in the rural countryside of metropolitan counties or adjacent nonmetropolitan counties. Development of new housing on large parcels of land is growth with a different character than that occurring at the city's edge. Instead of relatively dense development of 4–6 houses per acre, exurban development consists of scattered single houses on large parcels (often 10 acres or more). Rural large-lot development is not a new phenomenon, although it may be getting more attention than in the past. Growth in the area used for housing rose steadily throughout the last century (. . . Peterson and Branagan, 2000).

Large-lot categories dominate this process, and growth in large-lot development has accelerated with periods of prosperity and recession since 1970. The largest lot size category (10–22 acres) accounted for 55 percent of the growth in housing area since 1994, and lots greater than 1 acre accounted for over 90 percent of land for new housing. About 5 percent of the acreage used by houses built between 1994 and 1997 is for existing farms, and 16 percent is in existing urban areas within Metropolitan Statistical Areas (MSAs) defined by the Bureau of the Census. Thus, nearly 80 percent of the acreage used for recently constructed housing—about 2 million acres—is land outside urban areas or in non-metropolitan areas. Almost all of this land (94 percent) is in lots of 1 acre or larger, with 57 percent on lots of 10 acres or larger.

The people who move into these new houses may be pioneers moving from cities that once seemed distant. They may be pioneers in another sense: Areas experiencing this kind of development may be just starting on a gradual process of infill and expansion that will gradually transform the once-rural countryside into suburban and urban settlements resembling the existing urban fringe.

Critical Thinking

1. What is "urban sprawl" and how does it "consume" land?

2. Why does sprawl happen and who is responsible for it?

3. What do some people see as the benefits of low-density housing developments?

4. What are some of the costs of rural development? Who should be responsible for those costs?

5. Is continued worth at the urban fringe and into rural land a sustainable form of development? Can it be? Explain/discuss.

From *Economic Research Service, U.S. Department of Agriculture. Agricultural Economic Report No. 803*, 2001. Copyright © 2001 by United States Department of Agriculture. Reprinted by permission.

The End of a Myth

We thought the sea was infinite and inexhaustible. It is not. Exploring a new vision for Earth's greatest wilderness.

JULIA WHITTY

At dawn on a remote beach along Mexico's Baja California Peninsula, in a world of pink and red desert and forests of *cardón* cactus, I come upon an unlikely mass stranding of Portuguese man-of-wars, or bluebottles. They've come ashore in the night and their blue sails are as bright and shiny as living tissue, as if the beach were strewn with thousands of excised lungs. Their tentacles lie hopelessly tangled around them, reminiscent of dissected blood vessels, giving the otherwise peaceful morning the feel of the abattoir, as if many animals have been butchered and their larger parts consumed.

Vision is the art of seeing things invisible, wrote Jonathan Swift. And sometimes the invisible is a huge, dominant, virtually omnipotent presence in our world, the feature for which our planet should have been named, and may well be named by distant intelligent beings with the means to peer farther than we can at present. The ocean is our blind spot: a deep, dark, distant, and complex realm covering 70.8 percent of Earth's surface. We have better maps of the surface of Mars than of our own sea floor. Yet under our skin, we're a plasma ocean, so entwined with the outer seas that we can't easily know either ourselves or our water world.

This ocean is the largest wilderness on Earth, home to wildlife in staggering multispecies aggregations, and with a lineage of life three billion years older than anything above sea level. Its three-dimensional realm comprises 99 percent of all habitable space and is so embedded with life as to be largely composed of life, with an ounce of seawater home to as many as 30 billion microorganisms—and counting. Honing our technological eyesight, we begin to observe what was once too small to be seen, in an exercise that mirrors infinity.

For most of our time on Earth, most of what we've known of the ocean has been its dead oddities on the beach. The Portuguese man-of-wars must once have seemed the leftovers of immortals, stamped with the teethmarks of Oceanus, Varuna, or Tangaroa. Modern explanations are likewise riddled with paradox, since science reveals the man-of-war to be not a jellyfish but a siphonophore, not a single body but a collection of bodies, a colony of as many as 1,000 polyps. The blue bottle bobbing at the surface, the pneumatophore, functions as an upside-down sailboat keeping the colony afloat, its inflatable sail rigged to navigate wind and waves. The nearly invisible tentacles, the dactylozooids, trail scores of feet below the boat and fish for prey with built-in stinging harpoons. The digestive polyps, the gastrozooids, cook up the catch and serve it to the pneumatophore, the dactylozooids, and the last members of the colony, the gonozooids, whose job it is to reproduce the man-of-war. Combined, the team is as integrated as a single animal. Yet absent of leadership, insight, or foresight, these strange conglomerates—we don't know if they function as one or many beings—beach, often in blue fleets by the dozens or hundreds. As we might if our legs were separate entities from our heads, our stomachs and sex organs separate again.

But we are not jellyfish. And we see the alarms, the messages inside the man-of-wars' blue bottles joining a host of other distress signals washing ashore these days from oil spills, fish kills, slain cetaceans, washed-up seabirds composed in part of lethally indigestible plastic. Some are silent sirens: the disappearing seashells and horseshoe crabs, the missing sea turtles and their eggs, the vanished egg casings of sharks and rays, the lost coral debris, the dwindling beach-spawning grunion, and the anguillid eels, those long-distance travelers that migrate thousands of miles from ocean to river and back again, and appear to be evaporating from the face of the earth.

Only of late have we learned to see the ocean's surprising vulnerabilities. That it's neither infinite nor inexhaustible. That it's the beleaguered terminus of all our downstream pollutants, part of a dynamic system intensely interactive with land and atmosphere and everything we do there. Only in the past decade has science discovered the ocean to be fragile in the way only really enormous things are fragile: with resilience teetering on the brink of collapse. Yet our behavior lags far behind our understanding, and the ocean awaits our enlightened action.

One in seven people on Earth depend on food from the sea as their primary source of protein. Yet one of the more optimistic assessments calculates that we've depleted up to a third of all the world's fisheries, with 7 percent to 13 percent of stocks collapsed, perhaps never to recover. These declines happen in our lifetime: bluefin tuna, once cheap, becomes exorbitant; species once scorned become market favorites.

The ocean is Earth's last frontier and its fish stocks are the bison we're currently obliterating with trawlers, longliners, purse seiners, and gillnets. Modern fisheries target not only the ocean's herbivores but also its carnivores—the apex predators such as tuna, sharks, and billfish, the marine equivalent of wolves, mountain lions, and grizzlies. Of late, we've begun large-scale hunting of the field mice of the sea, the forage fish, such as sardines and anchovies, which form the ecological backbone of many marine food webs and which, as we've learned from examples off California, Peru, Japan, and Namibia, collapse catastrophically whenever the stresses of climate change intersect with the stresses of overfishing. Worse, we're turning our guns on the ocean's primary consumers, like krill, which, one or three trophic links removed, feed most everything else that lives in or makes its living from the sea, including us.

It's not easy to calculate the magnitude of our appetites. We visit the ocean and look forward to eating the food of the sea, even those of us who would not in our wildest dreams consider eating elephants, lions, or leopards on our visit to a dry wilderness in Africa or India. Yet it's about more than the cost of eating wildlife. It's also about suffering. We hook, bludgeon, net, drown, and drag to death seabirds, sea turtles, seals, dolphins, and whales in the course of hunting seafood. We blindly assume that fish feel no pain—though many scientists firmly believe otherwise—and leave unquestioned the inhumane business of slaughtering them by the billions in the wild with truly cold-blooded detachment.

Life cycles in the sea are more complex than those of terrestrial life. Virtually all species that spend their adulthood anchored to the sea floor, such as corals, oysters, and sponges, spend their youth adrift: a two-part life history that allows them to disperse before becoming immobile. Most swimming species, such as swordfish and tuna, spend their embryo-like larval lives as translucent plankton-pickers no bigger than fingernail clippings afloat on currents. This fundamental difference in strategy between marine and terrestrial development magnifies our impacts in frightening ways. We hunt adults in one part of the ocean and destroy the nursery grounds of their larvae in another part and poison the habitat of their juveniles somewhere else while tangentially depleting the species they depend on for food at each stage.

Consider the Atlantic bluefin tuna, an endangered species on the Red List of the International Union for Conservation of Nature (IUCN). Overfishing contributes to its decline, yet pollutants inflict an insidious, often invisible toll. In 2010, tuna spawning partially converged in time and space with the *Deepwater Horizon* catastrophe in the Gulf of Mexico, one of its only two known breeding areas. We don't know the effects of 206 million gallons of oil and almost two million gallons of chemical dispersant on adult fish that came to spawn that year, or will come the next, or the next. We don't know the outcome of unprecedented levels of pollution on delicate eggs, and therefore on entire generations of an endangered species. We may not know for years, decades, or ever. Add to that ongoing

debacle the other ongoing calamities in the Gulf: the disappearing wetlands; the channelization of the Mississippi River; the overfertilization of America's breadbasket, which downriver fuels the world's second-largest oceanic dead zone; the laying of 36,000 miles of offshore pipeline; the drilling of 52,000 offshore wells; the thousands of rigs left abandoned. Most of what undermines the Gulf has been done with little or no consideration of its waters and wildlife, in marked contrast to our developing attitudes toward the land. In this, the battered Gulf of Mexico is a microcosm of the global ocean.

Meanwhile the international body of the IUCN estimated that Atlantic bluefin tuna has declined by as much as 51 percent in only three tuna generations as a result of overfishing of adult fish by longlines, drift nets, traps, bait boats, and purse seines. Its demise may also be hastened by unregulated recreational fisheries that impose no limits on the catch of juvenile tuna, as well as, increasingly, by their capture for tuna farms. We don't know much about the status of Atlantic bluefin tuna as larvae, when it dwells among zooplankton communities in warm surface waters. We do know that the mid-ocean gyres are now rife with plastic debris that breaks down at sea into tiny pieces, which are eaten and passed up the food web, starting with zooplankton, to endure seemingly forever, the modern immortals. We do know that phytoplankton—the microscopic organisms fueling zooplankton—which produce half of all plant matter on Earth, have declined an average of 1 percent a year since 1950, for a staggering total of 40 percent worldwide, according to one study. If nearly half of all wild and cropped plants disappeared from the terrestrial world, all animals, including humans, would likewise suffer severe depopulations. Humble phytoplankton. These are the microscopic organisms that perform on a global scale what we can't even manage on a village scale, namely, turning sunlight into food. They are the keystone players not just of the sea but of all the planet. Through the process of photosynthesis, they produce half the oxygen we breathe, mitigate the carbon dioxide we unleash into the air, and support all marine life. Their demise stems from a different kind of pollution, that of atmospheric carbon dioxide and other greenhouse gases, which produce a warming, more stratified ocean, with less mixing between layers and less stirring up of nutrients from the depths to support phytoplankton on the surface. In a real sense, the ocean begins to petrify and phytoplankton to starve.

A warming ocean also threatens to reroute the travels of drifting species, like the Portuguese man-of-war, as well as all that follow them, like endangered leatherback and loggerhead turtles. Man-of-wars meander tropical and subtropical seas, adrift on wind-driven currents. But changes in global temperature can send more rain and fresh meltwater into the ocean, altering the saltiness of seawater and therefore its density and changing the force of the powerful underwater rivers embedded in the depths. In one of the most exciting scientific investigations of the twentieth century, oceanographers mapped these saltwater rivers to discover they form a system connecting all the oceans of the world into one World Ocean. They called this system the ocean conveyor because it transports the rainfall from one ocean basin through subsurface currents around the world. More important, the conveyor also carries heat, drawing warmth from equatorial waters to the high latitudes—a crucial global thermostat. But now a warmer ocean, freshened by melting icecaps and glaciers, threatens to disrupt the oceanic rivers and upend one of our most critical climate regulators.

That's not the worst of it. We know that the ocean currently sequesters about a third of our atmospheric carbon dioxide emissions—yet another seemingly inexhaustible service performed by these seemingly inexhaustible waters. As we tally the true cost of this mitigation we see what was largely invisible even a decade ago—the other CO_2 problem, of rising carbon dioxide in the ocean unleashing a complex series of chemical reactions that make seawater more acidic. The ocean has, since the onset of the Industrial Revolution, become about 30 percent more acidic—a gargantuan change in chemistry that has also reduced carbonate ion concentrations in seawater by 16 percent. Carbonate ions are needed for marine life to make their shelters, their reefs and shells, which means that rising acidity threatens the survival of entire ecosystems, starting with phytoplankton and including coral communities, Antarctic systems reliant on sea urchins, and many human food webs, from oyster beds to salmon fisheries. Nor will mobile species escape. Pteropods, or sea butterflies, the swimming snails that flap through open waters and are important fuel for temperate herring, salmon, whales, and seabirds, are threatened with an acidic extinction even before the forecast rise of an ocean that will be two and a half times more acidic by 2100. The latest research shows that fish eggs and larvae are far more vulnerable to rising acidity than scientists had imagined.

Seen through a myopic lens, changes in the ocean's pH are a threat to our dinner plates and wallets. Seen through the telescope of history, acidification may have been one of the primary "kill mechanisms" behind the grimmest of all extinctions, 252 million years ago, when nearly all life on Earth—in the seas and on the land—perished. Recovery from the Great Dying took an astonishing 30 million years and required a near-total reboot of life. In our modern ocean, beset by dwindling phytoplankton, warming waters, melting ice, rising acidity, corroding reefs, dying shellfish beds, and collapsing food webs, life itself threatens to sputter out.

Underwater off a beach along Australia's Great Barrier Reef I come upon a living Portuguese man-of-war drifting into the winds on the crimped wonton of its blue sail. The retractable fishing line, the dactylozooid, dangles beneath, bouncing with the waves to connect the chattery surface with the silent parade of floating zooplankton 15 feet below. Tucked fringelike under the sail are the digestive polyps and the reproductive polyps. They are such a pale blue as to be nearly invisible and are as reactive as exposed nerves, flinching at things I can't see.

And then I can see, attached to one squirming polyp, the strange creature known as a blue dragon—a floating sea slug that makes its living eating venomous man-of-wars. It's a fantastically beautiful creature, about an inch and a half long, that looks something like a swimming lizard with winglike legs tipped with feathery toes. It lives upside down at the surface, buoyed by a swallowed air bubble, and not only endures the stinging tentacles that would send you or me screaming from the water but also eats the most venomous of them, which it doesn't digest but stores inside special sacs in its "fingertips." Armed by its own prey—as if you or I could eat pork and grow boar's tusks—blue dragons are themselves formidably venomous creatures.

As far as we know, the man-of-war has no choice but to suffer this nibbling thief that steals its limbs and its weapons. Onward it meanders, an ethereally lethal self-contained society of captain, fishermen, harpoonists, cooks, and procreators, adrift in seemingly perpetual orbit through a blue emptiness. Yet even as it's being eaten, it eats, reeling in on its fishing line a nearly transparent juvenile fish. Still alive, but paralyzed by venom, the tiny victim stares unblinkingly at the approaching sail of the bluebottle, at the digestive polyps wriggling like blind tongues as they reach out to latch on, to form little mouths that combine into one large mouth. When the fish is enveloped, the man-of-war begins to digest its meal in a bath of corrosive enzymes.

Blind and rudderless, the enmeshed trio drifts, prodded by waves building into surf that somersaults against the beach. The current sweeps under them. The wind tips the bluebottle's sail landward. Since I am not a jellyfish, I can gaze both underwater and topside toward shore, where the future of their all-consuming promenade seems clear. Yet not one of them appears to anticipate the invisible shoal ahead, the shipwreck onto a world without ocean.

Critical Thinking

1. What is the magnitude of our appetite for the ocean's resources? How much of it have we eaten so far? Can that rate of consumption be continued? Discuss.

2. List the threats to our oceans the author discusses. Given what you have read in this Environment 13/14, which of these four threats do you think will become the most critical and difficult to address in the next 50 years? Explain.

3. Who should be most responsible and bear most of the cost burden that will be necessary to protect and preserve our oceans? (*Hint:* Construct a concept map of the ecosystem services the oceans provide and make linkages to the geographical locations of the human groups that depend on the services.)

Land-Use Choices: Balancing Human Needs and Ecosystem Function

Ruth S. DeFries, Jonathan A. Foley, and Gregory P. Asner

Conversion of land to grow crops, raise animals, obtain timber, and build cities is one of the foundations of human civilization. While land use provides these essential ecosystem goods, it alters a range of other ecosystem functions, such as the provisioning of freshwater, regulation of climate and biogeochemical cycles, and maintenance of soil fertility. It also alters habitat for biological diversity. Balancing the inherent trade-offs between satisfying immediate human needs and maintaining other ecosystem functions requires quantitative knowledge about ecosystem responses to land use. These responses vary according to the type of land-use change and the ecological setting, and have local, short-term as well as global, long-term effects. Land-use decisions ultimately weigh the need to satisfy human demands and the unintended ecosystem responses based on societal values, but ecological knowledge can provide a basis for assessing the trade-offs.

Front Ecol Environ 2004; 2(5): 249–257

In a nutshell:

- Land-use change to provide food, fiber, timber, and space for settlements is one of the foundations of human civilization
- There are often unintended consequences, including feedbacks to climate, altered flows of freshwater, changes in disease vectors, and reductions in biodiversity
- Land-use decisions ultimately weigh the inherent trade-offs between satisfying immediate human needs and unintended ecosystem consequences, based on societal values
- Ecological knowledge to assess these ecosystem consequences is a prerequisite to assessing the full range of trade-offs involved in land-use decisions

P eople transform landscapes to obtain food, fiber, timber, and other ecosystem goods. This basic aspect of human existence holds true whether a subsistence farmer is growing food to feed his family from marginal lands in southern Africa or a multinational conglomerate is fertilizing and irrigating land in the midwestern US to export crops worldwide. The intended consequence of this land use is clear—to appropriate primary production for human consumption (Vitousek *et al.* 1986). The unintended consequences for the watershed, atmosphere, human health, and biological diversity often remain hidden. The implicit assumption is that the intended consequence of appropriating primary production for human consumption outweighs the unintended consequences for other ecosystem functions.

Land-use change is intricately related to both economic development and the ecological characteristics of the landscape. Within a particular region, land use potentially follows a series of transitions that parallel economic development—from wildlands with low human population densities, to frontier clearing and subsistence agriculture with the majority of the population employed in food production for local consumption, to intensive agriculture supporting mainly urban populations. . . . Regions might pass through these transitions rapidly over a period of years, or slowly over a period of centuries. In some cases, a particular region may never complete the full transition if economic conditions do not enable the infrastructure for fertilizer, irrigation, or transport; if consumer demand for products from intensive agriculture is too weak; or if arable and accessible land is not available.

We can see examples of different land-use transitions occurring throughout the world. Parts of the Amazon basin, for example, are currently experiencing a rapid transition from wildlands to intensive agriculture within a period of years, as forests that were initially cleared for pasture are now being converted to intensive agriculture in areas where infrastructure and ecological conditions are conducive to this type of usage (Laurance *et al.* 2001). Other regions, such as the Indian subcontinent and China, experienced frontier clearing many thousands of years ago, and much of the agricultural activity is still for local subsistence (Ellis and Wang 1997), except in pockets such as the Punjab where fertile soils and infrastructure inputs resulting from the Green Revolution have completed the transition. Other parts of the world remain in the subsistence stage, with few near-term prospects for moving through the transition to intensive agriculture. In sub-Saharan Africa, for instance, the vast majority of the population obtains food from subsistence farming or pastoralism.

At the other end of the transition scale, much of the cropland in the eastern US is returning to forest (Williams 1989; Foster 1992) while intensive agriculture in the Midwest provides food for consumption in distant locations. Puerto Rico has witnessed a transition from having 10% of its land covered with forest in the 1940s to more than 40% today, with a shift from agriculture to manufacturing and migration from rural to urban areas (Grau et al. 2003). Similarly, forest cover has increased in parts of the Ecuadorian Amazon, with a shift from cattle ranching to crops for urban and export markets (Rudel et al. 2002). Yet other regions experience cycles as they pass through these stages in land-use transition; for example, the Central American Mayan forests which are wildlands today once supported dense human populations during the peak of that civilization (Turner II et al. 2003), and the Amazon Basin, where large-scale transformation of landscapes existed around 1200 to 1600 AD in places now considered to be "pristine" (Heckenberger et al. 2003).

The land area within each stage of the land-use transition varies from continent to continent and changes over time with economic development, population growth, technological capabilities, and many other factors. Few wildlands remain unaffected by human presence, roads, or other infrastructure. Sanderson et al. (2002) estimate that 83% of the land surface is either directly or indirectly affected by human influences. The remaining large tracts of wildlands are located where it is too cold (boreal forests of Canada and Russia, and the Arctic tundra), too hot (desert regions of Africa and Central Australia), or too inaccessible (the Amazon Basin).

Today, approximately 33% of the land surface is under agricultural use, either as cropland (12%) or pasture (21%). . . . In arid and semi-arid regions, managed grazing—the single largest form of land use—takes place in areas often considered to be "wild open spaces" or wildlands (Goldewijk and Battjes 1997; Asner et al. in press a). The transition from wildlands to agricultural use over the past several hundred years has reduced previously forested lands by 20 to 50% (Matthews et al. 2000), and over 25% of grasslands have been converted to cropland (White et al. 2000). Urban areas cover only a small percentage of the landscape, but the resident populations generate a food demand that alters land use over a much larger area.

Subsistence agriculture, including both cultivation of crops and grazing domestic animals on rangelands, supports about half the world's population (Marsh and Grossa Jr 1996). In the coming decades, the expansion of cropland areas is likely to be a major factor in sub-Saharan Africa and Latin America, but elsewhere the mainstay of food production will come from intensification of agriculture through higher yields and more multiple cropping (Alexandratos 1999; Turner II 2002).

Ecosystem Consequences of Land-Use Change

The implications of these land-use transitions for ecosystem function are profound. In addition to providing food, fiber, fuelwood, and space for human settlement, ecosystems also perform a vast array of other functions necessary for life (Daily 1997; Millennium Ecosystem Assessment 2003). For example, ecosystems provide habitat for other plant and animal species, regulate climate by modulating energy and water flows to the atmosphere and sequester carbon that would otherwise reside in the atmosphere as a greenhouse gas, maintain the flow of freshwater into streams, and regulate vectors that transmit disease (Table 1). Land-use change enhances the share of primary production for human consumption, but decreases the share available for other ecosystem functions.

These land-use transitions offer a general framework for identifying different ecosystem consequences characteristic of different stages. As forests are cleared in the frontier stage, for example, carbon dioxide is emitted to the atmosphere as a result of clearing, burning, and vegetation decay. During the intensive stage in the transition, however, the major consequence for greenhouse gas emissions is likely to be nitrous oxide emissions from fertilizer use; carbon dioxide emissions will be much less important, since the initial biomass has already been removed (Galloway et al. in press). Other ecosystem responses such as disease regulation and biodiversity also follow characteristic responses, depending on stage in the transition. . . . Habitat fragmentation alters biodiversity in the initial clearing stage, but in the subsistence stage there are substantial declines in mammal populations associated with bushmeat hunting (Newmark et al. 1994).

Ecosystem responses to land-use change also vary in different ecological settings, even for the same type of land-use transition. For example, conversion to agricultural land in temperate and boreal latitudes may lead to a cooling of the surface as a result of increased albedo (amount of incoming light reflected back to the atmosphere) associated with a brighter land surface (Bonan 1999). The same type of conversion in tropical latitudes has precisely the opposite warming effect, where surface temperature increases with reduced transpiration in crops and pastures compared with high-biomass tropical forests (Costa and Foley 2000; Bounoua et al. 2002; DeFries et al. 2002b). The consequences of urbanization also vary with ecological setting. Urbanization in forested systems probably increases the tendency to flash floods, as water runs off impervious surfaces rather than percolating through the soil. However, in arid systems, urbanization might reduce this tendency due to the greater number of lawns and other green spaces.

Grazing has opposing impacts on vegetation under different ecological conditions. Heavy grazing is now widely implicated in the expansion of woodland vegetation in formerly open grassland and savannas in temperate latitudes (Archer et al.

Table 1 Some ecosystem functions altered by land-use change

Ecosystem function	Role of landscape in providing function	Example of altered function with land-use change
Ecosystem goods (food, fiber, fuelwood)	Provides primary production for human appropriation	Conversion to cropland increases fraction appropriated for human consumption
Provision of freshwater	Regulates flow of water to streams	Urbanization increases tendency to flash floods from storm runoff
	Maintains water quality	Agricultural runoff increases nutrient loads in streams
Climate regulation	Sequesters greenhouse gases through biogeochemical cycling	Tropical deforestation releases carbon dioxide to the atmosphere
	Exchanges water, energy, and momentum with atmosphere	Forest removal increases albedo and cools surface
Disease regulation	Restricts habitat for disease vectors	Deforestation increases human—primate contacts and spreads zoonotic diseases
	Maintains healthy climate	Urbanization creates heat islands
Biological diversity (genetic resources, biochemicals, cultural benefits)	Provides habitat for plant and animal species	Forest conversion increases habitat fragmentation
Soil fertility	Replenishes soil nutrients	Increased soil erosion from clearing depletes fertility

1995; Asner *et al.* 2003); in the humid tropics, ranching systems decrease woodland and forest cover, by up to 15 000 km^2 per year in the Brazilian Amazon alone (Houghton *et al.* 2000). The ecological community's ability to quantify these varying responses to land-use change in different stages of the transition and in different ecological settings is only just beginning to emerge.

Previous efforts have identified ecological principles for guiding land-management decisions (Dale and Haeuber 2001). Such principles highlight the importance of varying time scales of ecological processes, local environmental factors such as topography, climate, and soils that constrain land-use decisions in a particular place, species that control ecological processes disproportionate to their abundance, disturbances such as fire and flood that affect ecosystems, and the spatial configuration of habitat fragments that influence ecosystem response. Applying these ecological principles to quantify the way in which different types of land use in different ecological settings alter ecosystem function is a prerequisite for assessing the trade-offs from land uses designed to satisfy immediate human needs.

Time and Space Scales of Ecosystem Responses

Ecosystem responses to land use vary in space and time, and assessing trade-offs associated with land-use decisions requires explicit recognition of the scale of analysis. The increased tendency of a stream to flash flood in response to urbanization, for example, occurs on short time scales of minutes to hours over relatively small areas, Climate responses to greenhouse

gas emissions, on the other hand, occur over decades to centuries on continental to global scales, and species extinction as a result of landscape fragmentation occurs over decades as population sizes dwindle.

Spatial scales also vary; for example, land-use change in uplands could have distant consequences in the form of flooding downstream (Costa *et al.* 2003). Deforestation in the tropical lowlands of Costa Rica alters convection of moisture to the atmosphere and leads to drier conditions for upland forests (Lawton *et al.* 2001). If the spatial scale of analysis does not encompass the larger area with sufficient spatial detail, ecosystem responses will be overlooked or predictions will be inaccurate.

Assessments of trade-offs associated with land-use change are only meaningful if the spatial and temporal scales of analysis are carefully defined and explicitly stated. Too narrow a definition of either can result in a misperception of the problem. If soil nutrients decline over time under agricultural use, or if deforestation affects climate or streamflow in distant locations, the perceived impact is crucially dependent on the time period chosen for analysis.

Long-term and far-away responses to land-use change are not always obvious. If land use reduced habitat and the abundance of a top predator declined, for example, the effect could cascade into population expansions for prey species at lower trophic levels (Carpenter and Kitchell 1993; Terborgh *et al.* 2001). A change in disturbance regime, such as the intensity or frequency of flood or fires, could alter habitats for maintaining biodiversity, change the fluxes of carbon and other greenhouse gases to the atmosphere, and affect a variety of other ecosystem functions. These effects might not be apparent for many years.

Non-Linearities and Thresholds in Ecosystem Responses

Ecosystems are complex, dynamic systems with interactions between nutrients, plants, animals, soils, climate, and many other components, A linear response to land-use change—for example, a decline in water quality from agricultural runoff in direct proportion to the area undergoing conversion to agriculture—is unlikely in such complex systems. The more common ecosystem response is non-linear, so that small changes in land use would have large ecosystem consequences, or vice versa, depending on the degree of land-use change.

One example of a non-linear response is the application of nitrogen fertilizer for intensive agriculture and nitrate leaching into the Mississippi River system, ultimately contributing to the "dead zone" in the Gulf of Mexico (Donner *et al.* 2002). The application of nitrogen fertilizer involves a trade-off between increasing crop yields and nitrate leaching. Balancing an objective of maximum crop yields with minimum damage to coastal fisheries requires knowledge of the response of nitrate leaching to fertilizer application. . . .

Even more complex is the non-linear interaction between processes occurring simultaneously but at different temporal and spatial scales. Many managed grazing systems faltered in the southwestern US during the 1950s, as a major regional-scale drought took hold and persisted for nearly a decade (Fredrickson *et al.* 1998). Areas which were once productive for cattle ranching continued to be grazed well into the drought period, denuding very large areas of grassland. Cattle were eventually removed, as ranching ceased in some regions and waned in many others, but when precipitation returned to the region, there was a relatively sudden (non-linear) increase in woody vegetation across many large tracts of land (Buffington and Herbel 1965). The proliferation of woody vegetation was a consequence of heavy grazing that had removed herbaceous species, but these southwestern arid and semi-arid regions were pushed across a bioclimatic threshold induced by the interaction of land management and climate variability.

Ecosystems may also respond to land use via thresholds, often perceived as "the straw that breaks the camel's back". Such is the case when population abundances dwindle to the point where persistence of the species is no longer viable, resulting in a population crash. Many examples of this process come from Hawaii, where endemic honeycreeper bird populations have gone extinct following large-scale habitat loss from human settlement and agricultural expansion. For many species, fragmented habitat persists only at higher (cooler and wetter) elevations (Benning *et al.* 2002). The smaller parcels of remaining habitat no longer carry the rarest species, and avian malaria is moving to higher elevations as a result of climatic warming, pushing bird populations below the number of individuals needed to maintain a viable population.

Understanding the non-linearities and thresholds in ecosystem responses to land-use change is challenging. For those responses where ecological knowledge is available, communicating the response of ecosystem function to land-use change is straightforward for one or a few ecosystem functions. With a larger number of ecosystem functions responding at multiple scales, portraying the responses to land-use change becomes unwieldy (Heal *et al.* 2001). Quantifying and portraying this complexity to be useful for land-use decisions is a major challenge to the ecological community.

Assessing Trade-Offs from Land-Use Decisions

Land-use decisions must ultimately balance competing societal objectives based on available information about the intended and unintended ecosystem consequences. Objectives might include increasing crop production or revenue from timber on the one hand and maintenance of biodiversity, watershed protection, or any number of other ecosystem services on the other hand. Ecological knowledge to quantify the ecosystem responses in physical units underpins our ability to assess the trade-offs. Decision makers can only take the full range of consequences into account if the consequences are identified and quantified to the extent possible. . . .

Decision makers have been hindered from taking account of the full range of ecosystem consequences, partially because it is not possible with current scientific understanding to identify and quantify them over the appropriate temporal and spatial scales and to evaluate them in commensurate units. A number of case studies have quantified individual ecosystem consequences of land use, for example nitrogen pollution to the atmosphere and aquatic systems associated with intensive fertilizer use in the Yaquii Valley, Mexico (Matson *et al.* 1997; Turner II *et al.* 2003), loss of biodiversity from forest fragmentation (Pimm and Raven 2000), and carbon dioxide fluxes to the atmosphere from tropical deforestation (Achard *et al.* 2002; DeFries *et al.* 2002a). Fewer studies quantify the full suite of responses, and fewer still assess the consequences against the positive benefits for society.

We propose that the use of ecological knowledge as input to land-use decisions rests on the following premises.

Premise 1: Land Use Inherently Involves Trade-Offs

The positive benefit from appropriating ecosystem goods for human consumption is a valid objective even if other ecosystem consequences might be adversely affected. Using ecological knowledge as input to decision making implies highlighting not only the negative ecosystem consequences, but also the benefits in terms of increased yields, revenue from timber, or some other measure of societal advantages. At the same time, it is misleading to make land-use decisions based solely on the immediate societal benefits without weighing the consequences for ecosystem function and, ultimately, the impacts of changes in ecosystem function on society.

Table 2 Examples of measurable indicators appropriate for assessing short-term, local scale and long-term, global scale trade-offs associated with land-use change

Ecosystem function	Indicators appropriate for assessing short-term, local-scale trade-offs	Indicators appropriate for assessing long-term, global-scale trade-offs
Crop production	Calories/person for local consumption	Revenue from food export
Climate regulation	Rainfall from local convection	Greenhouse gas emissions altering global climate
Disease regulation	Incidence of disease locally	Spatial range of disease vectors
Biodiversity	Local species richness	Extinctions of endemic species
Soil fertility	Soil erosion in catchment area	Requirements for fertilizer to replenish fertility lost from long-term leaching
Freshwater	Groundwater recharged locally	Incidence of downstream flooding

Premie 2: Quantifying the Full Suite of Ecosystem Responses to Land-Use Change Is a Prerequisite to Assessing These Trade-Offs

Land-use decisions involve weightings placed on different ecosystem consequences according to their societal value or translation into monetary or other commensurate units, using techniques available from environmental economics (Pagiola *et al.* in press). Often, this process does not explicitly involve the full suite of ecosystem responses and only considers the positive benefits from land-use change. Ideally, the full suite of ecosystem responses at varying spatial and temporal scales should be considered explicitly. The ability to quantify the ecological response in physical units is a prerequisite, and facilitates the further step to assess trade-offs based on societal values. . . . A land-use decision—for example, converting forests to agricultural land—can account for ecosystem consequences such as biodiversity, watershed health, and climate feedbacks to the extent that these consequences can be estimated based on ecological knowledge.

A simple yet powerful approach to portraying these trade-offs is to use a "spider diagram" to depict changes in ecosystem function associated with different land-use alternatives. . . . While simple in concept, such a diagram raises a host of questions. Which ecosystem functions should be included? What are the appropriate units to measure them? Can they be quantified based on current scientific understanding? Should long-term consequences such as species extinction debts (Pimm and Raven 2000) and far-away consequences such as climate change from greenhouse gas emissions be included, or only the near-term and local consequences?

Understanding the full suite of ecosystem consequences requires quantifiable and measurable indicators for each of the ecosystem functions (NRC 2000; Millennium Ecosystem Assessment 2003; Table 2). Quantities depicted could be an absolute measure (eg tons of carbon stored), a measure relative to a previous quantity (percentage change in storage), or an amount relative to a "potential" or ideal "sustainable" amount (percentage difference from maximum). . . . In reality,

the ability to quantify the ecosystem responses and reduce them to commensurate units is not straightforward and is fraught with uncertainty. A conceptual structure for explicitly estimating responses and assessing trade-offs, however, can highlight the possible responses and reveal otherwise hidden assumptions.

To the extent that ecological knowledge can quantify the ecosystem responses at varying scales, a series of spider diagrams representing different scales can inform decision makers about the trade-offs. Appropriate units of analysis probably vary at different scales (Table 2). Crop production, for example, might be more usefully quantified in terms of calories for local consumption if land-use change is for the purpose of feeding local populations. Revenue from agricultural exports is a more applicable unit for analysis for national-level concerns. Likewise, the local response of climate regulation might be measured by changes in rainfall from altered local convection, while a longer time scale of analysis would include greenhouse gas emissions that alter global climate patterns. Explicit recognition of the different ecosystem responses at these varying spatial and temporal scales is fundamental to understanding the tradeoffs involved in land-use decisions.

Identifying the "Win-Win" and "Small Loss-Big Gain" Opportunities

Graphical depictions . . . are simple and readily communicable to decision makers, but it is difficult to portray non-linearities and thresholds in ecosystem responses to land-use change. Ideally, ecological analysis can identify the "win–win" approaches whereby societal benefit from land–use change increases as ecosystem function is preserved. An example of this is seen in the well known case of the cost-effective preservation of the watershed to maintain the quality of New York City's drinking water. In that case, preserving the watershed by limiting suburban development (a "win" for ecosystem function) was less costly than building a water-treatment plant to meet water quality standards for New York City's water supply (a "win" for immediate societal needs) (Daily and Ellison 2002 . . .).

A more common situation is likely to contain non-linear ecosystem responses, so that the challenge is to identify opportunities for "small loss–big gain" (a small compromise in immediate societal benefit for a big gain in ecosystem function, or conversely a big gain in social benefit with only a small loss in ecosystem function). For example, lowland riparian areas cover approximately 3% of the Greater Yellowstone ecosystem and are important breeding areas for maintaining populations of some bird species. Exurban development of rural homes around Yellowstone National Park is occurring in these same habitats (Hansen *et al.* 2002). Restricting development in this small percentage of the landscape (a small compromise in immediate societal benefit) will provide a disproportionate advantage to bird populations (big gain in ecosystem function).

Another example of "small loss–big gain" comes from the tropical timber industry. Conventional logging practices have traditionally been operations that leave the forest severely damaged and ecologically impoverished. The advent of reduced-impact logging (RIL) methods has led to demonstrably lower canopy damage levels, faster forest recovery rates, and reduced fire susceptibility (Pereira *et al.* 2002). However, RIL costs more, due to increased demand for technical expertise, and it leaves some high value species standing. The net outcome is less forest fragmentation over time (Asner *et al.* in press b) and possibly shorter harvest cycles (J Zweede pers comm) Small "losses" in cost and in leaving certain species results in "gains" as a result of the more sustainable timber practices, better forest condition following extraction, and thus a cascade of positive ecological repercussions.

Ecological analyses to identify such opportunities for "small loss–big gain" based on understanding of nonlinear responses of ecosystem function to land-use change can provide practical alternatives. Other, less desirable but possible relationships between land-use change to satisfy immediate human needs and ecosystem function can be categorized as "win–lose" and "big loss–small gain".

Conclusions

We propose a few simple principles that provide the building blocks for assessing the range of ecosystem responses to land use. Such principles underpin the ability of the ecological community to provide useful scientific information as input to land-use decisions.

1. Ecosystem responses to land use characteristically vary according to stage in the transition from frontier clearing to intensive human-dominated landscapes and according to the ecological setting. These characteristic responses provide the basis for understanding possible consequences of future land-use decisions.

2. Land-use decisions often result in trade-offs between intentional appropriation of ecosystem goods to satisfy human needs and unintended ecosystem responses. Balancing the trade-offs ultimately depends on societal values.

3. A prerequisite to assessing trade-offs inherent in land-use decisions is quantitative analysis of the full range of ecosystem responses, including both the intended ecosystem goods and unintended ecosystem consequences. Analyses need to explicitly consider the consequences at a range of temporal and spatial scales.

4. Ecosystems are likely to respond in a non-linear manner to land use. These non-linear responses offer the opportunity to identify land-use alternatives with small losses in satisfying immediate human needs but large gains in maintaining ecosystem function.

With increasing population and development pressures, there is little doubt that land-use change will continue over the coming decades, as food demand increases and urban areas expand in many parts of the world. These land-use changes will appropriate an increasing share of primary production to satisfy human needs while simultaneously altering many other functions of the landscape, ranging from regulation of disease vectors to species habitat to freshwater flows. Quantifying the trade-offs between meeting immediate societal needs for resources such as food, fiber, timber, and space and other less-intended consequences for ecosystem function is a major challenge for the ecological community.

References

Achard F, Eva H, Stibig HJ, *et al.* 2002. Determination of deforestation rates of the world's humid tropical forests. *Science* **297:** 999–1002.

Alexandratos N. 1999. World food and agriculture: outlook for the medium and longer term. *Proc Natl Acad Sci* **96:** 5908–14.

Archer S, Schimel DS, and Holland EA. 1995. Mechanisms of shrubland expansion: land use, climate, and CO_2. *Climatic Change* **29:** 91–99.

Asner GP, Archer S, Hughes RF, *et al.* 2003. Net changes in regional woody vegetation cover and carbon storage in Texas drylands, 1937–1999. *Global Change Biol* **9:** 316–35.

Asner GP, Elmore E, Martin RE, and Olander L. Grazing systems and global change. *Annu Rev Env Resour.* In press a.

Asner G, Keller M, and Silva JNM, Spatial and temporal dynamics of forest canopy gaps following selective logging in the eastern Amazon. *Global Change Biol.* In press b.

Benning TL, LaPointe D, Atkinson CT, and Vitousek PM. 2002. Interactions of climate change with biological invasions and land use in the Hawaiian Islands: modeling the fate of endemic birds using a geographic information system. *Proc Natl Acad Sci* **99:** 14246–49.

Bonan GB. 1999. Frost followed the plow: impacts of deforestation on the climate of the United States. *Ecol Applic* **9:** 1305–15.

Bounoua L, DeFries R, Collatz GJ, *et al.* 2002. Effects of land cover conversion on surface climate. *Climatic Change* **52:** 29–64.

Buffington LC and Herbel CH. 1965. Vegetational changes on a semidesert grassland range. *Ecol Monogr* **35:** 139–64.

Carpenter SR and Kitchell HF (Eds). 1993. The trophic cascade in lakes. Cambridge, UK: Cambridge University Press.

Costa MH, Botta A, and Cardille J. 2003. Effects of large-scale change in land cover on the discharge of the Tocantins River, Amazonia. *J Hydrol* **283:** 206–17.

Costa MH and Foley JA. 2000. Combined effects of deforestation and doubled atmospheric CO_2 concentration on the climate of Amazonia. *J Climate* **13:** 18–34.

Daily GC (Ed). 1997. Nature's services: societal dependence on natural ecosystems. Washington, DC: Island Press.

Daily GC and Ellison K. 2002. The new economy of nature: the quest to make conservation profitable. Washington, DC: Island Press.

Dale V and Haeuber R. 2001. Applying ecological principles to land management. New York, NY: Springer-Verlag.

DeFries R, Houghton RA, Hansen M, *et al.* 2002a. Carbon emissions from tropical deforestation and regrowth based on satellite observations for the 1980s and 90s. *Proc Natl Acad Sci* **99:** 14256–61.

DeFries RS, Bounoua L, and Collatz GJ. 2002b. Human modification of the landscape and surface climate in the next fifty years, *Global Change Biol* **8:** 438–58.

Donner SD, Coe MT, Lenters JD, *et al.* 2002. Modeling the impact of hydrological changes on nitrate transport in the Mississippi River Basin from 1955–1994. Global Biogeochemical Cycles: DOI:10.1029/2001GB001396, August 001397.

Ellis EC and Wang SM. 1997. Sustainable traditional agriculture in the Tai Lake Region of China. *Agr Ecosyst Environ* **61:** 177–93.

Foley J A, Costa MA, Delire C, *et al.* 2003. Green surprise? How terrestrial ecosystems could affect earth's climate. *Front Ecol Environ* **1:** 38–44.

Foley J A, Ramankutty N, and Leff B. Global land use changes. In: Our changing planet: a view from space. Cambridge, UK,: Cambridge University Press. Unpublished.

Foster DR. 1992. Land-use history (1730–1990) and vegetation dynamics in central New England, USA. *Ecol* **80:** 753–72.

Fredrickson E, Havstad KM, Estell R, and Hyder P. 1998. Perspectives on desertification: south-western United States. *J Arid Environ* **39:** 191–207.

Galloway JN, Dentener FJ, Capone DG, *et al.* Nitrogen cycles: past, present, and future. Biogechemistry. In press.

Goldewijk CGM and Battjes JJ. 1997. A hundred year (1890–1990) database for integrated environmental assessments (HYDE version 1.1). Report No. 422514002, National Institute of Public Health and the Environment (RIVM), Bilthoven, The Netherlands.

Grau HR, Aide TM, Zimmerman JK, *et al.* 2003. The ecological consequences of socioeconomic and land-use changes in postagriculture Puerto Rico. *BioScience* **53:** 1159–69.

Hansen AJ, Rasker R, Maxwell B, *et al.* 2002. Ecology and socioeconomics in the New West: A case study from Greater Yellowstone. *BioScience.* In press.

Heal G, Daily GC, Ehrlich PR, *et al.* 2001. Protecting natural capital through ecosystem service districts. *Stanford Environ Law J* **20:** 333–64.

Heckenberger MJ, Kuikuro A, Kuikuro UT, *et al.* 2003. Amazonia 1492: Pristine forest or cultural parkland? *Science* **301:** 1710–14.

Houghton RA, Skole DL, Nobre CA, *et al.* 2000. Annual fluxes of carbon from deforestation and regrowth in the Brazilian Amazon. *Nature* 403: 301–04.

Laurance WF, Cochrane MA, Bergen S, *et al.* 2001. Environment: The future of the Brazilian Amazon. *Science* **291:** 438–39.

Lawton RO, Nair RS, Pielke RAS, and Welch RM. 2001. Climatic impacts of tropical lowland deforestation on nearby montane cloud forests. *Science* **294:** 584–87.

Marsh WM and Grossa JM Jr. 1996. Environmental geography: science, land use, and earth systems. New York, NY: John Wiley and Sons.

Matson PA, Parton WJ, Power AG, and Swift MJ. 1997. Agricultural intensification and ecosystem properties. *Science* **277:** 504–08.

Matthews E, Rohweder M, Payne R, and Murray S. 2000. Pilot analysis of global ecosystems: forest ecosystems. Washington, DC: World Resources Institute.

Millennium Ecosystem Assessment. 2003. Ecosystems and human well-being: a framework for assessment. Washington, DC: Island Press.

Mustard J, DeFries R, Fisher T, and Moran EF (Eds). Land use and land cover change pathways and impacts. Dordrecht, The Netherlands: Kluwer Academic Publishers. In press.

Newmark WD, Manyanza DN, Gamassa DM, and Sariko HI. 1994. The conflict between wildlife and local people living adjacent to protected areas in Tanzania: human density as a predictor. *Conserv Biol* **8:** 249–55.

NRC. 2000. Ecological indicators for the Nation. Washington, DC: National Academy Press.

Pagiola S, Acharya G, and Dixon JA, Economic analysis of environmental impacts, London, UK: Earthscan. In press.

Pereira R, Zweede J, Asner GP, and Keller M. 2002. Forest canopy damage from conventional and reduced impact selective logging in Eastern Amazon. *Forest Ecol Manag* **168:** 77–89.

Pimm SL and Raven P. 2000. Extinction by numbers. *Nature* **403:** 843–45.

Rudel TK, Bates DM, and Machinguiashi R. 2002. A tropical forest transition? Agricultural change, out-migration, and secondary forests in the Ecuadorian Amazon. *Ann Assoc Am Geogr* **92:** 87–102.

Sanderson EW, Jaiteh M, Levy M, *et al* 2002. The human footprint and the last of the wild. *BioScience* **52:** 891–904.

Terborgh J, Lopez L, Nuñez P *et al.* 2001. Ecological meltdown in predator-free forest fragments. *Science* **295:** 1923–26.

Turner BL II. 2002, Toward integrated land-change science: Advances in 1.5 decades of sustained international research on land-use and land-cover change. In: Steffen W, Jager J, Carson D, and Bradshaw C (Eds). Challenges of a changing Earth: Proceedings of the Global Change Open Science Conference, Amsterdam, Netherlands, 10–13 July 2000. Heidelberg, Germany: Springer-Verlag.

Turner BL II, Matson PA, McCarthy J, *et al.* 2003. Illustrating the coupled human–environment system for vulnerability analysis: three case studies. *Proc Natl Acad Sci* **100;** 8080–85.

Vitousek P, Ehrlich PR, Ehrlich AH, and Matson PA. 1986. Human appropriation of the products of photosynthesis. *BioScience* **36:** 368–73.

White RP, Murray S, and Rohweder M. 2000. Pilot analysis of global ecosystems: grassland ecosystems. Washington, DC: World Resources Institute.

Williams M. 1989. Americans and their forests: a historical geography. New York, NY: Cambridge University Press.

Critical Thinking

1. What are some of the ecosystem functions altered by land use changes resulting from human consumption of the environment?

2. What are the benefits of understanding the ecological "trade-offs" inherent in any land-use decisions before the changes are made?

3. What are the similarities in this article's focus on "assessing trade-offs from land-use decisions" and Article 17: "Full Cost Accounting in Environmental Decision-Making"?

4. What are the differences in these approaches to land-use consumption decisions?

5. With regard to the approaches to land-use decision making in both Article 36 and 17, what obstacles to you see with any attempts to balance human needs and ecosystem functions?

Economic Report into Biodiversity Crisis Reveals Price of Consuming the Planet

Species losses around the world could really cost us the Earth with food shortages, floods and expensive clean up costs

JULIETTE JOWIT

In every corner of the globe the evidence of the global biodiversity crisis is now impossible to ignore.

In the UK, a third of high priority species and two thirds of habitats are declining, according to government figures that emerged today on the UK's Biodiversity Action Plan. Since 1994 despite the extra attention provided by the plan, 5% of the species it covered are thought to have gone extinct.

Around the world the picture is as bad or worse: the International Union for the Conservation of Nature believes one in five mammals, one in three amphibians and one in seven birds are extinct or globally threatened, and other species groups still being assessed are showing similar patterns.

Simon Stuart, a senior IUCN scientist, has warned that for the first time since the dinosaurs humans are driving plants and animals to extinction faster than new species can evolve.

For decades, nature lovers have watched the fens being drained, or noticed the decline of cuckoos in spring and butterflies in summer. But until recently these changes have been overshadowed by growing fears about the impact of climate change.

However, as the impact of these species losses around the world have mounted—riots over food shortages, costly floods and landslides, expensive bills for cleaning polluted water, and many more disasters—attention has finally started to turn to the impact of human beings literally consuming the planet's natural resources.

So it was in 2007, just months after the British government made global waves with the biggest ever report on the economics of climate change by Lord Stern, that world governments met in Potsdam, in Germany, and asked the leading economist and senior banker Pavan Sukhdev to do the same for the natural world.

The study—called The Economics of Ecosystems and Biodiversity (TEEB) shows that on average one-third of Earth's habitats have been damaged by humans—with, for example, 85% of seas and oceans and more than 70% of Mediterranean shrubland affected. It also warns that in spite of growing awareness of the danger of natural destruction it will "still continue on a large scale".

Following an interim report last year, the study group will publish its final findings this summer, in advance of the global Convention On Biological Diversity conference in Japan in October, marking 2010 as the International Year of Biodiversity.

Based on a host of academic and expert studies, the TEEB report is expected to say that the ratio of costs of conserving ecosystems or biodiversity to the benefits of doing so range from 10:1 to 100:1. "Our studies found ranges of 1:10, 1:25, 1:60," said Sukhdev. "The point is they are all big ratios: I'd do business on those ratios . . . I'm fine with 1:10."

One report estimated the cost of building and maintaining a more comprehensive network of global protected areas—increasing it from the current 12.5%-14% to 15% of all land and from 1% to 30% of the seas—would be $45bn a year, while the benefits of preserving the species richness within these zones would be worth $4-5tn a year. Another unpublished report for the UN by UK-based consultants Trucost claimed the combined cost of damage to the environment by the world's 3,000 biggest companies was $2.2tn in 2008.

Echoing Lord Stern's famous description of climate change as "the greatest and widest-ranging market failure ever", Sukhdev—who supports action on climate change as well—said the destruction of the natural world was "a landscape of market failures", because the services of nature were nearly always provided for free, and so not valued until they were gone.

"The earth and its thin surface is our only home, and there's a lot that comes to us from biodiversity and ecosystems: we get food, fuel, fibre; we get the ability to have clean air and fresh water; we get a stable micro-climate where we live; if we wander into forests and wildernesses we get enjoyment, we get recreation, we get spiritual sustenance; all kinds of things—which in many cases are received free, and I think that's perhaps the nub of the problem," said Sukhdev who was [visiting] the UK as a guest of science research and education charity, the Earthwatch Institute.

"We fail to recognise the extent to which we are dependent on natural ecosystems, and not just for goods and services, but also for the stability of the environment in which we survive—there's an element of resilience that's been built into our lives, the ability of our environment to withstand the shocks to which we expose it . . . the more we lose, the less resilience there is to these shocks, and therefore we increase the risk to society and risk to life and livelihoods and the economy," he added. Sukhdev is a senior banker at Deutsche Bank, adviser to the UN Environment Programme. He also owns a rainforest restoration and eco-tourism project in Australia and an organic farm in south India.

The final reports, in five sections covering the economic methods and advice to policy makers, administrators, businesses and citizens, will make a series of recommendations for how to use economic values for different parts of nature, such as particular forests, wetlands, ocean habitats like coral reefs or individual species (one example given is paying farmers to tolerate geese wintering in Scotland), into ways to protect them.

One of the most immediate changes could be reform of direct and indirect subsidies, such as tax exemptions, which encourage over-production even when there is clear destruction of the long-term ability of the environment to provide what is needed, and below-cost pricing which leads to wasteful use and poor understanding of the value of the products. "Particularly worrying" are about $300bn of subsidies to agriculture and fishing; subsidies of $500bn for energy, $238–$306bn for transport and $67bn for water companies are also singled out.

Although the report is likely to argue some subsidies should be reformed rather than axed, an example of the huge potential impact was given by Sukhdev [during] a meeting in New York this week. [Sukhdev indicated] that to stop the global collapse of fish stocks, more than 20 million people employed in the industry may need to be taken out of service and retrained over the next 40 years.

Other suggested reforms include stricter limits on extraction and pollution; other environmental regulations such as restrictions on fishing net sizes or more damaging agricultural practices; higher penalties for breaking the limits, reform of taxes to encourage better practices; better public procurement; public funds for restoring damaged ecosystems such as reedbeds or heathlands; forcing companies that want to develop an area of land to restore or conserve another piece of land to "offset" the damage; and paying communities for the use of goods and services from nature—such as the proposed Redd international forestry protection scheme. Money raised by some policies could pay for others, says the report.

Sukhdev's team also wants companies and countries to adopt new accounting systems so alongside their financial accounts of income, spending, profits and capital, they also publish figures showing their combined impact on environmental or natural capital, and also social capital, such improvements in workers' skills or national education levels.

"We're in a society where more is better, where we tend to reward more production and more consumption . . . GDP tends to get associated with progress, and that's not necessarily the case."

Critical Thinking

1. Summarize briefly the current state of our biodiversity crisis.

2. What are some of the primary drivers responsible for biodiversity loss?

3. While this is only a summary report, are the changes and policies advocated regarding the "price of consuming the earth" aimed at the consumption behavior of individuals, or at the consumption of the environment by groups of humans? (*Hint:* Refer to Article 44.)

4. With reference to question 3, what are the pros and cons of the two different approaches (individual consumption behaviors and consumption of the environment by human groups) to dealing with the biodiversity issue?

Putting People in the Map: Anthropogenic Biomes of the World

ERLE C. ELLIS AND NAVIN RAMANKUTTY

Humans have fundamentally altered global patterns of biodiversity and ecosystem processes. Surprisingly, existing systems for representing these global patterns, including biome classifications, either ignore humans altogether or simplify human influence into, at most, four categories. Here, we present the first characterization of terrestrial biomes based on global patterns of sustained, direct human interaction with ecosystems. Eighteen "anthropogenic biomes" were identified through empirical analysis of global population, land use, and land cover. More than 75% of Earth's ice-free land showed evidence of alteration as a result of human residence and land use, with less than a quarter remaining as wildlands, supporting just 11% of terrestrial net primary production. Anthropogenic biomes offer a new way forward by acknowledging human influence on global ecosystems and moving us toward models and investigations of the terrestrial biosphere that integrate human and ecological systems.

Front Ecol Environ 2008; 6(8); 439–447, doi:
10.1890/070062

Humans have long distinguished themselves from other species by shaping ecosystem form and process using tools and technologies, such as fire, that are beyond the capacity of other organisms (Smith 2007). This exceptional ability for ecosystem engineering has helped to sustain unprecedented human population growth over the past half century, to such an extent that humans now consume about one-third of all terrestrial net primary production (NPP; Vitousek *et al.* 1986; Imhoff *et al.* 2004) and move more earth and produce more reactive nitrogen than all other terrestrial processes combined (Galloway 2005; Wilkinson and McElroy 2007). Humans are also causing global extinctions (Novacek and Cleland 2001) and changes in climate that are comparable to any observed in the natural record (Ruddiman 2003; IPCC 2007). Clearly, *Homo sapiens* has emerged as a force of nature rivaling climatic and geologic forces in shaping the terrestrial biosphere and its processes.

In a Nutshell:

- Anthropogenic biomes offer a new view of the terrestrial biosphere in its contemporary, human-altered form
- Most of the terrestrial biosphere has been altered by human residence and agriculture
- Less than a quarter of Earth's ice-free land is wild; only 20% of this is forests and >36% is barren
- More than 80% of all people live in densely populated urban and village biomes
- Agricultural villages are the most extensive of all densely populated biomes and one in four people lives in them

Biomes are the most basic units that ecologists use to describe global patterns of ecosystem form, process, and biodiversity. Historically, biomes have been identified and mapped based on general differences in vegetation type associated with regional variations in climate (Udvardy 1975; Matthews 1983; Prentice *et al.* 1992; Olson *et al.* 2001; Bailey 2004). Now that humans have restructured the terrestrial biosphere for agriculture, forestry, and other uses, global patterns of species composition and abundance, primary productivity, land-surface hydrology, and the biogeochemical cycles of carbon, nitrogen, and phosphorus, have all been substantially altered (Matson *et al.* 1997; Vitousek *et al.* 1997; Foley *et al.* 2005). Indeed, recent studies indicate that human-dominated ecosystems now cover more of Earth's land surface than do "wild" ecosystems (McCloskey and Spalding 1989; Vitousek *et al.* 1997; Sanderson *et al.* 2002, Mittermeier *et al.* 2003; Foley *et al.* 2005).

It is therefore surprising that existing descriptions of biome systems either ignore human influence altogether or describe it using at most four anthropogenic ecosystem classes (urban/built-up, cropland, and one or two cropland/natural vegetation mosaic(s); classification systems include IGBP, Loveland *et al.* 2000; "Olson Biomes", Olson *et al.* 2001; GLC 2000, Bartholome and Belward 2005; and GLOBCOVER, Defourny *et al.* 2006). Here, we present an alternate view of the terrestrial biosphere, based on an empirical analysis of global patterns of sustained direct human interaction with ecosystems, yielding a global map of "anthropogenic biomes". We then examine the potential of anthropogenic biomes to serve as a new global framework for ecology, complete with testable hypotheses, that can advance research, education, and conservation of the terrestrial biosphere as it exists today—the product of intensive reshaping by direct interactions with humans.

Human Interactions with Ecosystems

Human interactions with ecosystems are inherently dynamic and complex (Folke *et al.* 1996; DeFries *et al.* 2004; Rindfuss *et al.* 2004); any categorization of these is a gross oversimplification. Yet there is little hope of understanding and modeling these interactions at a global scale without such simplification. Most global models of primary productivity, species diversity, and even climate depend on stratifying the terrestrial surface into a limited number of functional types, land-cover types, biomes, or vegetation classes (Haxeltine and Prentice 1996; Thomas *et al.* 2004; Feddema *et al.* 2005).

Human interactions with ecosystems range from the relatively light impacts of mobile bands of hunter-gatherers to the complete replacement of pre-existing ecosystems with built structures (Smil 1991). Population density is a useful indicator of the form and intensity of these interactions, as increasing populations have long been considered both a cause and a consequence of ecosystem modification to produce food and other necessities (Boserup 1965, 1981; Smil 1991; Netting 1993). Indeed, most basic historical forms of human-ecosystem interaction are associated with major differences in population density, including foraging (<1 person km^{-2}), shifting (>10 persons km^{-2}), and continuous cultivation (>100 persons km^{-2}); populations denser than 2500 persons km^{-2} are believed to be unsupportable by traditional subsistence agriculture (Smil 1991; Netting 1993).

In recent decades, industrial agriculture and modern transportation have created new forms of human-ecosystem interaction across the full range of population densities, from low-density exurban developments to vast conurbations that combine high-density cities, low-density suburbs, agriculture,

and even forested areas (Smil 1991; Qadeer 2000; Theobald 2004). Nevertheless, population density can still serve as a useful indicator of the form and intensity of human-ecosystem interactions within a specific locale, especially when populations differ by an order of magnitude or more. Such major differences in population density help to distinguish situations in which humans may be considered merely agents of ecosystem transformation (ecosystem engineers), from situations in which human populations have grown dense enough that their local resource consumption and waste production form a substantial component of local biogeo-chemical cycles and other ecosystem processes. To begin our analysis, we therefore categorize human-ecosystem interactions into four classes, based on major differences in population density: high population intensity ("dense", >100 persons km^{-2}), substantial population intensity ("residential", 10 to 100 persons km^{-2}), minor population ("populated", 1 to 10 persons km^{-2}), and inconsequential population ("remote", <1 person km^{-2}). Population class names are defined only in the context of this study.

Identifying Anthropogenic Biomes: An Empirical Approach

We identified and mapped anthropogenic biomes using the multi-stage empirical procedure . . . outlined below, based on global data for *population* (urban, non-urban), *land use* (percent area of pasture, crops, irrigation, rice, urban land), and *land cover* (percent area of trees and bare earth); data for NPP, IGBP land cover, and Olson biomes were obtained for later analysis. . . . Biome analysis was conducted at 5 arc minute resolution (5′ grid cells cover ~ 86 km^2 at the equator), a spatial resolution selected as the finest allowing direct use of high-quality land-use area estimates. First, "anthropogenic" 5′ cells were separated from "wild" cells, based on the presence of human populations, crops, or pastures. Anthropogenic cells were then stratified into the population density classes described above ("dense", "residential", "populated", and "remote"), based on the density of their non-urban population. We then used cluster analysis, a statistical procedure designed to identify an optimal number of distinct natural groupings (clusters) within a dataset (using SPSS 15.01), to identify natural groupings within the cells of each population density class and within the wild class, based on non-urban population density and percent urban area, pasture, crops, irrigated, rice, trees, and bare earth. Finally, the strata derived above were described, labeled, and organized into broad logical groupings, based on their populations, land-use and land-cover characteristics, and their regional distribution, yielding the 18 anthropogenic biome classes and three wild biome classes. . . .

A Tour of the Anthropogenic Biomes

When viewed globally, anthropogenic biomes clearly dominate the terrestrial biosphere, covering more than three-quarters of Earth's ice-free land and incorporating nearly 90% of terrestrial

NPP and 80% of global tree cover.... About half of terrestrial NPP and land were present in the forested and rangeland biomes, which have relatively low population densities and potentially low impacts from land use (excluding residential rangelands....) However, one-third of Earth's ice-free land and about 45% of terrestrial NPP occurred within cultivated and substantially populated biomes (dense settlemgnts, villages, croplands, and residential rangelands)....

Of Earth's 6.4 billion human inhabitants, 40% live in dense settlements biomes (82% urban population), 40% live in village biomes (38% urban), 15% live in cropland biomes (7% urban), and 5% live in rangeland biomes (5% urban; forested biomes had 0.6% of global population).... Though most people live in dense settlements and villages, these cover just 7% of Earth's ice-free land, and about 60% of this population is urban, living in the cities and towns embedded within these biomes, which also include almost all of the land we have classified as urban (94% of 0.5 million km^2, although this is probably a substantial underestimate; Salvatore *et al.* 2005)....

Village biomes, representing dense agricultural populations, were by far the most extensive of the densely populated biomes, covering 7.7 million km^2, compared with 1.5 million km^2 for the urban and dense settlements biomes. Moreover, village biomes house about one-half of the world's non-urban population (1.6 of ~ 3.2 billion persons). Though about one-third of global urban area is also embedded within these biomes, urban areas accounted for just 2% of their total extent, while agricultural land (crops and pasture) averaged >60% of their area. More than 39% of densely populated biomes were located in Asia, which also incorporated more than 60% of that continent's total global area, even though this region was the fifth largest of seven regions ... Village biomes were most common in Asia, where they covered more than a quarter of all land. Africa was second, with 13% of village biome area, though these covered just 6% of Africa's land. The most intensive land-use practices were also disproportionately located in the village biomes, including about half the world's irrigated land (1.4 of 2.7 million km^2) and two-thirds of global rice land (1.1 of 1.7 million km^2....)

After rangelands, cropland biomes were the second most extensive of the anthropogenic biomes, covering about 20% of Earth's ice-free land. Far from being simple, crop-covered landscapes, cropland biomes were mostly mosaics of cultivated land mixed with trees and pastures.... As a result, cropland biomes constituted only slightly more than half of the world's total crop-covered area (8 of 15 million km^2), with village biomes hosting nearly a quarter and rangeland biomes about 16%. The cropland biomes also included 17% of the world's pasture land, along with a quarter of global tree cover and nearly a third of terrestrial NPP. Most abundant in Africa and Asia, residential, rainfed mosaic was by far the most extensive cropland biome and the second most abundant biome overall (16.7 million km^2), providing a home to nearly 600 million people, 4 million km^2 of crops, and about 20% of the world's tree cover and NPP—a greater share than the entire wild forests biome.

Rangeland biomes were the most extensive, covering nearly a third of global ice-free land and incorporating 73% of global pasture (28 million km^2), but these were found primarily in arid and other low productivity regions with a high percentage of bare earth cover (around 50%....) As a result, rangelands accounted for less than 15% of terrestrial NPP, 6% of global tree cover, and 5% of global population.

Forested biomes covered an area similar to the cropland biomes (25 million km^2 versus 27 million km^2 for crop-lands), but incorporated a much greater tree-covered area (45% versus 25% of their global area). It is therefore surprising that the total NPP of the forested biomes was nearly the same as that of the cropland biomes (16.4 versus 16.0 billion tons per year). This may be explained by the lower productivity of boreal forests, which predominate in the forested biomes, while crop-land biomes were located in some of the world's most productive climates and soils.

Wildlands without evidence of human occupation or land use occupied just 22% of Earth's ice-free land in this analysis. In general, these were located in the least productive regions of the world; more than two-thirds of their area occurred in barren and sparsely tree-covered regions. As a result, even though wildlands contained about 20% cover by wild forests (a mix of boreal and tropical forests ...), wild-lands as a whole contributed only about 11% of total terrestrial NPP.

Anthropogenic Biomes Are Mosaics

It is clear from the biome descriptions above, from the land-use and land-cover patterns ..., and most of all, by comparing our biome map against high-resolution satellite imagery ..., that anthropogenic biomes are best characterized as heterogeneous landscape mosaics, combining a variety of different land uses and land covers. Urban areas are embedded within agricultural areas, trees are interspersed with croplands and housing, and managed vegetation is mixed with semi-natural vegetation (eg croplands are embedded within rangelands and forests). Though some of this heterogeneity might be explained by the relatively coarse resolution of our analysis, we suggest a more basic explanation: that direct interactions between humans and ecosystems generally take place within heterogeneous landscape mosaics (Pickett and Cadenasso 1995; Daily 1999). Further, we propose that this heterogeneity has three causes, two of which are anthropogenic and all of which are fractal in nature (Levin 1992), producing similar patterns across spatial scales ranging from the land holdings of individual households to the global patterning of the anthropogenic biomes.

We hypothesize that even in the most densely populated biomes, most landscape heterogeneity is caused by natural variation in terrain, hydrology, soils, disturbance regimes (eg fire), and climate, as described by conventional models of ecosystems and the terrestrial biosphere (eg Levin 1992; Haxeltine and Prentice 1996; Olson *et al.* 2001). Anthropogenic enhancement of natural landscape heterogeneity represents a secondary cause of heterogeneity within anthropogenic biomes, explained in part by the human tendency to seek out and use the most productive lands first and to work and populate these lands most intensively (Huston 1993). At a global scale, this process may explain why wildlands are most common in those parts of the biosphere with the least potential for agriculture (ie polar regions, mountains, low fertility tropical soils . . .) and why, at a given percentage of tree cover, NPP appears higher in anthropogenic biomes with higher population densities (compare NPP with tree cover, especially in wild forests versus forested biomes, . . .). It may also explain why most human populations, both urban and rural, appear to be associated with intensive agriculture (irrigated crops, rice), and not with pasture, forests, or other, less intensive land uses. . . . Finally, this hypothesis explains why most fertile valleys and floodplains in favorable climates are already in use as croplands, while neighboring hillslopes and mountains are often islands of semi-natural vegetation, left virtually undisturbed by local populations (Huston 1993; Daily 1999). The third cause of landscape heterogeneity in anthropogenic biomes is entirely anthropogenic: humans create landscape heterogeneity directly, as exemplified by the construction of settlements and transportation systems in patterns driven as much by cultural as by environmental constraints (Pickett and Cadenasso 1995).

All three of these drivers of heterogeneity undoubtedly interact in patterning the terrestrial biosphere, but their relative roles at global scales have yet to be studied and surely merit further investigation, considering the impacts of landscape fragmentation on biodiversity (Vitousek *et al.* 1997; Sanderson *etal.* 2002).

A Conceptual Model for Anthropogenic Biomes

Given that anthropogenic biomes are mosaics—mixtures of settlements, agriculture, forests and other land uses and land covers—how do we proceed to a general ecological understanding of human–ecosystem interactions within and across anthropogenic biomes? Before developing a set of hypotheses and a strategy for testing them, we first summarize our current understanding of how these interactions pattern terrestrial ecosystem processes at a global scale using a simple equation:

$$Ecosystem\ processes = f(population\ density,\ land\ use,\ biota,\ climate,\ terrain,\ geology)$$

Those familiar with conventional ecosystem-process models will recognize that ours is merely an expansion of these, adding human population density and land use as parameters to explain global patterns of ecosystem processes and their changes. With some modification, conventional land-use and ecosystem-process models should therefore be capable of modeling ecological changes within and across anthropogenic biomes (Turner *et al.* 1995; DeFries *et al.* 2004; Foley *et al.* 2005). We include population density as a separate driver of ecosystem processes, based on the principle that increasing population densities can drive greater intensity of land use (Boserup 1965, 1981) and can also increase the direct contribution of hyumans to local econsystem processes (eg resource consumption, combustion,

excretion; Imhoff *et al.* 2004). For example, under the same environmental conditions, our model would predict greater fertilizer and water inputs to agricultural land in areas with higher population densities, together with greater emissions from the combustion of biomass and fossil fuel.

Some Hypotheses and Their Tests

Based on our conceptual model of anthropogenic biomes, we propose some basic hypotheses concerning their utility as a model of the terrestrial biosphere. First, we hypothesize that anthropogenic biomes will differ substantially in terms of basic ecosystem processes (eg NPP, carbon emissions, reactive nitrogen . . .) and biodiversity (total, native) when measured across each biome in the field, and that these differences will be at least as great as those between the conventional biomes when observed using equivalent methods at the same spatial scale. Further, we hypothesize that these differences will be driven by differences in population density and land use between the biomes . . ., a trend already evident in the general tendency toward increasing cropped area, irrigation, and rice production with increasing population density. . . . Finally, we hypothesize that the degree to which anthropogenic biomes explain global patterns of ecosystem processes and biodiversity will increase over time, in tandem with anticipated future increases in human influence on ecosystems.

The testing of these and other hypotheses awaits improved data on human-ecosystem interactions obtained by observations made within and across the full range of anthropogenic landscapes. Observations within anthropogenic landscapes capable of resolving individually managed land-use features and built structures are critical, because this is the scale at which humans interact directly with ecosystems and is also the optimal scale for precise measurements of ecosystem parameters and their controls (Ellis *et al.* 2006). Given the considerable effort involved in making detailed measurements of ecological and human systems across heterogeneous anthropogenic landscapes, this will require development of statistically robust stratified-sampling designs that can support regional and global estimates based on relatively small landscape samples within and across anthropogenic biomes (eg Ellis 2004). This, in turn, will require improved global data, especially for human populations and land-use practices. Fortunately, development of these datasets would also pave the way toward a system of anthropogenic ecore-gions capable of serving the ecological monitoring needs of regional and local stakeholders, a role currently occupied by conventional ecoregion mapping and classification system (Olson *et al.* 2001).

Are Conventional Biome Systems Obsolete?

We have portrayed the terrestrial biosphere as composed of anthropogenic biomes, which might also be termed "anthromes" or "human biomes" to distinguish them from conventional biome systems. This begs the question: are conventional biome systems obsolete? The answer is certainly "no". Although we have proposed a basic model of ecological processes within and across anthropogenic biomes, our model remains conceptual, while existing models of the terrestrial biomes, based on climate, terrain, and geology, are fully operational and are useful for predicting the future state of the biosphere in response to climate change (Melillo *et al.* 1993; Cox *et al.* 2000; Cramer *et al.* 2001).

On the other hand, anthropogenic biomes are in many ways a more accurate description of broad ecological patterns within the current terrestrial biosphere than are conventional biome systems that describe vegetation patterns based on variations in climate and geology. It is rare to find extensive areas of any of the basic vegetation forms depicted in conventional biome models outside of the areas we have defined as wild biomes. This is because most of the world's "natural" ecosystems are embedded within lands altered by land use and human populations, as is apparent when viewing the distribution of IGBP and Olson biomes within the anthropogenic biomes. . . .

Ecologists Go Home!

Anthropogenic biomes point to a necessary turnaround in ecological science and education, especially for North Americans. Beginning with the first mention of ecology in school, the biosphere has long been depicted as being composed of natural biomes, perpetuating an outdated view of the world as "natural ecosystems with humans disturbing them". Although this model has long been challenged by ecologists (Odum 1969), especially in Europe and Asia (Golley 1993), and by those in other disciplines (Cronon 1983), it remains the mainstream view. Anthropogenic biomes tell a completely different story, one of "human systems, with natural ecosystems embedded within them". This is no minor change in the story we tell our children and each other. Yet it is necessary for sustainable management of the biosphere in the 21st century.

Anthropogenic biomes clearly show the inextricable intermingling of human and natural systems almost everywhere on Earth's terrestrial surface, demonstrating that interactions between these systems can no longer be avoided in any substantial way. Moreover, human interactions with ecosystems mediated through the atmosphere (eg climate change) are even more pervasive and are dis-proportionately altering the areas least impacted by humans directly (polar and arid lands; IPCC 2007. . .) . Sustainable ecosystem management must therefore be directed toward developing and maintaining beneficial interactions between managed and natural systems, because avoiding these interactions is no longer a practical option (DeFries *et al.* 2004; Foley *et al.* 2005). Most importantly, though still at an early stage of development, anthropogenic biomes offer a framework for incorporating humans directly into global ecosystem models, a capability that is both urgently needed and as yet unavailable (Carpenter *et al.* 2006).

Ecologists have long been known as the scientists who travel to uninhabited lands to do their work. As a result, our understanding of anthropogenic ecosystems remains poor when compared with the rich literature on "natural" ecosystems. Though much recent effort has focused on integrating humans into ecological research (Pickett *et al.* 2001; Rindfuss *et al.* 2004 . . .) and support for this is increasingly available from the US National Science Foundation (www.nsf.gov; eg HERO, CNH, HSD programs), ecologists can and should do more to "come home" and work where most humans live. Building ecological science and education on a foundation of anthropogenic biomes will help scientists and society take ownership of a biosphere that we have already altered irreversibly, and moves us toward understanding how best to manage the anthropogenic biosphere we live in.

Conclusions

Human influence on the terrestrial biosphere is now pervasive. While climate and geology have shaped ecosystems and evolution in the past, our work contributes to the growing body of evidence demonstrating that human forces may now outweigh these across most of Earth's land surface today. Indeed, wildlands now constitute only a small fraction of Earth's land. For the foreseeable future, the fate of terrestrial ecosystems and the species they support will be intertwined with human systems: most of "nature" is now embedded within anthropogenic mosaics of land use and land cover. While not intended to replace existing biome systems based on climate, terrain, and geology, we hope that wide availability of an anthropogenic biome system will encourage a richer view of human-ecosystem interactions across the terrestrial biosphere, and that this will, in turn, guide our investigation, understanding, and management of ecosystem processes and their changes at global and regional scales.

References

Bartholome E and Belward AS. 2005. GLC2000: a new approach to global land cover mapping from Earth observation data. *Int J Remote Sens* **26**: 1959–77.

Boserup E. 1965. The conditions of agricultural growth: the economics of agrarian change under population pressure. London, UK: Allen and Unwin.

Boserup E. 1981. Population and technological change: a study of long term trends. Chicago, IL: University of Chicago Press.

Carpenter SR, DeFries R, Dietz T, *et al.* 2006. Millennium Ecosystem Assessment: research needs. *Science* **314**: 257–58.

Cox PM, Betts RA, Jones CD, *et al.* 2000. Acceleration of global warming due to carbon-cycle feedbacks in a coupled climate model. *Nature* **408**: 184–87.

Cramer W, Bondeau A, Woodward FI, *et al.* 2001. Global response of terrestrial ecosystem structure and function to CO_2 and climate change: results from six dynamic global vegetation models. *Global Change Bioi* **7**: 357–73.

Cronon W. 1983. Changes in the land: Indians, colonists, and the ecology of New England. New York, NY: Hill and Wang.

Daily GC. 1999. Developing a scientific basis for managing Earth's life support systems. *Conserv Ecol* **3**: 14.

DeFries RS, Foley JA, and Asner GP. 2004. Land-use choices: balancing human needs and ecosystem function. *Front Ecol Environ* **2**: 249–57.

Defourny P, Vancutsem C, Bicheron P, et al. 2006. GLOBCOVER: a 300 m global land cover product for 2005 using Envisat MERIS time series. In: Proceedings of the ISPRS Commission VII mid-term symposium, *Remote sensing: from pixels to processes;* 2006 May 8–11; Enschede, Netherlands.

Ellis EC. 2004. Long-term ecological changes in the densely populated rural landscapes of China. In: DeFries RS, Asner GP, and Houghton RA (Eds). Ecosystems and land-use change. Washington, DC: American Geophysical Union.

Ellis EC, Wang H, Xiao HS, et al. 2006. Measuring long-term ecological changes in densely populated landscapes using current and historical high resolution imagery. *Remote Sens Environ* **100**: 457–73.

Feddema JJ, Oleson KW, Bonan GB, et al. 2005. The importance of land-cover change in simulating future climates. *Science* **310**: 1674–78.

Folke C, Holling CS, and Perrings C. 1996. Biological diversity, ecosystems, and the human scale. *Ecol Appl* **6**: 1018–24.

Foley JA, DeFries R, Asner GP, et al. 2005. Global consequences of land use. *Science* **309**: 570–74.

Friedl MA, McIver DK, Hodges JCF, et al. 2002. Global land cover mapping from MODIS: algorithms and early results. *Remote Sens Environ* **83**: 287–302.

Galloway JN. 2005. The global nitrogen cycle. In: Schlesinger WH (Ed). Treatise on geochemistry. Oxford, UK: Pergamon.

Golley FB. 1993. A history of the ecosystem concept in ecology: more than the sum of the parts. New Haven, CT: Yale University Press.

Haxeltine A and Prentice IC. 1996. BIOME3; an equilibrium terrestrial biosphere model based on ecophysiological constraints, resource availability, and competition among plant functional types. *Global Biogeochem Cy* **10**: 693–710.

Huston M. 1993. Biological diversity, soils, and economics. *Science* **262**: 1676–80.

Imhoff ML, Bounoua L, Ricketts T, et al. 2004. Global patterns in human consumption of net primary production. *Nature* **429**: 870.

IPCC (Intergovernmental Panel on Climate Change). 2007. Climate change 2007: the physical science basis. Summary for policy makers. A report of Working Group I of the Intergovernmental Panel on Climate Change. Geneva, Switzerland: IPCC.

Levin SA. 1992. The problem of pattern and scale in ecology. *Ecology* **73**: 1943–67.

Loveland TR, Reed BC, Brown JF, et al 2000. Development of a global land-cover characteristics database and IGBP DISCover from 1 km AVHRR data. *Int J Remote Sens* **21**: 1303–30.

Matson PA, Parton WJ, Power AG, and Swift MJ. 1997. Agricultural intensification and ecosystem properties. *Science* **277**: 504–09.

Matthews E. 1983. Global vegetation and land use: new high-resolution databases for climate studies. *J Clim Appl Meteorol* **22**: 474–87.

McCloskey JM and Spalding H. 1989. A reconnaissance level inventory of the amount of wilderness remaining in the world. *Amino* **18**: 221–27.

Melillo JM, McGuire AD, Kicklighter DW, et al. 1993. Global climate change and terrestrial net primary production. *Nature* **363**: 234–40.

Mittermeier RA, Mittermeier CG, Brooks TM, et al. 2003. Wilderness and biodiversity conservation. *P Natl Acad Sci USA* **100**: 10309–13.

Netting RM. 1993. Smallholders, householders: farm families and the ecology of intensive sustainable agriculture. Stanford, CA: Stanford University Press.

Novacek MJ and Cleland EE. 2001. The current biodiversity extinction event: scenarios for mitigation and recovery. *P Natl Acad Sci USA* **98**: 5466–70.

Odum EP. 1969. The strategy of ecosystem development. *Science* **164**: 262–70.

Olson DM, Dinerstein E, Wikramanayake ED, et al. 2001. Terrestrial ecoregions of the world: a new map of life on Earth. *Bioscience* **51**: 933–38.

Pickett STA and Cadenasso ML. 1995. Landscape ecology: spatial heterogeneity in ecological systems. *Science* **269**: 331–34.

Pickett STA, Cadenasso ML, Grove JM, et al. 2001. Urban ecological systems: linking terrestrial ecological, physical, and socioeconomic components of metropolitan areas. *Annu Rev Ecol Syst* **32**: 127–57.

Qadeer MA. 2000. Ruralopolises: the spatial organisation and residential land economy of high-density rural regions in South Asia. *Urban Stud* **37**: 1583–1603.

Rindfuss RR, Walsh SJ, Turner ll BL, et al. 2004. Developing a science of land change: challenges and methodological issues. *P Natl Acad Sci USA* **101**: 13976–81.

Ruddiman WF. 2003. The anthropogenic greenhouse era began thousands of years ago. *Climatic Change* **61**: 261–93.

Salvatore M, Pozzi F, Ataman E, et al. 2005. Mapping global urban and rural population distributions. Rome, Italy: UN Food and Agriculture Organisation. Environment and Natural Resources Working Paper 24.

Sanderson EW, Jaiteh M, Levy MA, et al. 2002. The human footprint and the last of the wild. *BioScience* **52**: 891–904.

Smil V. 1991. General energetics: energy in the biosphere and civilization, 1st edn. New York, NY: John Wiley & Sons.

Smith BD. 2007. The ultimate ecosystem engineers. *Science* **315**: 1797–98.

Theobald DM. 2004. Placing exurban land-use change in a human modification framework. *Front Ecol Environ* **2**: 139–44.

Thomas CD, Cameron A, Green RE, et al. 2004. Extinction risk from climate change. *Nature* **427**: 145–48.

Turner II BL, Skole D, Sanderson S, et al. 1995. Land-use and land cover change: science/research plan. Stockholm, Sweden: International Geosphere-Biosphere Ptrogramme. IGBP Report no 35.

Vitousek PM, Mooney HA, Lubchenco J, and Melillo JM. 1997. Human domination of Earth's ecosystems. *Science* **277**: 494–99.

Wilkinson BH and McElroy BJ. 2007. The impact of humans on continental erosion and sedimentation. *Geol Soc Am Bull* **119**: 140–56.

Critical Thinking

1. How does "putting people on the map" helps us address the consuming the earth issue?

2. What empirical analyses did the authors use to construct their map of anthropogenic biomes of the world?

3. One section of the article is "Ecologist Go Home!" Explain why the authors would say this. Do you agree with their argument? Why or why not?

4. How does this article support the validity and importance of including geography and place (i.e., the Approaching Environmental Issues paradigm) in any discourse regarding our consumption of the planet?

Acknowledgements—ECE thanks S Gliessman of the Department of Environmental Studies at the University of California, Santa Cruz, and C Field of the Department of Global Ecology, Carnegie Institute of Washington at Stanford, for graciously hosting his sabbatical. P Vitousek and his group, G Asner, J Foley, A Wolf, and A de Bremond provided helpful input. T Rabenhorst provided much-needed help with cartography. Many thanks to the Global Land Cover Facility (www.landcover.org) for providing global land-cover data and to C Monfreda for rice data.

UNIT 5

The Consumption-Sustainability Conundrum: Is There an Answer?

Unit Selections

Learning Outcomes

After reading this unit, you should be able to:

- Explain why the concept of sustainability has become the environment mantra for the 21st century.

- Assess the idea of "collaborative consumption" and explain how this idea is argued to move society to new realms of consumerism.

- Offer insights on the roles that cultural values and worldviews play in the environmentally destructive behaviors affluent societies so often engage in.

- Summarize the misconceptions that the "cornucopians" have regarding sustainability.

- Explain why the greatest threat to global and environmental stability may come from a distorted belief that we can consume the Earth at the rate we are and still avoid the consequences of this appetite.

Student Website
www.mhhe.com/cls

Internet References

Alliance for Global Sustainability (AGS)
http://globalsustainability.org

The Earth Institute at Columbia University
www.earth.columbia.edu

Earth Pledge Foundation
www.earthpledge.org

Energy Justice Network
www.energyjustice.net/peak

EnviroLink
http://envirolink.org

Global Footprint Network (GFN)
www.footprintnetwork.org

Going Green
www.goinggreen.com

The Green Guide
http://environment.nationalgeographic.com/environment/green-guide/

Grass-roots.org
www.grass-roots.org
National Center for Electronics Recycling
www.electronicsrecycling.org/Public/default.aspx
Resources
www.etown.edu/vl

Sustainable Communities Online
www.sustainable.org
Terrestrial Sciences
www.cgd.ucar.edu/tss
YouSustain
www.yousustain.com/footprint/actions

Where there is great doubt, there will be great awakening; small doubt, small awakening; no doubt, no awakening.

Zen Master Dōgen, circa 1230 A.D.

This edition of *Environment 13/14* concludes with six articles in Unit 5 exploring the question of how can we continue to consume the earth—which we must—without destroying it and ourselves in the process. Currently, the answer and guiding light to our species survival has been the mantra of *sustainability.* And, it would seem that all members of the human race would intuitively agree that it makes sense to work toward building a sustainable relationship with our planet. However, when we put the two words together—*sustainable consumption*—coming to a global consensus as to what that actually means, not to mention a consensus on how we implement it and encourage (enforce) it, humanities' intuitive faculties get a little muddled.

To begin, some critics feel the term *sustainability* has become so widely used that it may be in danger of becoming meaningless. Others criticize the term for being applied to all manner of non-sustainable consumption behaviors in the hope of "greening" them. Can a "green" factory making "green" Barbie dolls really be referred to as sustainable? Are the continuing production and consumption of material artifacts of status and self-centered pleasure sustainable? Can the amount of truly unessential material stuff we produce through the transformation of earth's natural resources and purchase on Black Friday really be a sustainable way to measure the nation's economic health and wealth?

Still other critics say the definition of sustainability is so vague that it is open to any and all interpretations. *Sustainability: development that meets the needs of the present without compromising the ability of future generations to meet their own needs.* Words such as *development, meet, needs, compromise, ability,* and *future generations* will mean different things to different income groups, religious groups, ethnic groups, agricultural or urban economies, geographical regions, and nations. Article 39 by Eric Zencey examines some of these issues in "Theses on Sustainability," which provides 18 "theses" regarding the term sustainability.

On the other hand, some commentators believe that examining our resource consumption behaviors rather than engaging in continued discourse about sustainability will bear more fruit for understanding. They argue that we are steeped in excessive materialism and this obsession is threatening the earth and global stability. Yet some authors in this section see new social technologies and environmental imperatives moving us to a new realm of consumerism. But one point is agreed upon: The formulation

and implementation of policies for the greening of lifestyles and consumption policies will not be an easy task. Ultimately, most critical observers engaged in trying to find a resolution to the consumption-sustainability conundrum believe that humanity's overconsumption of the earth's natural resources must be understood at the individual level and validated at the global level.

Article 40, "Collaborative Consumption: Shifting the Consumer Mindset" by Rachael Botsman, and Article 41, "Toward a New Consciousness: Values to Sustain Human and Natural Communities" by Anthony Leiserowitz and Lisa Fernandez, argue that changes in social technologies, cultural values, and world views can change our current destructive consumption patterns. The first steps to these changes, however, must begin in the affluent societies.

Writer Jeremy Seabrook expresses concerns in Article 43, "Consuming Passions," that the wealthy in particular cannot continue to pursue the high levels of material consumption that characterizes the rich. He argues that the wealthy have a distorted kind of mysticism that we can consume the earth at our current rates and avoid the consequences of this appetite.

The last article of this section offers suggestions on how we can reduce this overconsumption. In Article 44, "What Can Be Done to Reduce Consumption," Paul Brown and Linda Cameron examine closely the overconsumption issues at the individual consumer level. The authors explore the problems associated with identifying what overconsumption is and offer approaches that can be used to reduce the problem.

To conclude this edition of *Environment 13/14,* I would like to end with an observation and a question. In the nonsocioeconomic-geopolitical world of biology and ecology, the term *sustainability* is really more of a *result* of a functioning life system than a *way to achieve* a functioning life system. In other words, living systems (cells, organisms, Earth) have the ability to regulate internal conditions, usually by a system of feedback controls, so as to stabilize health and functioning, regardless of the outside changing conditions. This is known as *homeostasis.* Thus, it is through homeostasis that an organism *achieves life sustainability.*

Is it possible for the human race to behave as a unified organism and employ this biological principle of homeostasis as the *modus operandi* to maintain the constancy of our *external* environment and thus achieve our *sustainable relationship* with Earth?

Theses on Sustainability

A Primer

Eric Zencey

[1]

THE TERM HAS BECOME so widely used that it is in danger of meaning nothing. It has been applied to all manner of activities in an effort to give those activities the gloss of moral imperative, the cachet of environmental enlightenment. "Sustainable" has been used variously to mean "politically feasible," "economically feasible," "not part of a pyramid or bubble," "socially enlightened," "consistent with neoconservative small-government dogma," "consistent with liberal principles of justice and fairness," "morally desirable," and, at its most diffuse, "sensibly far-sighted."

[2]

NATURE WILL DECIDE what is sustainable; it always has and always will. The reflexive invocation of the term as cover for all manner of human acts and wants shows that sustainability has gained wide acceptance as a longed-for, if imperfectly understood, state of being.

[3]

AN ACT, PROCESS, OR STATE of affairs can be said to be economically sustainable, ecologically sustainable, or socially sustainable. To these three some would add a fourth: culturally sustainable.

[4]

NATURE IS MALLEABLE and has enormous resilience, a resilience that gives healthy ecosystems a dynamic equilibrium. But the resiliency of nature has limits and to transgress them is to act unsustainably. Thus, the most diffuse usage, "sensibly far-sighted," is the usage that contains and properly reflects the strict ecological definition of the term: a thing is ecologically sustainable if it doesn't destroy the environmental preconditions for its own existence.

[5]

ECONOMIC SUSTAINABILITY describes the point at which a less-developed economy no longer needs infusions of capital or aid in order to generate wealth. This definition is misleading: for many of those who use it (including traditional economists and many economic aid agencies), "economic sustainability" means "sustainable within the general industrial program of using fossil fuels to generate wealth and produce economic growth," a program that is, of course, not sustainable.

[6]

SOCIAL SUSTAINABILITY describes a state in which a society does not contain any dynamics or forces that would pull it apart. Such a society has sufficient cohesion to overcome the animosities that arise from (for instance) differences of race, gender, wealth, ethnicity, political or religious belief; or from differential access to such boons as education, opportunity, or the nonpartisan administration of justice. Social sustainability can be achieved by strengthening social cohesion (war is a favorite device), through indoctrination in an ideology that bridges the disparities that strain that cohesion, or through diminishing the disparities themselves. (Or all three.)

[7]

CULTURAL SUSTAINABILITY asks that we preserve the opportunity for nonmarket or other nonindustrial cultures to maintain themselves and to pass their culture undiminished to their offspring.

[8]

HUMAN CIVILIZATION has been built on the exploitation of the stored solar energy found in four distinct carbon pools: soil, wood, coal, petroleum. The latter two pools represent antique, stored solar energy, and their stock is finite. Since agriculture and forestry exploit current solar income, civilizations built on the first two pools—soil and wood—had the opportunity to be sustainable. Many were not.

[9]

THE 1987 UN BRUNDTLAND REPORT offered one widely accepted definition of what sustainability means: "meet[ing] the needs of the present without compromising the ability of future generations to meet their own needs." This definition contains within it two key concepts. One is the presumption of a distinction between needs and wants, a distinction that comes into sharp relief when we compare the consumption patterns of people in rich and in poor nations: rich nations satisfy many of their members' wants—indeed, billions of dollars are spent to stimulate those wants—even as poor nations struggle to satisfy human needs. Two: we face what Brundtland called "limitations imposed by the state of technology and social organization on the environment's ability to meet present and future needs."

[10]

THAT A DISTINCTION can usefully be drawn between wants and needs seems obvious. Mainstream economics, however, refuses to countenance such a distinction. (Marxist economics does, which, from the viewpoint of an ecologically enlightened economics, is one of the few ways in which it is distinguishable from its neoclassical alternative.) The work of Wilfred Pareto was crucial to this refusal. His contribution to economic theory marks a turning point in the evolution (some would say devolution) of nineteenth-century political economy into the highly mathematized discipline of economics as we know it today. Pareto's novel idea: because satisfactions and pleasures are subjective—because no one among us can say with certainty, "I like ice cream more than you do"—there is no rational way to compare the degree of pleasure that different people will gain by satisfying desires. All we can do is assert that if an economic arrangement satisfies *more* human wants, it is objectively better than an arrangement that satisfies *fewer* human wants. This seems commonsensical until we unpack that caveat "all we can do." An economic arrangement achieves Pareto Optimality if, within it, no one can be made better off (in his own estimation) without making someone else worse off (in her own estimation). Economic science, in its desire to be grounded on rational, objective principles, thus concludes that were we to take a dollar from a billionaire and give it to a starving man to buy food, we can't know for certain that we have improved the sum total of human satisfaction in the world. For all we know, the billionaire might derive as much pleasure from the expenditure of his billionth dollar as would a starving man spending a dollar on food. All we can do—all!—is promote the growth of income; and if we care about that starving man, we must work to produce two dollars' worth of goods where before there was only one, so that both the billionaire and the starving man can satisfy their wants.

[11]

THUS WAS neoclassical economic theory, putatively value-free and scientific, made structurally dependent on a commitment to infinite economic growth, a value-laden, unscientific, demonstrably unsustainable commitment if ever there was one.

[12]

THE BRUNDTLAND assertion that we face "limitations imposed by the state of technology and social organization on the environment's ability to meet present and future needs" can be read as both acknowledging ecological limits to human activity and as sidestepping the major issue that those ecological limits have brought to the fore. Can humans, through technological development, solve any problem brought on by resource scarcity and the limited capacity of ecosystems to absorb our acts and works? When all is said and done, can we enlarge the economy's ecological footprint forever in order to create wealth? Gradually, we are coming to recognize that the answer is no.

[13]

AN ECONOMY CAN BE MODELED as an open thermodynamic system, one that exchanges matter and energy across its border (that mostly conceptual, sometimes physical line that separates culture from its home in nature). An economy sucks up valuable low-entropy matter and energy from its environment, uses these to produce products and services, and emits degraded matter and energy back into the environment in the form of a high-entropy wake. (Waste heat. Waste matter. Dissipated and degraded matter: yesterday's newspaper, last year's running shoes, last decade's dilapidated automobile.) An economy has ecological impact on both the uptake and emission side. The laws of thermodynamics dictate that this be so. "You can't make something from nothing; nor can you make nothing from something," the law of conservation of matter and energy tells us. With enough energy we could recycle all the matter that enters our economy—even the molecules that wear off the coins in your pocket. But energy is scarce: "You can't recycle energy," says the law of entropy. Or, in a colloquial analogy: Accounts must balance and bills must be paid. To operate our economic machine we pay an energy bill; we must ever take in energy anew.

[14]

ESTABLISHING an ecologically sustainable economy requires that humans accept a limit on the amount of scarce low entropy that we take up from the planet (which will also, necessarily, limit the amount of degraded matter and energy that we emit). An effective approach would be to use market mechanisms, such as would occur if we had an economy-wide tax on low-entropy uptake (the extraction of coal and oil, the cutting of lumber). The tax rate could be set to ensure that use doesn't exceed a limit—the CO_2 absorption capacity of the planet, the regenerative ability of forests. Producers and consumers would have freedom under the cap brought about by the tax. With such a tax, the tax on workers' income could be abandoned. (As the slogan says, we should "tax bads, not goods." Work is good. Uptake of scarce resources is bad.)

[15]

FOR DECADES environmentalism has been primarily a moral vision, with principles susceptible to being reduced to fundamentalist absolutes. Pollution is wrong; it is profanation. *We have no right,* environmentalism has said, to cause species extinction, to destroy habitat, to expand the dominion of culture across the face of nature. True enough, and so granted. But even Dick Cheney agreed that environmentalism is essentially, merely, a moral vision. ("Conservation," he said, on his way to giving oil companies everything they wanted, "may be a personal virtue, but it is not a sufficient basis for a sound, comprehensive energy policy.") The time has long since passed for the achievement of sustainability to be left to simple moral admonition, to finger-wagging in its various forms. It's time to use the power of the market—the power of self-interest, regulated and channeled by wise policy—to do good. Environmentalism must become an economic vision.

[16]

ACCEPTING A LIMIT on the economy's uptake of matter and energy from the planet does not mean that we have to accept that history is over, that civilization will stagnate, or that we cannot make continual improvements to the human condition. A no-growth economy is not a no-development economy; there would still be invention, innovation, even fads and fashions. An economy operating within ecological limits will be in dynamic equilibrium (like nature, its model): just as ecosystems evolve, so would the economy. Quality of life (as it is measured by the Index of Sustainable Economic Welfare, an ecologically minded replacement for GDP) would still improve. If a sustainable economy dedicated to development rather than growth were achieved through market mechanisms, consumers would still reign supreme over economic decision making, free to pursue satisfactions—and fads and fashions—as they choose.

[17]

OUR CHALLENGE is to create something unprecedented in human history: an ecologically sustainable civilization that offers a high standard of living widely shared among its citizens, a civilization that does not maintain itself through more-or-less hidden subsidies from antique solar income, or from the unsustainable exploitation of ecosystems and peoples held in slavery or penury, domestically or in remote regions of the globe. The world has never known such a civilization. Most hunting-and-gathering tribes achieved a sustainable balance with their environments, living off current solar income in many of its forms rather than on the draw-down of irreplaceable stocks, but we can't say that any of them achieved a high standard of material well-being. Medieval western Europe lived in balance with its soil community, achieving a form of sustainable agriculture that lasted until the invention of coal- and steam-propelled agriculture a few centuries ago, but few of us would trade the comforts and freedoms we enjoy today for life as a serf on a baronial estate, or even for the pre-electricity, pre-petroleum life of a mid-nineteenth-century farmer.

[18]

NO, THERE IS NO PRECEDENT for what we are struggling to create. We have to make it up ourselves.

Critical Thinking

1. What other issues regarding sustainability can you extrapolate from Thesis 9 that we will have to address in the near future?

2. Why might the idea of "sustainable economy" (Thesis 5) be misleading?

3. Make a list of what you think are "wants" and "needs" (10 each). After that, imagine you make less than $2 a day and live in Sub-Saharan Africa. Compare and contrast the lists.

4. What does the author mean that "a no-growth economy is not a no-development economy"?

5. Describe how you incorporate the biology principle of "homeostasis" into the socioeconomic idea of sustainability.

Collaborative Consumption: Shifting the Consumer Mindset

Collaborative consumption is organized sharing, bartering, lending, trading, renting, gifting and swapping through online and real-world communities. "What's Mine Is Yours: The Rise of Collaborative Consumption" explores this invigorating shift from an unfettered zeal for individual getting and spending toward a rediscovery of collective good.

RACHEL BOTSMAN AND ROO ROGERS

There is something sad about all this stuff we work so hard to buy, can't live with, but inevitably can't bear to part with. In the same way that we focus on where to bury our waste, not where the waste came from, we also spend inordinate amounts of energy and money storing excess stuff rather than asking the hard truths of why we have so much in the first place. The comedian George Carlin riffed on this in his classic stand-up routine about stuff: "The whole meaning of life has become trying to find a place to put your stuff. . . . Have you ever noticed how other people's shit is shit and your stuff is stuff?" The controversial David Fincher movie *Fight Club* struck a painful chord with viewers who have ever experienced that addictive feeling of always wanting more, regardless of how much they have. Most people remember two lines from the movies: "The first rule of Fight Club—you do not talk about Fight Club" and "The things you own end up owning you."

Tyler and Jack, the two main characters in the movie, seem to represent the stark choice that modern consumerism offers, best summarized by esteemed German social psychologist Erich Fromm as "To Have or to Be." Jack (Ed Norton), is a stereotypical 30-year-old insomniac yuppie who keeps trying to fill his emotional voids and feel "complete" with the things he acquires. "I flip through catalogues and wonder what kind of dining set defines me as a person." But no matter what Jack buys, he's never satisfied. That's before he meets Tyler (Brad Pitt), who, throughout the movie, takes anticonsumerist jabs such as, "You are not the clothes you wear. You are not the contents of your wallet. . . . You are not your grande latte. You are not the car you drive. You are not your fucking khakis. You're the all-singing, all-dancing crap of the world." Tyler shows Jack that acquiring more and more stuff is a meaningless pursuit devoid of purpose and fulfillment. "Goddamn it. . . . Advertising has us chasing cars and clothes, working jobs we hate so we can buy shit we don't need." The main theme of *Fight Club* runs counter to much of what consumer advertising preys on—we won't find happiness or the meaning of our lives in the shopping mall or in the click of a mouse.

Research has proved that people who can afford to buy and hold on to more material goods are not necessarily more satisfied with their lives. Indeed, the reverse is often true. Economist Richard Layard has researched the relationship between growth, hyper-consumerism and happiness. His findings are illustrated by a graph on which one line represents a soaring increase of income and personal consumption per capita since 1950 (it has more than doubled) and the other line, marking Americans and Britons that describe themselves as "very happy" in an annual Gallup survey, remains flat. In fact, the number of people reporting to be "very happy" peaked in 1957 just as the conspicuous cycle of "work and spend," and a revolution of rising materialistic expectations, began. Happiness became an elusive moving target. Nothing was ever enough.

Telling societal indicators paint a vivid picture of this decrease in well-being. Since 1960 the divorce rate has doubled in the United States; teen suicide rates have tripled; violent crime has quadrupled; the prison population has quintupled; and the percentage of babies born to unmarried parents has sextupled. Not exactly indicators of a satisfied consumer society. And it is only getting worse, as indicated by the massive increase in depression, anxiety, insomnia, heart disease, and obesity since the 1980s. As political scientist Robert Lane comments in *The Loss of Happiness in Market Democracies,* "The appetite of our present materialism depends upon stirring up our wants—but not satisfying them." Economists describe this emotional phenomenon as the "hedonic treadmill." We work hard to acquire more stuff but feel unfulfilled because there is always something better, bigger, and faster than in the present. The distance between what we have and what we want, the "margins of discontent," widens as the number of things we own increases. In other words, the more we have, the more we want.

We are taught to dream and desire new things from an early age, when asked frequently "What do you want for Christmas?" or "What do you want for your birthday?" Susan Fournier and Michael Guiry, former associate marketing professors at Harvard Business School, conducted a study called Consumption Dreaming Activity. They asked participants, "What things would you like to own or do someday?" Contrary to the researchers' expectations, the lists varied little regardless of sex, income, education or standard or living. Generally speaking, lists were full of desires for material possessions; almost half the sample (44 percent) mentioned new cars; more than one in four (29 percent) listed luxury items such as yachts, antiques, jewelry, and designer clothes; and 16 percent just asked for the money—enough to buy anything they could possibly want. Where the study gets most interesting is not just the type of items respondents wrote down but the level of detail and elaboration they included; 42 percent of all things listed were described vividly. One participant wrote down not just wanting a car but "an emerald green Jaguar."

As the professors noted, "This level of detail and elaboration could reflect that consumers have 'perfect things' in mind when they formulate wish lists." Here we see the amount of time and headspace most of us give to future purchases. Not only do the things we own fill up our closets and our lives, but they also fill our minds.

Changing the Consumer Mindset

The ideological debate between those who believe in self-interest as the purest way to maximize production and those who believe that it operates as an affront to the collective good and equality has dominated our political, economic and philosophical discourse for centuries. But while we've debated, the world has continued undistracted down a path of self-destructive growth. It is through the fog of anxiety that Collaborative Consumption has emerged with a simple consumer proposition. It meets all the same consumer needs as the old model of mass consumption but helps address some of our most worrying economic and environmental issues. While it is complex to audit and project the entire environmental impact of Collaborative Consumption, it reduces the number of new products and raw materials consumed and it creates a different consumer mindset.

When Jonathon Porritt was chairman of the Ecology Party in the U.K. from 1978 to 1984, he and his colleagues struggled with what became known as the Great Washing Machine Debate. Porritt, a leading environmental-thought leader in the U.K., serves as an adviser to many entities, from Marks & Spencer to Prince Charles to the sustainability think tank Forum for the Future. In the 1980s, when he was still active in the Ecology Party, which would later be renamed the Green Party, he was faced with the problem of what to do with the deterioration of one of the first mainstream product service systems: the laundromat. At the time, masses of people were going to shopping centers to purchase personal home washing machines, either for the first time, for upgrade or for replacement, resulting in what Porritt calls a "staggering increase in the number of personally owned machines." Between 1964 and 1992, the percentage of homes in the U.K. alone with washing machine rose from 53 percent to 88 percent. At the same time, 50 percent of all launderettes closed. Given that the average home washing machine was only used four to five times a week and consumed more than 21.7 percent of our personal water usage, and that each year around 2 million used washing machines were being discarded, Porritt and his colleagues were concerned. The move away from collective services and toward a self-service society had serious environmental implications.

The Ecology Party considered two choices: Lobby for some form of governmental taxation and incentives, or undertake a strong messaging program to change the consumer mindset back toward using laundromats. Neither option was attractive. Government was slow and infatuated with economic growth over sustainable causes. And a strong messaging campaign would only alienate the consumer. "By being over-prescriptive you become your own worst enemy and force people into even more defensive and negative behavior," Porritt said.

The Great Washing Machine Debate was just one example of a larger struggle taking place in the environmental movement. How do you address the public and inspire sustainable behavior without being negative or dogmatic? According to Porritt, this issue is still huge in the environmental movement today, which recognizes the inefficacy of trying to guilt people into a more sustainable choice but nonetheless struggles to find an alternative.

Fast-forward 20 years, and there is a different answer to the Great Washing Machine Debate and the conflict between enticing self-interest and ensuring social good. It exists at 122 Folsom St. in San Francisco. Brainwash is a laundromat founded in 1999 by Jeffrey Zalles. Zalles admits that his primary concern isn't being green. What he's done is figured out how to make Laundromats cool again. Brainwash woos customers with additional offerings of a café, happy hours, live music, stand-up comedy nights, pinball machines, free Wi-Fi, and even a place to do your homework. The space is bright and modern, with indoor and outdoor seating, cool music playing, funky artwork on the walls and helpful, friendly staff—a little different from the dark and dingy experience associated with most laundromats.

A big part of Brainwash's success is based on a simple insight: Customers need something to do while waiting for their laundry to finish, and it needs to be better than what they would do at home. That's why the sense of community that Zalles has built through cultural events and Meetups is so smart and critical. "Everyone who comes here could afford to get their own home washing machine . . . but where is the fun in that!" says Zalles. Indeed, the demand for Brainwash is so overwhelming that he is looking to open more franchises this year.

The idea behind Brainwash is simple, but the behavioral impact is significant. And by doing so it achieves seemingly opposing outcomes: clean clothes, fun, friends, affordability and environmental responsibility. Instead of forcing consumers to sacrifice personal convenience and comfort for the doing the right thing, Jeff makes the right thing more attractive. By diversifying the motivations and putting the emphasis on consumer experience over a prescriptive sense of obligation, Brainwash achieves Porritt's goal with barely a whimper of activist politics. Brainwash hardly identifies itself with its purpose. Does it exist to provide a cheap alternative to get your clothes washed? Is it a cool café and culture club where you can hobnob and hang out? Or is a powerful green statement? The answer, of course, is all of the above.

The key difference in Brainwash's approach to the Great Washing Machine Debate is that instead of trying to change consumers, the system itself has changed to accommodate needs and wants in a more sustainable and appealing way, with little burden on the individual. In this respect, Collaborative Consumption actually enables an entitled, self-interested consumer who is so well-served he doesn't even realize he is doing something different or "good." By taking an indirect, open-ended approach, Collaborative Consumption enables consumers to break down the stereotypes of collectivism or environmentalism and simply do what works the best for them. It is so intuitive to our basic needs that consumers often fall into it by accident. One could argue that it doesn't matter whether the system leads to a change in mindset as long as it converts our consumption into positive outcomes: fewer products, more efficient usage, less material consumed, reduced waste and more social capital.

We've seen certain consistent and specific motivations for participating in Collaborative Consumption: cost savings, coming together, convenience, and being more socially conscious and sustainable. The fact that it attracts new consumers based on traditional self-interested motivation, including money,

value and time, and that it converts this into positive social and environmental outcomes, should not distract from its overall impact on consumer behavior.

When someone enters Collaborative Consumption through one particular door—a clothing exchange, a car-sharing scheme or a [laundromat]—they become more receptive to other kinds of collective or community-based solutions. Over time, these experiences create a deep shift in consumer mindset. Consumption is no longer an asymmetrical activity of endless acquisition but a dynamic push and pull of giving and collaborating in order to get what you want. Along the way, the acts of collaboration and giving become an end in itself. Collaborative Consumption shows consumers that their material wants and needs do not need to be in conflict with the responsibilities of a connected citizen. The idea of happiness being epitomized by the lone shopper surrounded by stuff becomes absurd, and happiness becomes a much broader, more iterative process.

Critical Thinking

1. What is "collaborative consumption"?
2. How is/isn't the idea compatible with the ideas of sustainability?
3. Identify some similarities between the points made in this article and Article 4 regarding consumption.
4. Do you think the idea of collaborative consumption would work better in "developing countries" with a growing middle class than nations with small, powerful, wealthy elites?

From *Mother Earth News*, September/October 2008, pp. 1–6. Copyright © 2008 by Ogden Publications, Inc. Reprinted by permission.

Toward a New Consciousness: Values to Sustain Human and Natural Communities

Anthony A. Leiserowitz and Lisa O. Fernandez

O ur world, our only habitat, is a biotic system under such stress it threatens to fail in fundamental and irreversible ways. Major change is required to stabilize and restore its functional integrity. Examine any of the great environmental challenges confronting us—climate change, biodiversity loss, pollution, resource depletion—and a similar pattern emerges. A modest number of people know a great deal about these afflictions and unfolding tragedies, but their messages have difficulty overcoming public apathy, political denial, or entrenched opposition. Most of all, these messages rarely spur responsive public action, basic shifts in values and attitudes, or the behavioral change needed at the scale or within the time frame required. The result is what is commonly referred to as a failure of political will, but this phrase fails to capture the depth of the cultural void or social malfunction involved.

At its deepest level, if we are to address the linked environmental, social, and even spiritual crises, we must address the wellsprings of human caring, motivation, and social identity. Many have concluded that what we need is a major shift in our core values and dominant culture—in effect, the evolution of a new consciousness. Aldo Leopold wrote to a friend in 1944 that little could be done in conservation "without creating a new kind of people."[1] Peter Senge and his colleagues have similarly argued, "When it is all said and done, the only change that will make a difference is the transformation of the human heart."[2]

To explore these themes, the Yale School of Forestry & Environmental Studies convened an esteemed group of leaders representing diverse disciplines, including the natural sciences, social sciences, philosophy, communications, education, religion, ethics, public policy, business, philanthropy, history, the creative arts, and the humanities.[3] The conference focused on the role of cultural values and worldviews in environmentally destructive behavior within affluent societies—patterns that are being adopted throughout the world, including the rising centers of Western-style affluence in the developing world. The conference was intented to help catalyze further investigation of the critical role of cultural values and worldviews in the global environmental crisis and the implementation of concrete initiatives to accelerate a paradigm shift in human values, attitudes, and behaviors toward the natural world.

Diagnoses

The failure of the developed world to fully comprehend or confront the size, severity, and urgency of the global environmental crisis requires a deep examination of the prevailing worldviews, structures and institutions, and norms and beliefs within modern society that maintain and reinforce a self-destructive relationship with the natural world.

Worldviews

*   *Anthropocentrism, materialism, and alienation from nature.* The anthropocentric notion that humans stand "above" and independent of nature, rather than "within" and interdependent with it, has deep cultural and historical roots, some argue, dating back at least to the biblical cosmology of Genesis. Further, since the Enlightenment, the reigning scientific worldview has held that matter is dead and inert, encouraging human beings to believe that they can manipulate and rearrange the material world any way they like, with few moral or ethical constraints, duties, or obligations. One result is that members of modern societies are increasingly physically, psychologically, and culturally separated from the natural world. We live in a system that has severed or rendered invisible many of our connections to nature. The packaged chicken in the grocery store has been cleaned, sanitized, and presented in a way that disguises that it was once a living, breathing animal, that inhabited a particular place (a factory farm), was bred, fattened, pumped with growth hormones and antibiotics, and slaughtered by migrant workers. The cell phone is an

assemblage of literally hundreds of material elements, mined, milled, and gathered from around the world, manufactured, assembled, distributed, and disposed of by faceless people in unknown places, with unknown environmental consequences. The entire edifice of the global economic system is constructed upon this underlying worldview and accompanying detachment of products from their natural origins.

As a result, there are few daily reminders of the natural world as the foundation on which civilization stands. People, especially children, are spending less and less time outside in natural settings, which some have called the "extinction of nature experience."[4] Human contact with other species and wild nature is increasingly mediated through the television, constrained within the safe confines of the rectangular screen. There seems to be a growing societal blindness to the beauty, succor, and necessity of the more-than-human world.

Surveys do find that people around the world strongly profess environmental values,[5] yet these values are increasingly less rooted in actual experience and interaction with nature and thus begin to float free, untethered, unintegrated into everyday behavior. The well-documented gap between our professed environmental values and actual behavior stems in part from this increasing detachment from the natural world.

- *Reductionism.* The prevailing scientific worldview seeks understanding by breaking complex objects of study into smaller and smaller parts, with the assumption that complex behavior is the simple result of the interaction of these parts. Thus, if we can just understand and model the behavior of each piece, we will understand the behavior of the whole.[6] Over the centuries, this approach has generated tremendous advances in scientific knowledge, leading to the establishment of disciplinary fields of expertise. At the same time, however, this approach has led to hyperspecialization within science, where entire subdisciplines and entire careers are spent investigating smaller and smaller twigs on the "tree of knowledge." As a result, many researchers can no longer understand the breadth of their own discipline, much less how their discipline might intersect with others.

  This approach, however, has been recently challenged by the findings of systems and complexity theory, which demonstrate the existence of emergent properties unpredictable from the interaction of their constituent parts in systems ranging in size from microscopic to cosmological, in disciplines as diverse as chemistry, ecology, and astronomy. Likewise, interdisciplinary research has received increasing attention and funding, as scientists and funders have recognized the importance of holistic and systems perspectives at play in natural and social phenomena and the environmental crisis.

- *Binary and dichotomous thinking.* Good versus evil, humans versus nature, economy versus environment: binary or dichotomous thinking is often problematic, as it separates the world into simplistic, separate, and opposing categories, while privileging one of the two. Lost is the potential for gray areas of difference, "win-win" solutions, or the possibility of an interdependent relationship between the two. For example, there are tremendous opportunities to protect the environment while growing the economy—for instance, through green jobs and renewable energy technologies. On a deeper level, the dualistic separation of humans and nature reinforces the false notion that humans are outside and above nature and natural processes, instead of emergent from and inextricably interconnected to them.

- *Radical individualism.* American society often privileges competition over collaboration and individualism over community, equity, or social justice. Meanwhile, studies have demonstrated that radical individualists are less likely to believe environmental problems exist and more likely to oppose environmental policies and programs.[7] Taken to an extreme, individualism privileges personal autonomy at the expense of what is best for communities or society as a whole. While individualism remains a core value, it also needs to be balanced with other core values, such as equality, fairness, and justice.

- *Economism.* Just as all cultures have a complex of myths about nature and the proper human relation to nature, so do we have a complex of myths about the economy, which can collectively be referred to as economism. Given the privileged place economic analysis holds in policymaking and the acquiescence of other disciplines to the rules of economic discourse, many individual decisions, some with deep moral implications, are now determined primarily by income and prices. We increasingly perceive and understand "reality" from our particular position in the economic system and perceive the value of others and nature through an economic lens. Our dreams for the future are often dominated by portrayals of economic and material progress. The field of economics makes a number of often unquestioned, flawed assumptions, such as the belief in a direct and consistent relationship between income and human well-being, an autonomous, rational-actor model of human decisionmaking and behavior, that the economy is independent of ecology, and that perpetual economic growth is possible on a planet of finite resources. Meanwhile, the implementation of these ideas in the real world is a major driver of the environmental crisis.

- *Cornucopianism and technological optimism.* For centuries, the bounty of nature seemed unlimited. In the twentieth century, however, the world witnessed an explosion in scientific knowledge and technology and an accompanying exponential increase in human beings' power to exploit nature. While science and technology have unquestionably improved human health and well-being, technologies invented to solve one problem have often had unanticipated and negative human or ecological consequences—for example, the pesticide DDT. Further, science and technology do not operate in a vacuum—scientific and technological advances are mediated and inflected through existing social structures, norms, and values. In turn, outside forces like venture capital drive much scientific research and lead to the development of certain technologies and not others, based on market values. Finally, science and technology have vastly increased the human impact on the natural world, ranging from individual environmental disasters, like Chernobyl and the *Exxon Valdez* oil spill, to large-scale problems like climate change and the ozone hole. We have now entered the Anthropocene era, in which human beings are one of the dominant forces of change on the planet. This rate and scale of the human impact is radically new and is due in large part to the exponential increase in the human ability to manipulate the world. Finally, while environmental science and green technologies will certainly be important contributors to the effort to find solutions to global environmental problems, such as climate change, overfishing, biodiversity extinctions, and ocean acidification, they alone are insufficient to solve these problems, which are also rooted in politics, economics, social relations, and culture.

Structures and Institutions

- *Media: balance and compartmentalization.* "Balanced" and "objective" reporting are core values of the news media. Perversely, however, the implementation of these values has led to misleading news coverage of critical environmental issues. "Balance" has often been interpreted as meaning that each side of a debate merits equal mention. Thus many news stories have, in the interest of "balance," placed the views of the overwhelming majority of scientists on a level playing field with a small minority of dissenters, leading to the false impression that there is more scientific controversy about an issue than actually exists.

  Likewise, too many environmental news stories frame environmental issues only in terms of natural science or politics. For example, many environmental stories describe human impacts on the natural world, without necessarily connecting these impacts back to human beings. Meanwhile, stories about environmental justice—the disproportionate environmental harms imposed on the poor, people of color, and the disempowered—often fail to get adequate attention. Even climate change has often been described in terms of its impacts on nonhuman nature, such as glaciers or polar bears, with inadequate attention to the potential impacts on human beings or the implications for global environmental justice.

- *Academia: disciplinary silos.* Disciplines within academia (natural and social sciences and the humanities) are often isolated from one another. More broadly, too many academics talk only to each other, using language and jargon incomprehensible to even the educated layperson. The traditional disciplinary

structure, along with the reward system of academia (status, tenure, and promotion) all constrain the holistic, integrated, and interdisciplinary research and teaching required to address environmental problems.

- *Humanities: an anthropocentric focus.* The humanities, as evidenced by their very name, continue to retain an almost exclusive focus on human beings and their affairs, often treating the natural world as a mere backdrop to human history and culture. There is a burgeoning genre of nonfiction nature writing; however, it remains marginalized within the study of literature. Meanwhile, this genre itself has historically been dominated by "cabin" and "wilderness" narratives of lone individuals confronting and reflecting upon the natural world.[8] Many culturally, racially, and ethnically diverse voices are now emerging, often challenging deeply held conceptions of the human-nature relationship.[9]

- *Environmentalism: an inadequate reach.* Some argue that environmentalism largely remains a reform movement committed to the assumption that the environmental crisis can be solved within the current political and economic system, without challenging underlying values or questioning contemporary lifestyles.[10] For 40 years, the environmental movement has worked to develop new policies, regulations, and legislation to protect the environment and relied on large expert bureaucracies and the judicial system to enforce these rules and regulations. Likewise, many environmentalists today are working to promote green thinking and practice within corporations and consumer markets. Working within the system, environmentalists have tended to be pragmatic and incrementalist, often focused on solving individual problems rather than addressing deeper underlying causes. Environmentalism needs to sharpen its critique of contemporary culture, economics, and politics; reach out and form alliances with other social movements; invest in the intellectual development of core concepts, ideals, and values; and wage effective campaigns to win hearts and minds.

- *Policy: dysfunctional political systems.* Many political systems are crippled by cronyism, revolving doors, corporate influence, lobbyists, special interests, gerrymandering, scandal, and a lack of inspired leadership. The local level, however, is proving to be fertile ground for transformative action. Cities, counties, states, and other local groups have taken bold action to address both local and global issues, such as climate change. While serving as the inspiration and testing ground for new ideas and approaches, however, these local solutions ultimately have to be scaled up to the national and international levels if we are to successfully deal with our global environmental challenges.

- *Philanthropy: a lack of holistic, strategic, and systems thinking.* The philanthropic sector often invests in projects to fix pressing environmental and social problems. Philanthropic organizations have become very good at describing what they are against, but rarely do they invest in projects that help articulate what they are for—detailed, concrete, and positive visions of a better world and roadmaps to help get us there. Thus, much of philanthropic giving has been relatively tactical and piecemeal, not strategic. This tendency is reinforced by the corporatization of foundations, with increasing emphasis placed on quantifiable, short-term results.

Norms and Beliefs

- *Environmental issues lack urgency.* Many political leaders and members of the public in the United States have not yet comprehended the urgency of the environmental crisis. While the sense of urgency about climate change has grown recently, it still is underappreciated, and we are running out of time to avoid the worst consequences. Meanwhile, climate change is just one of many global environmental stressors that have potentially disastrous consequences, yet barely register on the radar screens of leaders (including ocean acidification, nitrogen pollution, overfishing, patterns of consumption, and population growth). Although the broad public professes positive environmental attitudes and expresses concern about the state of the world's environment,[11] a very large gap between individual and societal attitudes and behaviors clearly remains.

- *Scientists should not advocate.* Many scientific disciplines are currently struggling with the proper role of science and the scientist in society. Some argue that scientists should focus only on the production of scientific facts and leave value judgments to policymakers and the public. They further argue that when scientists speak out as advocates for action, they diminish the public perception that scientists provide objective truth, debase scientific credibility, and reduce scientists to just another special interest group prone to making up, selecting, or distorting facts to fit a preestablished subjective agenda.

  In response, other scientists argue that science—through the scientific method and rigorous empiricism—has identified and described a wide array of human factors currently tearing ecosystems apart, degrading human health and well-being, and destroying the life-support systems of the planet, in rapid and irreversible succession. Given these pervasive and dangerous impacts, these scientists argue that to stand by and say nothing, especially given scientists' unique understanding of what is happening, is problematic at best and immoral at worst.

- *Environmental behavior is an individual responsibility.* The prototypical environmental act today is recycling—primarily an individual behavior. Likewise, individuals are told they should buy green products, turn down the thermostat, buy compact fluorescent light bulbs, drive less, buy more fuel efficient cars, eat organic, and eat local. Meanwhile, relatively little attention is focused on the vital need for systemic changes in collective behavior. Political action, carbon pricing, government incentives and subsidies for clean energy development, and increased regulation of polluters are all examples of social policies and behaviors that are required to deal with the environmental crisis. Individual consumption and conservation, while important on many levels, are simply inadequate to address the scale and scope of our current challenges.

- *Consumerism as the basis of self-identity.* The desire for and expression of individual identity has become a major force in modern culture and societies. These desires have been amplified and exploited by marketers to sell products. Individuals now adopt distinct "lifestyles" or particular ensembles of material products, homes, color schemes, and hobbies that become both sources of individual identity and the means by which these identities are signified to others. This process helps to fuel consumerism, which is the primary engine of many developed economies. These economies in turn drive much of the increasing exploitation and degradation of the global environment. As personal identity becomes further entangled with consumer behavior, it becomes harder and harder to challenge existing patterns of consumption.

Prescriptions

After diagnosis comes the difficult but critical challenge of searching for cures. We must ask ourselves what kind of a world we want to live in, what kind of world we want our descendents to live in, and how we can get there. Given the enormity of the task, the following proposals certainly do not exhaust the realm of possibilities. Effecting a mass change in public environmental values, priorities, and behavior will require the concerted efforts of millions of committed individuals and organizations seeking a better and more sustainable world—a movement already well under way. These proposals are intended to spark a broader conversation about ways to catalyze deep change and inspire others to search for, create, and implement their own answers to these fundamental questions.

New Narratives

New narratives are needed to help guide and inspire social transformation and changes in the practice of science and education, religion and ethics, and policy and economics. Narratives ranging from sacred texts to national myths to individual life stories give meaning, order, and direction to the lives of individuals and entire societies. It is vital that we create new narratives that:

- Vividly depict the kind of world we are for, not just the problems we are against.
- Raise fundamental questions: How should individuals and societies measure success: ever higher incomes, growing GNP, greater material consumption? What truly makes individuals happy?

- Re-envision "the good life" and alter the trajectory of ever-greater material consumption: "Rich lives instead of lives of riches."
- Articulate ecocentric and biophilic ways of thinking. In this view, humanity is understood as coexisting within nature—a community that includes land, water, air, and biota. The central challenge is for humans to conceptualize ourselves as existing as part of and because of the biosphere. Our ecological niche is now the entire planet, but cultural evolution has not yet caught up to this new fact. We must now adapt to this global scale by reconceptualizing our relationship to nature.
- Emphasize themes of health and wellness. The global environmental crisis is part of a broader set of enormous challenges to human physical and mental health, the health and viability of other species, and planetary health.
- Reclaim the word "sacrifice." Human beings have long been willing to sacrifice their comfort, possessions, and even their lives for freedom, for equality, for God, for country. How can we reinvigorate and harness this force for the common good?
- Invoke the language of faith and spirituality. The discourses of science and policy, while necessary, are not sufficient to motivate mass changes in values and behavior. The work in world religions and ecology has important contributions to make in this regard. Many people will be more motivated to save the planet if the sacredness of creation is included in the conservation message. The sense of an enchanted, awe-inspiring universe and creation can reawaken a commitment to the Earth.
- Embed the human story in the story of the universe. A deep understanding of modern cosmology places human beings within the grand narrative of the universe—from the Big Bang, to the formation of galaxies, the coalescing of Earth and the solar system, and the origins and evolution of life. This narrative reminds us that human beings are not separate from nature and its processes—we emerged from it, we are the descendents of a vast, complex, ancient, and beautiful universe, inhabitants of an incredibly precious planetary home, and kin, genetically, to all other life on Earth. These ideas and this story fundamentally challenge our traditional understandings of what it means to be human in relation to the natural world.

Conference participants suggested several ways to promote and disseminate these new narratives, including the development of films and television programs and organizing a national conversation on "the good life" and the new American dream.

For example, serial melodramas have been used to promote mass changes in social values and behavior in the developing world regarding issues such as HIV/AIDs, infant mortality, and women's rights. These projects start with in-depth social science research to identify key target audiences in a society and the barriers preventing them from adopting the new behavior. Screenwriters then create stories with characters that represent the target audience, confronting the same barriers they confront, but finding ways to overcome them. Research has found that viewers and listeners strongly identify with these characters and their struggles and are inspired to change their own lives through the example of these role models.[12]

Likewise, it is vital that we track, catalogue, and broadcast real-world examples of the changes in behavior and ethical lifestyle we are trying to promote. What does a two-tons-carbon-per-year lifestyle actually look like, and what would it take to get there? Can we demonstrate that this way of living can be fun, meaningful, and more fulfilling than current lifestyles?

Finally, a series of structured dialogues in cities across the United States could be organized to help local communities and the country at large confront the global ecological crisis and provide a forum to deliberate the meaning of the American Dream in the twenty-first century. Such a forum should provide the opportunity to reflect on the meaning of "the good life" and our deepest values, goals, and aspirations as individuals, families, and communities, as well as to question the current trajectories of material consumption, environmental and social degradation, and the current meaning of the "pursuit of human happiness."[13]

Science and Education

Support and promote sustainability science. Sustainability science (also known as "boundary science") occurs in the "ecotones"[14] between basic and applied research. Sustainability science focuses on theoretically important questions that also have real-world applications. It seeks to understand the drivers of sustainability—economic growth, wealth and distribution, environmental protection, and human development and security—and often partners with real-world decisionmakers to answer their pressing questions and needs.[15] This often interdisciplinary research requires significantly more long-term support from funders, including the National Science Foundation, philanthropies, and scientific organizations. Further, the traditional structures of academia, funding, and reward systems remain major obstacles. Interdisciplinary research inherently takes longer to conduct as scientists must integrate different fields, methodologies, and theories in the effort to understand the complex, interconnected reality of major environmental and social problems, which cannot be understood solely from the standpoint of any one discipline.

- *Produce an Intergovernmental Panel on Climate Change (IPCC)–like assessment of global sustainability values, attitudes, and behavior.* Our empirical understanding of the current state, trajectories, and drivers of sustainability values, attitudes, and behaviors around the world is very limited. Collaborative research to identify, measure, and explain the trends in sustainability values, attitudes, and behaviors over time is critically needed. This research should integrate survey, ethnographic, historical, and experimental methods leading to both global-scale surveys repeated at regular time intervals, and local-scale, intensive studies to identify and overcome critical barriers to sustainable behavior. As a first step, an international workshop could be convened to gather the lessons learned from past studies of global values, attitudes, and sustainability behaviors and develop a collaborative research program.[16]

- *Construct and convey a range of possible futures.* Scientists can help support change by constructing empirically based scenarios, illustrating a range of potential futures for policymakers and the public to consider, evaluate, and choose among. These scenarios should describe both the potential futures that we desire and those we do not, extrapolating from current trends and trajectories and the key decisions that individuals, governments, companies, and civil society will be making over the next several decades.[17]

- *Encourage greater engagement of scientists in societal decisionmaking.* Scientists need to be encouraged and supported to participate in education, outreach, and policymaking. The engagement of science and scientists will be absolutely necessary (although insufficient) to achieve a global transition toward a sustainable world. Courses to teach scientists how to speak publicly about their research and about the policy implications should be integrated into graduate school science programs.[18] Reward systems within science and academia should be developed to encourage scientists to engage in (two-way) discussions with different audiences outside the lab and the ivory tower.

- *Create a national center for environmental education.* This organization would develop environmental science and studies curricula, materials, and teaching plans; train teachers; and integrate environmental science and studies into state standards, advanced placement courses, and local curricula for grades K–12, based on several curriculum principles. These include:
- Promoting environmental education as part of the core curriculum, not just the occasional event or field trip.
- Developing interdisciplinary, integrative, and theme-based approaches to environmental education.
- Teaching about both local and global environmental change and the connections between these scales.
- Developing courses, readers, and curricula on worldviews and nature.[19]
- Providing place-based experiential learning and exploration of local ecological processes and problems.

Religion and Ethics

- *Revitalize reverence for the Earth.* Spirituality, ritual, and scripture are all vital resources to help accelerate this moment of transition. Religions are among the oldest of human wisdom traditions and have shaped the human-nature relationship in cultures around the world. Though embedded in worldviews, religions can also form and transform those views, as the Quaker rejection of slavery in the nineteenth century and the role of religion in the U.S. civil rights movement make clear. Indeed, reverence for nature can be found in most of the world's religions; this moral force is beginning to awaken to the environmental crisis.

- *Revitalize the sense of the sacred Earth.* The Western humanities and culture often have dismissed or marginalized the sacred by placing it in the realm of the transcendent instead of the "here and now." For many, the sacred is limited to notions of the afterlife, or specific locations and buildings like churches, mosques, and synagogues. The sense of the sacred, however, can also enhance the human experience of connection and inextricable embeddedness in nature. The humanities and the world's religions can provide powerful symbols and language to reinvigorate a sense of sacred interconnection and interdependence with the natural world.

- *Convene a dialogue on cosmology.* Each religious tradition has emerged from a different set of cosmological frameworks, scriptures, and practices. At the same time, science now offers its own large-scale cosmological story. While there are certainly fundamental differences in these cosmological worldviews and epistemologies, there is also tremendous opportunity for a dialogue between science and religion to discuss the deeper significance of these scientific findings and how science and religion can work together to address the interlinked global environmental and human crises of sustainability.

- *Revitalize the Golden Rule.* "Treat others as you would have them treat you" is a fundamental principle of human ethics that can be found in many of the world's greatest religions. How do we reinvigorate this precept in our relations with each other, especially with regard to the great questions of environmental justice between the haves and have-nots both within and between countries? How might it be expanded to include ethical consideration of the natural world within the human community and vice versa?

- Promote ecological ethics as integral to social ethics and vice-versa. Environmental ethics has for too long been focused solely on the ethics of human behavior toward the nonhuman world. Likewise, social ethics have rarely incorporated a consideration of human moral duties and responsibilities toward the natural world. These two domains need to be interconnected, as it has become increasingly evident that the health and functioning of the environment affect the health and functioning of society, and vice versa. Environmental quality should be a human right.

- *Endorsement and adoption of the Earth Charter.* The Earth Charter, the result of a global, six-year participatory consultation process, presents four general-level values (community of life; ecological integrity; social and economic justice; and democracy, nonviolence, and peace).[20] These are elaborated with 16 intermediate-level principles and an additional 61 specific-level values. Since its release in 2000, the charter has been endorsed by more than 13,000 individuals and organizations. This soft-law document for a global ethics remains open for endorsement by other organizations and communities.

Policy and Economics

Policy analysts cannot create a movement by themselves. But they can help prepare the ground so that when a movement coalesces, policy tools and leaders are ready with a clear sense of which goals to pursue and paths to take. Likewise, it is imperative that environmentalism become more than another special interest. What is required is a systems shift, a new holistic view of the world we live in. A powerful, inspiring vision of a better world, not just a critique of the status quo is needed. If widely accepted, the policy changes will follow. Policymakers and analysts can help to develop the social and political capital and policy tools for the movement that is emerging in response to the ecological, social, and economic challenges of the present and future.

Use policy to encourage behavior change along with a change in values. The late Senator Daniel Patrick Moynihan (D-NY) argued that, "The central conservative truth is that culture, not politics, determines the success of a society. The central liberal truth is that politics can change a culture and save it from itself." Sociologists have found that the engrained routinization of behavior, over time, can lead to sea changes in values. Focusing solely on changing values first may miss the opportunity to engrain new behaviors, which may themselves lead to new values. Part of the importance of policy is that laws and regulations can require changes in behavior, whether or not citizens and companies currently hold the values that would lead to those behaviors without regulation.

Democratic governments, however, cannot govern without the consent of the governed and often cannot adequately enforce changes in individual behavior. Thus, policy instruments and value changes need to support each other, creating synergies and positive feedbacks that lead to large-scale changes in human behavior. Changes in smoking, seat-belt use, and drunk driving are all recent examples of the mutually reinforcing impacts of shifts in public values and attitudes on the one hand and changes in government policies on the other.

- *Prepare for the opportunities inherent in future crises.* There is often opportunity in crisis, and the policy domain needs to be prepared to act when it occurs. Crises like Pearl Harbor, Three Mile Island, and 9/11 resulted in rapid and fundamental shifts in public priorities and institutions. As global environmental conditions continue to deteriorate, there will be inevitable surprises, shocks, and disasters. How can leaders be prepared not only to better respond to the damage and destruction of these events, but also to take advantage of these "teachable moments"? We need to prepare for future ecological crises by creating institutions, systems, and roadmaps for change so that negative responses, such as authoritarianism, do not seize the day.

- *Reconnect people with nature.* A movement to bring the land back to the city is already quietly building in the form of Community-Supported Agriculture programs, farmers markets, efforts to source school lunches locally, and conversion of abandoned properties and brownfields into community gardens. A concerted effort is needed to amplify these innovations and explore other ways—such as parks and greenways—to reconnect people to nature within urban settings, while at the same time revitalizing communities and building social resilience.

- *Establish a U.S. Federal "Land Service" or "Green Corps" modeled on the Peace Corps.* Volunteers could work within the United States or internationally to help conserve, preserve, or restore natural environments and processes, or address global environmental challenges, such as climate change, loss of biodiversity, and water scarcity.

- *Develop better measures of societal progress and well-being than Gross Domestic Product.* Many economists have argued that GDP does not adequately measure the current state of either the economy or social progress and well-being. For example, many social and ecological "bads" are mischaracterized as positive economic benefits. An oil spill may generate millions of dollars in cleanup costs, which count as an increase in GDP. Meanwhile, the environmental and social costs, such as killed birds, fish, and animals, lost livelihoods, and lost communities are often not accounted for—they become "externalities." Attempts are under way to design measures of economic progress that internalize these environmental and social "externalities."[21] Meanwhile, others are calling for new measures of human well-being, as better indicators of changes in social welfare than simplistic and misleading measures like GDP.

Coming at a critical moment in human and natural history, the conference participants' collective efforts on behalf of environmental protection and social justice are important and inspiring. Many ideas described in this report represent themes that have been the subject of enormous scholarship and debate, and we encourage the interested reader to further investigate these rich research traditions. One place to begin is at the conference Web site, which includes links to related resources, organizations' and efforts.

Notes

1. A. Leopold, letter to Douglas Wade, 23 October 1944, Leopold Papers 10-8, 1, University of Wisconsin–Madison.

2. P. M. Senge, C. O. Scharmer, J. Jaworski, and B. S. Flowers, *Presence: An Exploration of Profound Change in People, Organizations, and Society* (New York: Currency Doubleday, 2005), 26.

3. A list of conference participants can be found at www.environment.yale.edu/newconsciousness.

4. R. M. Pyle, "The Extinction of Experience," *Horticulture* 56 (1978): 64–67.

5. A. Leiserowitz, R. Kates, and T. Parris, "Do Global Attitudes and Behaviors Support Sustainable Development?" *Environment* 47, no. 9 (2005): 22–38.

6. While the origins of this worldview have deep cultural roots, it was greatly crystallized in the thought of Descartes, who described the universe as a giant "clockwork" with individual mechanical parts, and Newtonian physics, which described the universe as the interaction of billiard ball–like objects.

7. A. Leiserowitz, "Climate Change Risk Perception and Policy Preferences: The Role of Affect, Imagery, and Values," *Climatic Change* 77 (2006): 45–72.

8. See H. D. Thoreau, *Walden, Or, Life in the Woods* (Boston, MA: Ticknor and Fields, 1854); H. Beston, *Outermost House: A Year of Life on the Great Beach of Cape Cod* (New York: Doubleday, Doran & Co., 1928); A. Leopold, *A Sand County Almanac* (New York: Oxford University Press, 1949); and A. Dillard, *Pilgrim at Tinker Creek* (New York: Bantam, 1975).

9. The following describe both alternative cultural approaches to understanding the natural world and wrenching experiences of environmental and social change: N. S. Momaday, *House Made of Dawn* (New York: Harper & Row, 1968); S. Ortiz, *Woven Stone* (Tucson, AZ: University of Arizona Press, 1992); and K. D. Moore, K. Peters, T. Jojola, and A. Lacy, eds., *How It Is: The Native American Philosophy of V. F. Cordova* (Tucson, AZ: University of Arizona Press, 2008).

10. G. Speth, *The Bridge at the End of the World: Capitalism, the Environment, and Crossing from Crisis to Sustainability* (New Haven, CT: Yale University Press, 2008); and M. Dowie, *Losing Ground: American Environmentalism at the Close of the Twentieth Century* (Cambridge, MA: MIT Press, 1996).

11. A. Leiserowitz, R. Kates, and T. Parris, "Sustainability Values, Attitudes and Behaviors: A Review of Multi-national and Global Trends," *Annual Review of Environment and Resources* 31 (2006): 413–44.

12. A. Singhal, M. J. Cody, E. Rogers, and M. Sabido, eds., *Entertainment-Education and Social Change: History, Research, and Practice* (London: Lawrence Erlbaum, 2004).

13. One potential example is the National Conversation on Climate Action, a partnership between the Yale School of Forestry and Environmental Studies, ICLEI—Local Governments for Sustainability, and the Association of Science and Technology Centers. See www.climateconversation.org (accessed 10 July 2008).

14. An ecotone is a transition zone between two adjacent ecological communities, such as forest and grassland. It has some of the characteristics of each bordering community and often contains species not found in the overlapping communities. An ecotone may exist along a broad belt or in a small pocket, such as a forest clearing, where two local communities blend together. The influence of the two bordering communities on each other is known as the edge effect. An ecotonal area often has a higher density of organisms and a greater number of species than are found in either flanking community. "Ecotone," *Encyclopedia Britannica Online*, search.eb.can/eb/article_903194 (accessed 29 July 2008).

15. W. C. Clark, "Sustainability Science: A Room of One's Own," *Proceedings of the National Academy of Sciences* 104, no. 6 (6 February 2007): 1737–38.

16. Key questions include: What are the key factors that drive cultural evolution and social change? What can we learn from the analysis of past societal paradigm shifts? What universal and particular factors underlie each? What explains the differences in sustainability values, attitudes, and behaviors across different nations, regions, or levels of economic development? What value and lifestyle changes will be required to achieve a sustainable world? What can we learn from past successful and unsuccessful efforts to change public attitudes and behaviors (for instance, smoking and drunk driving)? What are the primary value, attitudinal, and structural barriers that constrain sustainable behavior in particular social, economic, political, cultural, and geographic contexts? Leiserowitz, Kates, and Parris, note 11; and P. R. Erlich and D. Kennedy, "Sustainability: Millennium Assessment of Human Behavior," *Science* 309, no. 5734 (22 July 2005): 562–63.

17. For example, see the work of the Great Transition Initiative, www.gtinitiative.org/.

18. The Aldo Leopold Leadership Program at Stanford University is one example of a successful effort to help scientists better communicate with journalists. See www.leopoldleadership.org (accessed 10 July 2008).

19. The Forum on Religion and Ecology, through its conferences, publications, and Web site, provides a rich set of resources. See www.yale.edu/religionandecology (accessed 10 July 2008).

20. The Earth Charter is available at earthcharterinaction.org/ec_splash/ (accessed 10 July 2008).

21. Redefining Progress and the International Forum on Globalization are two such organizations.

Critical Thinking

1. Do you agree or disagree with this statement: "At its deepest level, if we are to address the linked environmental, social, and even spiritual crises, we must address the wellsprings of human caring, motivation, and social identity." Explain.

2. To what extent do the research and findings in this study compare with points made in Articles 1, 2, and 3? Discuss.

3. What do the authors mean by *economism?*

4. What does the article say about "environmentalism" and the need for a critique of contemporary culture, economics and politics? Can you find any similarities between that statement and statements made in the Preface of this edition of Environment 13/14? Explain.

Consuming Passions

Everything That Can Be Done to Bring the Age of Heroic Consumption to Its Close Should Be Done

Jeremy Seabrook

The age of heroic consumption is surely drawing to a close. The inspiration of those whose principal virtue is the money that permits them to lay claim to a disproportionate share of the earth's resources is being by-passed in a world where a population of 9 billion must be accommodated by 2050.

The price tag on the possessions of the wealthy—their £10m mansions, £5m yachts, extravagant couture and priceless jewels, their private jets and lives apart from the great majority of humankind—are rapidly losing their power to enchant the rest of us. In an age when scientists, humanitarians and moral leaders are exhorting human beings to look to our impact upon the earth—and not solely in relation to the carbon footprint—it has become obsolete to gaze with breathless admiration upon individuals dedicated to the proposition that a whole world should be dying of consumption, and not just the 1.6m who perish from tuberculosis each year.

The greatest threat to global stability comes not from the poor but from the rich. This startling proposition runs directly into another received idea, which is that the risk of disorder is a result of excessive materialism. What we suffer from is not a surfeit of materialism, but a deficiency of it; for if we truly valued the material basis upon which all human systems depend, we would exhibit a far greater reverence for the physical world we inhabit. If materialism means respect for the elements that sustain life, then we are gravely wanting in it. What is sometimes referred to as "materialism" is actually something else: perhaps a distorted kind of mysticism which believes we can use up the earth and still avoid the consequences of our omnivorous appetites.

This is why the gross consumers of the age will be scorned as the pitiable destroyers of the sustenance, not only of the poor of today, but of everyone's tomorrow. It is natural for people to want to do the best for their children, but this is generally interpreted as leaving them a private monetary inheritance; but if the other side of this legacy is a befouled world, the enjoyment of today's privilege may become the curse of the future. In any case, there is a great deal of humbug in pious concern expressed for our children's children, since this rarely prevents those who give voice to such tender sentiments from living as though there were no tomorrow. "Live the dream" has become the cliche of the hour; although it requires no great wisdom to understand that dreams realised soon turn to ashes.

Everything that can be done to bring the age of heroic consumption to its close should be done. This means the promotion of a different understanding of wealth. The myriad aspects of a truly rich and fulfilled life should be rescued from the tyranny of money. Perhaps we have not entirely forgotten that the most joyful and exhilarating of human occupations derive from self-reliance, self-provisioning, not only in the basic goods that sustain life, but also in satisfactions that arise from the cost-free resourcefulness of ourselves and others.

This is why the A-listers, the celebs, the fat cats, the big spenders, the conspicuous consumers do not represent a "lifestyle" to be emulated at all costs, but serve as warning of the spectre of depletion and exhaustion awaiting us within a short space of time. When Thorstein Veblen wrote his Theory of the Leisure Class at the end of the 19th century, he saw "conspicuous waste and show" as a replacement for "earlier and more primitive displays of physical prowess". Even his caustic insights could not anticipate the degree to which the ornamental inutility of the very rich would lead them to become pioneers of planetary demolition.

Of course, downgrading the exploits of the major culprits in ransacking the earth is easier said than done. Cultures are not, as journalists and politicians sometimes suggest, to be discarded or "changed" at will. But sooner or later, a reduction in the abuse of the elements of life will be forced upon the world. If it proves impossible to take preventive action in this regard, we shall soon enough be overtaken by events—oil wars, water wars, even more brutal conflicts over land than we have already seen, food wars, social disruption, rioting and breakdown, such as the World Bank has already detected in some 37 countries in the last two years, will be the form in which the relentless plunder of the planet will resolve itself.

Just as the age of heroic labour—the Stakhanovite idea of selfless dedication to the building of Communism—perished, so heroic consumption—that equally selfless dedication to sustaining capitalism—has also had its day. Stakhanovites were so called after a coalminer in the Soviet Union in 1935 who exceeded his work quota by 14 times the fixed level, producing 102 tons of coal in six hours. This became a kind of "spontaneous" official policy in the construction of socialism.

How laughably old-fashioned this now sounds. And how swiftly things that appear immutable can change. It should be our ambition to ensure that the work of predatory individuals upon the fruits of the earth comes to appear as archaic and futile as the sacrifice of human energies in the Soviet Union to release the resources which, according to Marx, "slumbered in the lap of social labour".

Heroic consumption, unlike heroic labour, requires no official sponsorship. The incentive to get rich is so deeply embedded in capitalism, that it has been seen as an expression of human nature itself. The first task in achieving a decent security for all people on earth is to affirm the distinction between human nature and the nature of capitalism.

Human beings want, above all, to survive. The moral and social elevation of the wealthy and their profligacy suggests that they are prepared to sacrifice even this hitherto imperishable goal for the sake of transforming the beauty and value of the world into a wasteland. Enslavement to a reductive, diminished version of what it means to lead a rich life is still bondage; and when it must be protected by razor wire, guns, security guards and impregnable barriers, these become the very symbols of that unfreedom.

Critical Thinking

1. "What we suffer from is not a surfeit of materialism, but a deficiency of it." Explain what that statement means and how it supports the ideals of sustainability.

2. "Everything that can be done to bring the age of heroic consumption to its close should be done." Please explain what that means. If you were a multimillionaire, what would you think about that statement?

3. What arguments would the wealthiest people of the world's wealthiest nations and the cornucopians say about the validity of the statement in Question 2?

Reversal of Fortune

The formula for human well-being used to be simple: Make money, get happy. So why is the old axiom suddenly turning on us?

BILL MCKIBBEN

For most of human history, the two birds More and Better roosted on the same branch. You could toss one stone and hope to hit them both. That's why the centuries since Adam Smith launched modern economics with his book *The Wealth of Nations* have been so single-mindedly devoted to the dogged pursuit of maximum economic production. Smith's core ideas—that individuals pursuing their own interests in a market society end up making each other richer; and that increasing efficiency, usually by increasing scale, is the key to increasing wealth—have indisputably worked. They've produced more More than he could ever have imagined. They've built the unprecedented prosperity and ease that distinguish the lives of most of the people reading these words. It is no wonder and no accident that Smith's ideas still dominate our politics, our outlook, even our personalities.

But the distinguishing feature of our moment is this: Better has flown a few trees over to make her nest. And that changes everything. Now, with the stone of your life or your society gripped in your hand, you have to choose. It's More or Better.

Which means, according to new research emerging from many quarters, that our continued devotion to growth above all is, on balance, making our lives worse, both collectively and individually. Growth no longer makes most people wealthier, but instead generates inequality and insecurity. Growth is bumping up against physical limits so profound—like climate change and peak oil—that trying to keep expanding the economy may be not just impossible but also dangerous. And perhaps most surprisingly, growth no longer makes us happier. Given our current dogma, that's as bizarre an idea as proposing that gravity pushes apples skyward. But then, even Newtonian physics eventually shifted to acknowledge Einstein's more complicated universe.

[I] "We Can Do It If We Believe It": FDR, LBJ, and the Invention of Growth

It was the great economist John Maynard Keynes who pointed out that until very recently, "there was no very great change in the standard of life of the average man living in the civilized centers of the earth." At the utmost, Keynes calculated, the standard of living roughly doubled between 2000 B.C. and the dawn of the 18th century—four millennia during which we basically didn't learn to do much of anything new. Before history began, we had already figured out fire, language, cattle, the wheel, the plow, the sail, the pot. We had banks and governments and mathematics and religion.

And then, something new finally did happen. In 1712, a British inventor named Thomas Newcomen created the first practical steam engine. Over the centuries that followed, fossil fuels helped create everything we consider normal and obvious about the modern world, from electricity to steel to fertilizer; now, a 100 percent jump in the standard of living could suddenly be accomplished in a few decades, not a few millennia.

In some ways, the invention of the idea of economic growth was almost as significant as the invention of fossil-fuel power. But it took a little longer to take hold. During the Depression, even FDR routinely spoke of America's economy as mature, with no further expansion anticipated. Then came World War II and the postwar boom—by the time Lyndon Johnson moved into the White House in 1963, he said things like: "I'm sick of all the people who talk about the things we can't do. Hell, we're the richest country in the world, the most powerful. We can do it all. . . . We can do it if we believe it." He wasn't alone in thinking this way. From Moscow, Nikita Khrushchev thundered, "Growth of industrial and agricultural production is the battering ram with which we shall smash the capitalist system."

Yet the bad news was already apparent, if you cared to look. Burning rivers and smoggy cities demonstrated the dark side of industrial expansion. In 1972, a trio of MIT researchers released a series of computer forecasts they called "limits to growth," which showed that unbridled expansion would eventually deplete our resource base. A year later the British economist E. F. Schumacher wrote the best-selling *Small Is Beautiful.* (Soon after, when Schumacher came to the United States on a speaking tour, Jimmy Carter actually received him at the White House—imagine the current president making time for any economist.) By 1979, the sociologist Amitai Etzioni reported to President Carter that only 30 percent of Americans were "progrowth," 31 percent were "anti-growth," and 39 percent were "highly uncertain."

Such ambivalence, Etzioni predicted, "is too stressful for societies to endure," and Ronald Reagan proved his point. He convinced us it was "Morning in America"—out with limits, in with Trump. Today, mainstream liberals and conservatives compete mainly on the question of who can flog the economy harder. Larry Summers, who served as Bill Clinton's secretary of the treasury, at one point declared that the Clinton administration "cannot and will not accept any 'speed limit' on American economic growth. It is the task of economic policy to grow the economy as rapidly, sustainably, and inclusively as possible." It's the economy, stupid.

[2] Oil Bingeing, Chinese Cars, and the End of the Easy Fix

Except there are three small things. The first I'll mention mostly in passing: Even though the economy continues to grow, most of us are no longer getting wealthier. The average wage in the United States is less now, in real dollars, than it was 30 years ago. Even for those with college degrees, and although productivity was growing faster than it had for decades, between 2000 and 2004 earnings fell 5.2 percent when adjusted for inflation, according to the most recent data from White House economists. Much the same thing has happened across most of the globe. More than 60 countries around the world, in fact, have seen incomes per capita fall in the past decade.

For the second point, it's useful to remember what Thomas Newcomen was up to when he helped launch the Industrial Revolution—burning coal to pump water out of a coal mine. This revolution both depended on, and revolved around, fossil fuels. "Before coal," writes the economist Jeffrey Sachs, "economic production was

limited by energy inputs, almost all of which depended on the production of biomass: food for humans and farm animals, and fuel wood for heating and certain industrial processes." That is, energy depended on how much you could grow. But fossil energy depended on how much had grown eons before—all those billions of tons of ancient biology squashed by the weight of time till they'd turned into strata and pools and seams of hydrocarbons, waiting for us to discover them.

To understand how valuable, and irreplaceable, that lake of fuel was, consider a few other forms of creating usable energy. Ethanol can perfectly well replace gasoline in a tank; like petroleum, it's a way of using biology to create energy, and right now it's a hot commodity, backed with billions of dollars of government subsidies. But ethanol relies on plants that grow anew each year, most often corn; by the time you've driven your tractor to tend the fields, and your truck to carry the crop to the refinery, and powered your refinery, the best-case "energy output-to-input ratio" is something like 1.34-to-1. You've spent 100 Btu of fossil energy to get 134 Btu. Perhaps that's worth doing, but as Kamyar Enshayan of the University of Northern Iowa points out, "it's not impressive" compared to the ratio for oil, which ranges from 30-to-1 to 200-to-1, depending on where you drill it. To go from our fossil-fuel world to a biomass world would be a little like leaving the Garden of Eden for the land where bread must be earned by "the sweat of your brow."

And east of Eden is precisely where we may be headed. As everyone knows, the past three years have seen a spate of reports and books and documentaries suggesting that humanity may have neared or passed its oil peak—that is, the point at which those pools of primeval plankton are half used up, where each new year brings us closer to the bottom of the barrel. The major oil companies report that they can't find enough new wells most years to offset the depletion in the old ones; rumors circulate that the giant Saudi fields are dwindling faster than expected; and, of course, all this is reflected in the cost of oil.

The doctrinaire economist's answer is that no particular commodity matters all that much, because if we run short of something, it will pay for someone to develop a substitute. In general this has proved true in the past: Run short of nice big sawlogs and someone invents plywood. But it's far from clear that the same precept applies to coal, oil, and natural gas. This time, there is no easy substitute: I like the solar panels on my roof, but they're collecting diffuse daily energy, not using up eons of accumulated power. Fossil fuel was an exception to the rule, a one-time gift that underwrote a one-time binge of growth.

This brings us to the third point: If we do try to keep going, with the entire world aiming for an economy structured like America's, it won't be just oil that we'll run short of. Here are the numbers we have to contend with: Given current rates of growth in the Chinese economy, the 1.3 billion residents of that nation alone will, by 2031, be about as rich as we are. If they then eat meat, milk, and eggs at the rate that we do, calculates ecostatistician Lester Brown, they will consume 1,352 million tons of grain each year—equal to two-thirds of the world's entire 2004 grain harvest. They will use 99 million barrels of oil a day, 15 million more than the entire world consumes at present. They will use more steel than all the West combined, double the world's production of paper, and drive 1.1 billion cars—1.5 times as many as the current world total. And that's just China; by then, India will have a bigger population, and its economy is growing almost as fast. And then there's the rest of the world.

Trying to meet that kind of demand will stress the earth past its breaking point in an almost endless number of ways, but let's take just one. When Thomas Newcomen fired up his pump on that morning in 1712, the atmosphere contained 275 parts per million of carbon dioxide. We're now up to 380 parts per million, a level higher than the earth has seen for many millions of years, and climate change has only just begun. The median predictions of the world's climatologists—by no means the worst-case scenario—show that unless we take truly enormous steps to rein in our use of fossil fuels, we can expect average temperatures to rise another four or five degrees before the century is out, making the globe warmer than it's been since long before primates appeared. We might as well stop calling it earth and have a contest to pick some new name, because it will be a different planet. Humans have never done anything more profound, not even when we invented nuclear weapons.

How does this tie in with economic growth? Clearly, getting rich means getting dirty—that's why, when I was in Beijing recently, I could stare straight at the sun (once I actually figured out where in the smoggy sky it was). But eventually, getting rich also means wanting the "luxury" of clean air and finding the technological means to achieve it. Which is why you can once again see the mountains around Los Angeles; why more of our rivers are swimmable every year. And economists have figured out clever ways to speed this renewal: Creating markets for trading pollution credits, for instance, helped cut those sulfur and nitrogen clouds more rapidly and cheaply than almost anyone had imagined.

But getting richer doesn't lead to producing less carbon dioxide in the same way that it does to less smog—in fact, so far it's mostly the reverse. Environmental destruction of the old-fashioned kind—dirty air, dirty water—results from something going wrong. You haven't bothered to stick the necessary filter on your pipes, and so the crud washes into the stream; a little regulation, and a little money, and the problem disappears. But the second, deeper form of environmental degradation comes from things operating exactly as they're supposed to, just too much so. Carbon dioxide is an inevitable byproduct of burning coal or gas or oil—not something going wrong. Researchers are struggling to figure out costly and complicated methods to trap some CO_2 and inject it into underground mines—but for all practical purposes, the vast majority of the world's cars and factories and furnaces will keep belching more and more of it into the atmosphere as long as we burn more and more fossil fuels.

True, as companies and countries get richer, they can afford more efficient machinery that makes better use of fossil fuel, like the hybrid Honda Civic I drive. But if your appliances have gotten more efficient, there are also far more of them: The furnace is better than it used to be, but the average size of the house it heats has doubled since 1950. The 60-inch TV? The always-on cable modem? No need for you to do the math—the electric company does it for you, every month. Between 1990 and 2003, precisely the years in which we learned about the peril presented by global warming, the United States' annual carbon dioxide emissions increased by 16 percent. And the momentum to keep going in that direction is enormous. For most of us, growth has become synonymous with the economy's "health," which in turn seems far more palpable than the health of the planet. Think of the terms we use—the economy, whose temperature we take at every newscast via the Dow Jones average, is "ailing" or it's "on the mend." It's "slumping" or it's "in recovery." We cosset and succor its every sniffle with enormous devotion, even as we more or less ignore the increasingly urgent fever that the globe is now running. The ecological economists have an enormous task ahead of them—a nearly insurmountable task, if it were "merely" the environment that is in peril. But here is where things get really interesting. It turns out that the economics of environmental destruction are closely linked to another set of leading indicators—ones that most humans happen to care a great deal about.

[3] "It Seems That Well-Being Is a Real Phenomenon": Economists Discover Hedonics

Traditionally, happiness and satisfaction are the sort of notions that economists wave aside as poetic irrelevance, the kind of questions that occupy people with no head for numbers who had to major in liberal arts. An orthodox economist has a simple happiness formula: If you buy a Ford Expedition, then ipso facto a Ford Expedition is what makes you happy. That's all we need to know. The economist would call this idea "utility maximization," and in the words of the economic historian Gordon Bigelow, "the theory holds that every time a person buys something, sells something, quits a job, or invests, he is making a rational decision about what will . . . provide him 'maximum utility.' If you bought a Ginsu knife at 3 A.M. a neoclassical economist will tell you that, at that time, you calculated that this purchase would optimize your resources." The beauty of this principle lies in its simplicity. It is perhaps the central assumption of the world we live in: You can tell who I really am by what I buy.

Yet economists have long known that people's brains don't work quite the way the model suggests. When Bob Costanza, one of the fathers of ecological economics and now head of the Gund Institute at the University of Vermont, was first edging into economics in the early 1980s, he had a fellowship to study "social traps"—the nuclear arms race, say—in which "short-term behavior can get out of kilter with longer broad-term goals."

It didn't take long for Costanza to demonstrate, as others had before him, that, if you set up an auction in a certain way, people will end up bidding $1.50 to take home a dollar. Other economists have shown that people give too much weight to "sunk costs"—that they're too willing to throw good money after bad, or that they value items more highly if they already own them than if they are considering acquiring them. Building on such insights, a school of "behavioral economics" has emerged in recent years and begun plumbing how we really behave.

The wonder is that it took so long. We all know in our own lives how irrationally we are capable of acting, and how unconnected those actions are to any real sense of joy. (I mean, there you are at 3 A.M. thinking about the Ginsu knife.) But until fairly recently, we had no alternatives to relying on Ginsu knife and Ford Expedition purchases as the sole measures of our satisfaction. How else would we know what made people happy?

That's where things are now changing dramatically: Researchers from a wide variety of disciplines have started to figure out how to assess satisfaction, and economists have begun to explore the implications. In 2002 Princeton's Daniel Kahneman won the Nobel Prize in economics even though he is trained as a psychologist. In the book *Well-Being,* he and a pair of coauthors announce a new field called "hedonics," defined as "the study of what makes experiences and life pleasant or unpleasant. . . . It is also concerned with the whole range of circumstances, from the biological to the societal, that occasion suffering and enjoyment." If you are worried

that there might be something altogether too airy about this, be reassured—Kahneman thinks like an economist. In the book's very first chapter, "Objective Happiness," he describes an experiment that compares "records of the pain reported by two patients undergoing colonoscopy," wherein every 60 seconds he insists they rate their pain on a scale of 1 to 10 and eventually forces them to make "a hypothetical choice between a repeat colonoscopy and a barium enema." Dismal science indeed.

As more scientists have turned their attention to the field, researchers have studied everything from "biases in recall of menstrual symptoms" to "fearlessness and courage in novice paratroopers." Subjects have had to choose between getting an "attractive candy bar" and learning the answers to geography questions; they've been made to wear devices that measured their blood pressure at regular intervals; their brains have been scanned. And by now that's been enough to convince most observers that saying "I'm happy" is more than just a subjective statement. In the words of the economist Richard Layard, "We now know that what people say about how they feel corresponds closely to the actual levels of activity in different parts of the brain, which can be measured in standard scientific ways." Indeed, people who call themselves happy, or who have relatively high levels of electrical activity in the left prefrontal region of the brain, are also "more likely to be rated as happy by friends," "more likely to respond to requests for help," "less likely to be involved in disputes at work," and even "less likely to die prematurely." In other words, conceded one economist, "it seems that what the psychologists call subjective well-being is a real phenomenon. The various empirical measures of it have high consistency, reliability, and validity."

The idea that there is a state called happiness, and that we can dependably figure out what it feels like and how to measure it, is extremely subversive. It allows economists to start thinking about life in richer (indeed) terms, to stop asking "What did you buy?" and to start asking "Is your life good?" And if you can ask someone "Is your life good?" and count on the answer to mean something, then you'll be able to move to the real heart of the matter, the question haunting our moment on the earth: Is more better?

[4] If We're So Rich, How Come We're So Damn Miserable?

In some sense, you could say that the years since World War II in America have been a loosely controlled experiment designed to answer this very question. The environmentalist Alan Durning found that in 1991 the average American family owned twice as many cars as it did in

1950, drove 2.5 times as far, used 21 times as much plastic, and traveled 25 times farther by air. Gross national product per capita tripled during that period. Our houses are bigger than ever and stuffed to the rafters with belongings (which is why the storage-locker industry has doubled in size in the past decade). We have all sorts of other new delights and powers—we can send email from our cars, watch 200 channels, consume food from every corner of the world. Some people have taken much more than their share, but on average, all of us in the West are living lives materially more abundant than most people a generation ago.

What's odd is, none of it appears to have made us happier. Throughout the postwar years, even as the GNP curve has steadily climbed, the "life satisfaction" index has stayed exactly the same. Since 1972, the National Opinion Research Center has surveyed Americans on the question: "Taking all things together, how would you say things are these days—would you say that you are very happy, pretty happy, or not too happy?" (This must be a somewhat unsettling interview.) The "very happy" number peaked at 38 percent in the 1974 poll, amid oil shock and economic malaise; it now hovers right around 33 percent.

And it's not that we're simply recalibrating our sense of what happiness means—we are actively experiencing life as grimmer. In the winter of 2006 the National Opinion Research Center published data about "negative life events" comparing 1991 and 2004, two data points bracketing an economic boom. "The anticipation would have been that problems would have been down," the study's author said. Instead it showed a rise in problems—for instance, the percentage who reported breaking up with a steady partner almost doubled. As one reporter summarized the findings, "There's more misery in people's lives today."

This decline in the happiness index is not confined to the United States; as other nations have followed us into mass affluence, their experiences have begun to yield similar results. In the United Kingdom, real gross domestic product per capita grew two-thirds between 1973 and 2001, but people's satisfaction with their lives changed not one whit. Japan saw a fourfold increase in real income per capita between 1958 and 1986 without any reported increase in satisfaction. In one place after another, rates of alcoholism, suicide, and depression have gone up dramatically, even as we keep accumulating more stuff. Indeed, one report in 2000 found that the average American child reported higher levels of anxiety than the average child under psychiatric care in the 1950s—our new normal is the old disturbed.

If happiness was our goal, then the unbelievable amount of effort and resources expended in its pursuit since 1950

has been largely a waste. One study of life satisfaction and mental health by Emory University professor Corey Keyes found just 17 percent of Americans "flourishing," in mental health terms, and 26 percent either "languishing" or out-and-out depressed.

[5] Danes (and Mexicans, the Amish, and the Masai) Just Want to Have Fun

How is it, then, that we became so totally, and apparently wrongly, fixated on the idea that our main goal, as individuals and as nations, should be the accumulation of more wealth? The answer is interesting for what it says about human nature. Up to a certain point, more really does equal better. Imagine briefly your life as a poor person in a poor society—say, a peasant farmer in China. (China has one-fourth of the world's farmers, but one-fourteenth of its arable land; the average farm in the southern part of the country is about half an acre, or barely more than the standard lot for a new American home.) You likely have the benefits of a close and connected family, and a village environment where your place is clear. But you lack any modicum of security for when you get sick or old or your back simply gives out. Your diet is unvaried and nutritionally lacking; you're almost always cold in winter.

In a world like that, a boost in income delivers tangible benefits. In general, researchers report that money consistently buys happiness right up to about $10,000 income per capita. That's a useful number to keep in the back of your head—it's like the freezing point of water, one of those random figures that just happens to define a crucial phenomenon on our planet. "As poor countries like India, Mexico, the Philippines, Brazil, and South Korea have experienced economic growth, there is some evidence that their average happiness has risen," the economist Layard reports. Past $10,000 (per capita, mind you—that is, the average for each man, woman, and child), there's a complete scattering: When the Irish were making two-thirds as much as Americans they were reporting higher levels of satisfaction, as were the Swedes, the Danes, the Dutch. Mexicans score higher than the Japanese; the French are about as satisfied with their lives as the Venezuelans. In fact, once basic needs are met, the "satisfaction" data scrambles in mind-bending ways. A sampling of *Forbes* magazine's "richest Americans" have identical happiness scores with Pennsylvania Amish, and are only a whisker above Swedes taken as a whole, not to mention the Masai. The "life satisfaction" of pavement dwellers—homeless people—in Calcutta is among the lowest recorded, but it almost doubles when they move into a slum, at which point they are basically as satisfied with their lives as a sample of college students drawn from 47 nations. And so on.

On the list of major mistakes we've made as a species, this one seems pretty high up. Our single-minded focus on increasing wealth has succeeded in driving the planet's ecological systems to the brink of failure, even as it's failed to make us happier. How did we screw up?

The answer is pretty obvious—we kept doing something past the point that it worked. Since happiness had increased with income in the past, we assumed it would inevitably do so in the nature. We make these kinds of mistakes regularly: Two beers made me feel good, so ten will make me feel five times better. But this case was particularly extreme—in part because as a species, we've spent so much time simply trying to survive. As the researchers Ed Diener and Martin Seligman—both psychologists—observe, "At the time of Adam Smith, a concern with economic issues was understandably primary. Meeting simple human needs for food, shelter and clothing was not assured, and satisfying these needs moved in lockstep with better economics." Freeing people to build a more dynamic economy was radical and altruistic.

Consider Americans in 1820, two generations after Adam Smith. The average citizen earned, in current dollars, less than $1,500 a year, which is somewhere near the current average for all of Africa. As the economist Deirdre McCloskey explains in a 2004 article in the magazine *Christian Century,* "Your great-great-great-grandmother had one dress for church and one for the week, if she were not in rags. Her children did not attend school, and probably could not read. She and her husband worked eighty hours a week for a diet of bread and milk—they were four inches shorter than you." Even in 1900, the average American lived in a house the size of today's typical garage. Is it any wonder that we built up considerable velocity trying to escape the gravitational pull of that kind of poverty? An object in motion stays in motion, and our economy—with the built-up individual expectations that drive it—is a mighty object indeed.

You could call it, I think, the Laura Ingalls Wilder effect. I grew up reading her books—*Little House on the Prairie, Little House in the Big Woods*—and my daughter grew up listening to me read them to her, and no doubt she will read them to her children. They are the ur-American story. And what do they tell? Of a life rich in family, rich in connection to the natural world, rich in adventure—but materially deprived. That one dress, that same bland dinner. At Christmastime, a penny—a penny! And a stick of candy, and the awful deliberation about whether to stretch it out with tiny licks or devour it in an orgy of happy greed. A rag doll was the zenith of aspiration. My daughter likes dolls too, but her bedroom boasts a density of Beanie Babies that mimics the manic biodiversity of the deep rainforest. Another one? Really, so what? Its marginal utility, as an economist might say, is low. And so it is with all of us. We just haven't figured that out because the momentum of the past is still with us—we still imagine we're in that little house on the big prairie.

[6] This Year's Model Home: "Good for the Dysfunctional Family"

That great momentum has carried us away from something valuable, something priceless: It has allowed us to become (very nearly forced us to become) more thoroughly individualistic than we really wanted to be. We left behind hundreds of thousands of years of human community for the excitement, and the isolation, of "making something of ourselves," an idea that would not have made sense for 99.9 percent of human history. Adam Smith's insight was that the interests of each of our individual selves could add up, almost in spite of themselves, to social good—to longer lives, fuller tables, warmer houses. Suddenly the community was no longer necessary to provide these things; they would happen as if by magic. And they did happen. And in many ways it was good.

But this process of liberation seems to have come close to running its course. Study after study shows Americans spending less time with friends and family, either working longer hours, or hunched over their computers at night. And each year, as our population grows by 1 percent we manage to spread ourselves out over 6 to 8 percent more land. Simple mathematics says that we're less and less likely to bump into the other inhabitants of our neighborhood, or indeed of our own homes. As the *Wall Street Journal* reported recently, "Major builders and top architects are walling people off. They're touting one-person 'Internet alcoves,' locked-door 'away rooms,' and his-and-her offices on opposite ends of the house. The new floor plans offer so much seclusion, they're 'good for the dysfunctional family,' says Gopal Ahluwahlia, director of research for the National Association of Home Builders." At the building industry's annual Las Vegas trade show, the "showcase 'Ultimate Family Home' hardly had a family room," noted the *Journal.* Instead, the boy's personal playroom had its own 42-inch plasma TV, and the girl's bedroom had a secret mirrored door leading to a "hideaway karaoke room." "We call this the ultimate home for families who don't want anything to do with one another," said Mike McGee, chief executive of Pardee Homes of Los Angeles, builder of the model.

This transition from individualism to hyper-individualism also made its presence felt in politics. In the 1980s, British prime minister Margaret Thatcher asked, "Who is society? There is no such thing. There are individual men and women, and there are families." Talk about everything solid melting into air—Thatcher's maxim would have spooked Adam Smith himself. The "public realm"—things like parks and schools and Social Security, the last reminders of the communities from which we came—is under steady and increasing attack. Instead of contributing to the shared risk of health insurance, Americans are encouraged to go it alone with "health savings accounts." Hell, even the nation's most collectivist institution, the U.S. military, until recently recruited under the slogan an "Army of One." No wonder the show that changed television more than any other in the past decade was Survivor, where the goal is to end up alone on the island, to manipulate and scheme until everyone is banished and leaves you by yourself with your money.

It's not so hard, then, to figure out why happiness has declined here even as wealth has grown. During the same decades when our lives grew busier and more isolated, we've gone from having three confidants on average to only two, and the number of people saying they have no one to discuss important matters with has nearly tripled. Between 1974 and 1994, the percentage of Americans who said they visited with their neighbors at least once a month fell from almost two-thirds to less than half, a number that has continued to fall in the past decade. We simply worked too many hours earning, we commuted too far to our too-isolated homes, and there was always the blue glow of the tube shining through the curtains.

[7] New Friend or New Coffeemaker? Pick One

Because traditional economists think of human beings primarily as individuals and not as members of a community, they miss out on a major part of the satisfaction index. Economists lay it out almost as a mathematical equation: Overall, "evidence shows that companionship . . . contributes more to well-being than does income," writes Robert E. Lane, a Yale political science professor who is the author of *The Loss of Happiness in Market Democracies*. But there is a notable difference between poor and wealthy countries: When people have lots of companionship but not much money, income "makes more of a contribution to subjective well-being." By contrast, "where money is relatively plentiful and companionship relatively scarce, companionship will add more to subjective well-being." If you are a poor person in China, you have plenty of friends and family around all the time—perhaps there are four other people living in your room. Adding a sixth doesn't make you happier. But adding enough money so that all five of you can eat some meat from time to time pleases you greatly. By contrast, if you live in a suburban American home, buying another coffeemaker adds very little to your quantity of happiness—trying to figure out where to store it, or wondering if you picked the perfect model, may in fact decrease your total pleasure. But a new friend, a new connection, is a big deal. We have a surplus of individualism and a deficit of companionship, and so the second becomes more valuable.

Indeed, we seem to be genetically wired for community. As biologist Edward O. Wilson found, most primates live in groups and get sad when they're separated—"an isolated individual will repeatedly pull a lever with no reward other than the glimpse of another monkey." Why do people so often look back on their college days as the best years of their lives? Because their classes were so fascinating? Or because in college, we live more closely and intensely with a community than most of us ever do before or after? Every measure of psychological health points to the same conclusion: People who "are married, who have good friends, and who are close to their families are happier than those who are not," says Swarthmore psychologist Barry Schwartz. "People who participate in religious communities are happier than those who do not." Which is striking, Schwartz adds, because social ties "actually decrease freedom of choice"—being a good friend involves sacrifice.

Do we just think we're happier in communities? Is it merely some sentimental good-night-John-Boy affectation? No—our bodies react in measurable ways. According to research cited by Harvard professor Robert Putnam in his classic book *Bowling Alone,* if you do not belong to any group at present, joining a club or a society of some kind cuts in half the risk that you will die in the next year. Check this out: When researchers at Carnegie Mellon (somewhat disgustingly) dropped samples of cold virus directly into subjects' nostrils, those with rich social networks were four times less likely to get sick. An economy that produces only individualism undermines us in the most basic ways.

Here's another statistic worth keeping in mind: Consumers have 10 times as many conversations at farmers' markets as they do at supermarkets—an order of magnitude difference. By itself, that's hardly life-changing, but it points at something that could be: living in an economy where you are participant as well as consumer, where you have a sense of who's in your universe and how it fits together. At the same time, some studies show local agriculture using less energy (also by an order of magnitude) than the "it's always summer somewhere" system we operate on now. Those are big numbers, and it's worth thinking about what they suggest—especially since, between peak oil and climate change, there's no longer really a question that we'll have to wean ourselves of the current model.

So as a mental experiment, imagine how we might shift to a more sustainable kind of economy. You could use government policy to nudge the change—remove subsidies from agribusiness and use them instead to promote farmer-entrepreneurs; underwrite the cost of windmills with even a fraction of the money that's now going to protect oil flows. You could put tariffs on goods that travel long distances, shift highway spending to projects that make it easier to live near where you work (and, by cutting down on commutes, leave some time to see the kids). And, of course, you can exploit the Net to connect a lot of this highly localized stuff into something larger. By way of example, a few of us are coordinating the first nationwide global warming demonstration—but instead of marching on Washington, we're rallying in our local areas, and then fusing our efforts, via the website stepitup07.org, into a national message.

It's easy to dismiss such ideas as sentimental or nostalgic. In fact, economies can be localized as easily in cities and suburbs as rural villages (maybe more easily), and in ways that look as much to the future as the past, that rely more on the solar panel and the Internet than the white picket fence. In fact, given the trendlines for phenomena such as global warming and oil supply, what's nostalgic and sentimental is to keep doing what we're doing simply because it's familiar.

[8] The Oil-For-People Paradox: Why Small Farms Produce More Food

To understand the importance of this last point, consider the book *American Mania* by the neuroscientist Peter Whybrow. Whybrow argues that many of us in this country are predisposed to a kind of dynamic individualism—our gene pool includes an inordinate number of people who risked everything to start over. This served us well in settling a continent and building our prosperity. But it never got completely out of control, says Whybrow, because "the marketplace has always had its natural constraints. For the first two centuries of the nation's existence, even the most insatiable American citizen was significantly leashed by the checks and balances inherent in a closely knit community, by geography, by the elements of weather, or, in some cases, by religious practice."

You lived in a society—a habitat—that kept your impulses in some kind of check. But that changed in the past few decades as the economy nationalized and then globalized. As we met fewer actual neighbors in the course of a day, those checks and balances fell away. "Operating in a world of instant communication with minimal social tethers," Whybrow observes, "America's engines of commerce and desire became turbocharged."

Adam Smith himself had worried that too much envy and avarice would destroy "the empathic feeling and neighborly concerns that are essential to his economic model," says Whybrow, but he "took comfort in the fellowship and social constraint that he considered inherent in the tightly knit communities characteristic of the 18th century." Businesses were built on local capital investment, and "to be solicitous of one's neighbor was prudent insurance against future personal need." For the most part, people felt a little constrained about showing off wealth; indeed, until fairly recently in American history, someone who was making tons of money was often viewed with mixed emotions, at least if he wasn't giving back to the community. "For the rich," Whybrow notes, "the reward system would be balanced between the pleasure of self-gain and the civic pride of serving others. By these mechanisms the most powerful citizens would be limited in their greed."

Once economies grow past a certain point, however, "the behavioral contingencies essential to promoting social stability in a market-regulated society—close personal relationships, tightly knit communities, local capital investment, and so on—are quickly eroded." So re-localizing economies offers one possible way around the gross inequalities that have come to mark our societies. Instead of aiming for growth at all costs and hoping it will trickle down, we may be better off living in enough contact with each other for the affluent to once again feel some sense of responsibility for their neighbors. This doesn't mean relying on noblesse oblige; it means taking seriously the idea that people, and their politics, can be changed by their experiences. It's a hopeful sign that more and more local and state governments across the country have enacted "living wage" laws. It's harder to pretend that the people you see around you every day should live and die by the dictates of the market.

Right around this time, an obvious question is doubtless occurring to you. Is it foolish to propose that a modern global economy of 6 (soon to be 9) billion people should rely on more localized economies? To put it more bluntly, since for most people "the economy" is just a fancy way of saying "What's for dinner?" and "Am I having any?", doesn't our survival depend on economies that function on a massive scale—such as highly industrialized agriculture? Turns out the answer is no—and the reasons why offer a template for rethinking the rest of the economy as well.

We assume, because it makes a certain kind of intuitive sense, that industrialized farming is the most productive farming. A vast Midwestern field filled with high-tech equipment ought to produce more food than someone with a hoe in a small garden. Yet the opposite is true. If you are after getting the greatest yield from the land, then smaller farms in fact produce more food.

If you are one guy on a tractor responsible for thousands of acres, you grow your corn and that's all you can do—make pass after pass with the gargantuan machine across a sea of crop. But if you're working 10 acres, then you have time to really know the land, and to make it work harder. You can intercrop all kinds of plants—their roots will go to different depths, or they'll thrive in each other's shade, or they'll make use of different nutrients in the soil. You can also walk your fields, over and over, noticing. According to the government's most recent agricultural census, smaller farms produce far more food per acre, whether you measure in tons, calories, or dollars. In the process, they use land, water, and oil much more efficiently; if they have animals, the manure is a gift, not a threat to public health. To feed the world, we may actually need lots more small farms.

But if this is true, then why do we have large farms? Why the relentless consolidation? There are many reasons, including the way farm subsidies have been structured, the easier access to bank loans (and politicians) for the big guys, and the convenience for food-processing companies of dealing with a few big suppliers. But the basic reason is this: We substituted oil for people. Tractors and synthetic fertilizer instead of farmers and animals. Could we take away the fossil fuel, put people back on the land in larger numbers, and have enough to eat?

The best data to answer that question comes from an English agronomist named Jules Pretty, who has studied nearly 300 sustainable agriculture projects in 57 countries around the world. They might not pass the U.S. standards for organic certification, but they're all what he calls "low-input." Pretty found that over the past decade, almost 12 million farmers had begun using sustainable practices on about 90 million acres. Even more remarkably, sustainable agriculture increased food production by 79 percent per acre. These were not tiny isolated demonstration farms—Pretty studied 14 projects where 146,000 farmers across a broad swath of the developing world were raising potatoes, sweet potatoes, and cassava, and he found that practices such as cover-cropping and fighting pests with natural adversaries had increased production 150 percent—17 tons per household. With 4.5 million small

Asian grain farmers, average yields rose 73 percent. When Indonesian rice farmers got rid of pesticides, their yields stayed the same but their costs fell sharply.

"I acknowledge," says Pretty, "that all this may sound too good to be true for those who would disbelieve these advances. Many still believe that food production and nature must be separated, that 'agroecological' approaches offer only marginal opportunities to increase food production, and that industrialized approaches represent the best, and perhaps only, way forward. However, prevailing views have changed substantially in just the last decade."

And they will change just as profoundly in the decades to come across a wide range of other commodities. Already I've seen dozens of people and communities working on regional-scale sustainable timber projects, on building energy networks that work like the Internet by connecting solar rooftops and backyard windmills in robust mini-grids. That such things can begin to emerge even in the face of the political power of our reigning economic model is remarkable; as we confront significant change in the climate, they could speed along the same kind of learning curve as Pretty's rice farmers and wheat growers. And they would not only use less energy; they'd create more community. They'd start to reverse the very trends I've been describing, and in so doing rebuild the kind of scale at which Adam Smith's economics would help instead of hurt.

In the 20th century, two completely different models of how to run an economy battled for supremacy. Ours won, and not only because it produced more goods than socialized state economies. It also produced far more freedom, far less horror. But now that victory is starting to look Pyrrhic; in our overheated and underhappy state, we need some new ideas.

We've gone too far down the road we're traveling. The time has come to search the map, to strike off in new directions. Inertia is a powerful force; marriages and corporations and nations continue in motion until something big diverts them. But in our new world we have much to fear, and also much to desire, and together they can set us on a new, more promising course.

Critical Thinking

1. Why is trying to keep expanding the economy bad for our environment?

2. Explain what it means to have to pick between more or better. Is one selection more compatible with "environmental sustainability"? Why?

3. Locate on your text's world map, the region of the world's peoples who could "relate" to this article.

4. The author states: "Our single-minded focus on increasing wealth has succeeded in driving the planet's ecological systems to the brink of failure, even as it's failed to make

us happier." Beyond the author's evidence, what evidence supporting this statement do you see?

5. The author offers eight ideas to support his thesis. Locate one article in this text for each idea that you feel complements or conflicts with the author's view, and discuss.

6. Identify in this article three key terms, concepts, or principles that are used in your textbook (environmental science, economics, sociology, history, geography, etc.) or employed in the discipline you are currently studying. (Note: The terms, concepts, or principles may be implicit, explicit, implied, or inferred.)

7. Is it possible to be "happy" without being wealthy or owning material things? Explain. What do you need to be happy?

8. Do you think your definition of happiness is compatible with environmental sustainability? Discuss.

Test-Your-Knowledge Form

We encourage you to photocopy and use this page as a tool to assess how the articles in *Annual Editions* expand on the information in your textbook. By reflecting on the articles you will gain enhanced text information. You can also access this useful form on a product's book support website at www.mhhe.com/cls

NAME: DATE:

TITLE AND NUMBER OF ARTICLE:

BRIEFLY STATE THE MAIN IDEA OF THIS ARTICLE:

LIST THREE IMPORTANT FACTS THAT THE AUTHOR USES TO SUPPORT THE MAIN IDEA:

WHAT INFORMATION OR IDEAS DISCUSSED IN THIS ARTICLE ARE ALSO DISCUSSED IN YOUR TEXTBOOK OR OTHER READINGS THAT YOU HAVE DONE? LIST THE TEXTBOOK CHAPTERS AND PAGE NUMBERS:

LIST ANY EXAMPLES OF BIAS OR FAULTY REASONING THAT YOU FOUND IN THE ARTICLE:

LIST ANY NEW TERMS/CONCEPTS THAT WERE DISCUSSED IN THE ARTICLE, AND WRITE A SHORT DEFINITION:

NOTES